Driving Spaces

RGS-IBG Book Series

The *Royal Geographical Society (with the Institute of British Geographers) Book Series* provides a forum for scholarly monographs and edited collections of academic papers at the leading edge of research in human and physical geography. The volumes are intended to make significant contributions to the field in which they lie, and to be written in a manner accessible to the wider community of academic geographers. Some volumes will disseminate current geographical research reported at conferences or sessions convened by Research Groups of the Society. Some will be edited or authored by scholars from beyond the UK. All are designed to have an international readership and to both reflect and stimulate the best current research within geography.

The books will stand out in terms of:
* the quality of research
* their contribution to their research field
* their likelihood to stimulate other research
* being scholarly but accessible.

For series guides go to www.blackwellpublishing.com/pdf/rgsibg.pdf

Published

Driving Spaces: A Cultural-Historical Geography of England's M1 Motorway
Peter Merriman

Badlands of the Republic: Space, Politics and Urban Policy
Mustafa Dikeç

Geomorphology of Upland Peat: Erosion, Form and Landscape Change
Martin Evans and Jeff Warburton

Spaces of Colonialism: Delhi's Urban Governmentalities
Stephen Legg

People/States/Territories
Rhys Jones

Publics and the City
Kurt Iveson

After the Three Italies: Wealth, Inequality and Industrial Change
Mick Dunford and Lidia Greco

Putting Workfare in Place
Peter Sunley, Ron Martin and Corinne Nativel

Domicile and Diaspora
Alison Blunt

Geographies and Moralities
Edited by Roger Lee and David M. Smith

Military Geographies
Rachel Woodward

A New Deal for Transport?
Edited by Iain Docherty and Jon Shaw

Geographies of British Modernity
Edited by David Gilbert, David Matless and Brian Short

Lost Geographies of Power
John Allen

Globalizing South China
Carolyn L. Cartier

Geomorphological Processes and Landscape Change: Britain in the Last 1000 Years
Edited by David L. Higgitt and E. Mark Lee

Forthcoming

Politicizing Consumption: Making the Global Self in an Unequal World
Clive Barnett, Nick Clarke, Paul Cloke and Alice Malpass

Living Through Decline: Surviving in the Places of the Post-Industrial Economy
Huw Beynon and Ray Hudson

Swept up Lives? Re-envisaging 'the Homeless City'
Paul Cloke, Sarah Johnsen and Jon May

Climate and Society in Colonial Mexico: A Study in Vulnerability
Georgina H. Endfield

Resistance, Space and Political Identities
David Featherstone

Complex Locations: Women's Geographical Work and the Canon 1850–1970
Avril Maddrell

Geochemical Sediments and Landscapes
Edited by David J. Nash and Sue J. McLaren

Mental Health and Social Space: Towards Inclusionary Geographies?
Hester Parr

Domesticating Neo-Liberalism: Social Exclusion and Spaces of Economic Practice in Post Socialism
Adrian Smith, Alison Stenning, Alena Rochovská and Dariusz Świątek

Value Chain Struggles: Compliance and Defiance in the Plantation Districts of South India
Jeffrey Neilson and Bill Pritchard

Driving Spaces

A Cultural-Historical Geography of England's M1 Motorway

Peter Merriman

Blackwell
Publishing

BLACKWELL PUBLISHING
350 Main Street, Malden, MA 02148-5020, USA
9600 Garsington Road, Oxford OX4 2DQ, UK
550 Swanston Street, Carlton, Victoria 3053, Australia

First published 2007 by Blackwell Publishing Ltd

1   2007

Library of Congress Cataloging-in-Publication Data

Merriman, Peter.
    Driving spaces : a cultural-historical geography of England's M1 Motorway / Peter Merriman.
        p. cm.   –   (RGS-IBG book series)
    Includes bibliographical references and index.
    ISBN: 978-1-4051-3072-1 (pbk. : alk. paper)
    ISBN: 978-1-4051-3073-8 (hardcover : alk. paper)   1. M1 Motorway (England)–Design and construction–History.   2. Express highways–Social aspects–England–History.   3. Auto-mobile driving on highways–History.   4. England–Social life and customs.   5. Cultural landscapes–England.   I. Title.
    TE57.M47 2007
    388.1′220942–dc22

                                                                                                    2007011998

A catalogue record for this title is available from the British Library.

Set in 10 on 12pt Plantin
by SNP Best-set Typesetter Ltd., Hong Kong

For further information on
Blackwell Publishing, visit our website:
www.blackwellpublishing.com

To Mum and Dad

Contents

List of Figures   ix
Series Editors' Preface   xii
Acknowledgements   xiii

1 **Introduction: Driving Spaces**   1
   Mobilities   4
   Driving, Space, Social Relations   6
   Driving, Landscape, Visuality   12
   Geographies of the Modern Road   16
   Contents of the Book   20

2 **Envisioning British Motorways**   23
   Motoring and the Motor-Car Way, 1896–1930   24
   The German *Autobahnen*: The Politics and Aesthetics
      of a Nation's Roads   31
   Motorways for Britain? National Plans, National Defence   38
   Motorways, War and Reconstruction   43
   Motorways and the British Landscape   46

3 **Designing and Landscaping the M1**   60
   Legislating and Campaigning: Towards a National
      Motorway Network   61
   Locating the M1: Regional Planning, Local Protests
      and the Authority of the Engineer   67
   Landscape Architecture and the Post-war, Modern Road   73
   'A New Look at the English Landscape': Landscape
      Architecture, Movement and the Aesthetics of
      a Modern Motorway   83

Towards a Road Style: Service Areas in the Landscape   90
'Cutting Holes in the Landscape': Britain's Motorway Signs   97

4 **Constructing the M1**   103
'Operation Motorway': Constructing the M1 Motorway   104
Song of a Road: Folk Song, Working-Class Culture
   and the Labour of a Motorway   124

5 **Driving, Consuming and Governing the M1**   141
Motorway Driving, Embodiment, Competence   143
'Motorway Madness': Driving, Governing, Expertise   152
Motorway Modern: Consuming the M1   162
Motorway Service Areas and the Motorist-Consumer   178
Assessing the M1's Performance: Cost-Benefit Analysis,
   Scientific Experiments, Accidents   186

6 **Motorways and Driving since the 1960s**   200
The 'M1 Corridor'   202
Motorways and 'the Environment'   204
Dystopian and Marginal Landscapes?   208
Placeless Environments?   210
Placing the M1 in the Late Twentieth and
   Twenty-First Centuries   213

Appendix: Archival Sources   219
Notes   224
References   246
Index   285

Figures

2.1   An artist's impression of the Northern and Western
Motorway, commissioned by Lord Montagu of
Beaulieu, c.1923   28

2.2   'Suggested scheme of motorways' by the County
Surveyors' Society, May 1938   41

2.3   The front cover of George Curnock's *New Roads for
Britain*, published by the British Road Federation, 1944   58

2.4   Photomontage by Geoffrey Jellicoe of two motorway
carriageways passing alongside the banks of a river in
the Trough of Bowland, Lancashire, 1944   58

3.1   The Ministry of Transport's 1946 map of future national
routes, including motorways and reconstructed trunk roads   62

3.2   Cover of the British Road Federation's *The Case for
Motorways*, 1948, featuring a photo-montage by Geoffrey
Jellicoe   64

3.3   Map showing the location of Watford Gap in relation to
the communications corridors linking London,
the Midlands and the North-West, 1960   68

3.4   Map by A. J. Thornton showing the location of the first
sections of the London to Birmingham/Yorkshire
Motorway, 1959   72

3.5   Landscapes composed for appreciation by pedestrians
and motorists travelling at different speeds   75

3.6   *The Shadowed Road*, c.1808–10. Watercolour by
John Crome (1768–1821)   77

3.7   'The average landscape of the average motor-road in
Britain . . .'   78

3.8   'The Shadowed Road – modern style'   79

3.9   Motorway embankments in England and Germany   84
3.10   Two-span over-bridges spanning the M1 motorway   85
3.11   Two contrasting underpasses designed by Sir Owen
       Williams and Partners   88
3.12   Design by Clough Williams-Ellis of a footbridge for
       Watford Gap service area, 1959   92
3.13   *Hampton Court Bridge*, by Canaletto, c.1754   92
3.14   A maintenance depot loading hopper designed by
       Sir Owen Williams and Partners   95
4.1   Blackwood Hodge advertisement, 1959   107
4.2   Caterpillar advertisement, 1959   108
4.3   *London to Birmingham Motorway During Construction
       at Milton Near Northampton*. Oil painting by
       Terence Cuneo, 1958   114
4.4   Sir Robert Marriott and four Iranian journalists
       studying a model of the motorway in John Laing
       and Son Limited's news-room at construction
       headquarters, Newport Pagnell, September 1958   118
4.5   *Song of a Road*. Advertisement from the *Radio
       Times*, 1959   136
5.1   Cover of the Automobile Association's *Guide to
       the Motorway*, 1959   146
5.2   Cover of *Motorway* insert from *The Motor* magazine, 1959   148
5.3   India Tyres advertisement, 1959   150
5.4   Automotive Products Associated Limited
       advertisement, 1959   151
5.5   Automobile Association patrol vehicles, spotter plane
       and patrolmen on parade at junction 14 on 19
       October 1959   161
5.6   Monthly figures for the number of breakdown
       telephone calls received by the Automobile
       Association, November 1959 to October 1960   163
5.7   Cover of the sheet music for 'M1' by the Ted
       Taylor Four   165
5.8   The front cover of *The Mystery of the Motorway*, by
       Robert Martin   168
5.9   The front cover of *The Motorway Chase*, by Bruce Carter   170
5.10   British Industrial Plastics and Midland Red's
       Motorway Express coach   175
5.11   'The M.1. Motorway', black and white postcard, c.1960–3   183
5.12   'M1. Flyover, Dunstable Road, Luton'. Colour
       postcard of a bridge, embankment and roundabout
       at junction 11 of the M1   183

5.13   'Ye olde "A5" transport cafe'. Cartoon by Giles   184
5.14   The movements of vehicles involved in accidents on
       M1 in 1960 and 1961   195
5.15   'Approximate position of accidents at Park St.
       Terminal (A5/A405) in twelve months ending
       Oct. 31, 1960'   196

Series Editors' Preface

Like its fellow RGS-IBG publications, *Area*, the *Geographical Journal* and *Transactions*, the RGS-IBG Book Series only publishes work of the highest international standing. Its emphasis is on distinctive new developments in human and physical geography, although it is also open to contributions from cognate disciplines, such as anthropology, chemistry, geology and sociology, whose interests overlap with those of geographers. The Series places strong emphasis on theoretically-informed and empirically-strong texts. Reflecting the vibrant and diverse theoretical and empirical agendas that characterize the contemporary discipline, contributions are expected to inform, challenge and stimulate the reader. Overall, the RGS-IBG Book Series seeks to promote scholarly publications that leave an intellectual mark and that change the way readers think about particular issues, methods or theories.

For series guides go to www.blackwellpublishing.com/pdf/rgsibg.pdf

Kevin Ward (University of Manchester, UK) and
Joanna Bullard (Loughborough University, UK)
RGS-IBG Book Series Editors

Acknowledgements

A great many people have helped me during the research and writing of this book. The initial research was undertaken for my doctorate at the University of Nottingham, and I would like to thank staff and postgraduate students at the School of Geography for making it an enjoyable and invigorating place to research. I am particularly grateful to David Matless and Charles Watkins for their invaluable guidance and advice as my supervisors, and Stephen Daniels and Nigel Thrift for advising me to publish my research in book form. I have had the fortune of working in two supportive geography departments. At the University of Reading, Sophie Bowlby, Carl Cater, Erlet Cater, Steven Henderson, Sally Lloyd-Evans, JoAnn McGregor, Erika Meller, Gavin Parker, Rob Potter, Mike Raco, and the students I was fortunate enough to teach and supervise, all created a stimulating working environment and helped the book along in different ways. At the University of Wales, Aberystwyth I would like to thank staff and students in the Institute of Geography and Earth Sciences, including Pete Adey, Luke Desforges, Deborah Dixon, Bob Dodgshon, Bill Edwards, Kate Edwards, Gareth Hoskins, Martin Jones, Rhys Jones, Heidi Scott, Mark Whitehead and Mike Woods. Tim Cresswell and George Revill have provided a constant source of intellectual stimulation, as have members of the Historical Geography Research Group of the RGS-IBG.

I would like to thank Angela Cohen and Jacqueline Scott at Blackwell Publishing and Nick Henry and Kevin Ward, successive editors of the RGS-IBG series, for helping the smooth production of this book. Seminar and conference audiences in New York, Paris, Peckham . . . Aberdeen, Aberystwyth, Bristol, Cheltenham, Hull, Lancaster, Liverpool, London, Milton Keynes, Nottingham, Reading, Washington, DC, Windsor Great Park, and successive AAG and RGS-IBG annual conferences have provided useful and supportive feedback. Staff at the British Library, National

Library of Wales, Trinity College Dublin, British Library of Political and Economic Science, Imperial College and Science Museum Library, Landscape Institute Library, Automobile Association, Royal Automobile Club, Laing, Royal Institute of British Architects, Owen Williams, National Motor Museum, British Road Federation, Civic Trust and university libraries in Aberystwyth, Nottingham and Reading provided access to materials and invaluable help, as did archivists at The National Archives, Birmingham City Archives, the Institution of Civil Engineers, BBC Written Archives Centre, Hertfordshire County Record Office, and the Museum of English Rural Life at the University of Reading. Steve Biczysko, DI Evans and Bob Rogers provided me with invaluable information, and Michael May and Douglas Elbourne kindly agreed to talk to me about their work in the construction of the M1.

The author and publisher gratefully acknowledge the permission granted to reproduce the copyright material in this book. We would like to thank Hodder Arnold for permission to reprint an amended version of my article from *Cultural Geographies* as part of chapter 3. The first half of chapter 4 has been reprinted (in edited form) from *Journal of Historical Geography*, vol. 31, Peter Merriman, ' "Operation motorway": landscapes of construction on England's M1 motorway', pp. 113–33, Copyright (2005), with permission of Elsevier. The BBC Written Archives and Peggy Seeger kindly granted permission to reproduce extracts from the interviews and script of 'Song of a Road' in chapter 4. Copyright holders are acknowledged in the credit lines for individual figures. Every effort has been made to trace copyright holders and to obtain their permission for the use of copyright material. The publisher apologizes for any errors or omissions and would be grateful if notified of any corrections that should be incorporated in future reprints or editions of this book.

Finally, I would like to thank my friends and family, who have provided support and entertainment along this motorway journey.

Chapter One

Introduction: Driving Spaces

So, like earlier generations of English intellectuals who taught themselves Italian in order to read Dante in the original, I learned to drive in order to read Los Angeles in the original. . . . the freeway system in its totality is now a single comprehensible place, a coherent state of mind, a complete way of life, the fourth ecology of the Angeleno. . . . The freeway is where the Angelenos live a large part of their lives. . . . the actual experience of driving on the freeways prints itself deeply on the conscious mind and unthinking reflexes. As you acquire the special skills involved, the Los Angeles freeways become a special way of being alive. . . . (Banham 1971: 23, 213, 214)

The integrated [rural] freeway, married to its landscape, is an elegant composition in space, geared to high speed mobility. Its sculptural qualities can be enormous; it speaks of movement and the kinesthetic qualities of driving on it are vastly exciting. . . . It is further, a form of action calligraphy where the laws of motion generate a geometry which is part engineering, part painting, part sculpture, but mostly an exercise in choreography in the landscape. . . . At their best, these great ribbons of concrete, swirling through the land, give us the excitement of an environmental dance, where man can be in motion in his landscape theater. (Halprin 1966: 37)

. . . the Santa Monica/San Diego intersection is a work of art, both as a pattern on the map, as a monument against the sky, and as a kinetic experience as one sweeps through it. (Banham 1971: 89–90)

In the past decade, geographers have been drawing upon theories of mobility, embodiment, performance, materiality and practice in an attempt to provide increasingly nuanced understandings of the ways in which people more or less consciously and creatively inhabit and move through particular kinds of spaces, environments, places and landscapes.[1] Activities as diverse as dwelling in buildings, dancing, driving, walking and holiday-making are increasingly being examined in studies across the social sciences and humanities which are sensitive to the embodied inhabitation of, and movement through, particular spaces. Of course, few of these practices are new, and there is a fairly long history of critical commentaries, explorations and aesthetic interventions by writers, artists, landscape practitioners, engineers, dancers, musicians and film-makers, as well as academics and cultural commentators, who have explored the relations and tensions between landscape, movement, practice, perception and being. This is evident in the opening quotations (above) by the California landscape architect and environmental designer Lawrence Halprin, and the English architecture and design historian and cultural critic Reyner Banham.

In their focus on the motorist's embodied experience of the vernacular landscape, Reyner Banham and Lawrence Halprin's writings in the 1960s and early 1970s paralleled other well-known studies of the driving landscape – including Donald Appleyard, Kevin Lynch and John Myer's study of Boston's urban expressways in *The View from the Road*, Robert Venturi, Denise Scott Brown and Steven Izenour's architectural study of the Las Vegas strip, *Learning from Las Vegas*, and J. B. Jackson's extensive writings on the vernacular American landscape (Appleyard et al. 1964; Venturi et al. 1972; Jackson 1997). Banham and Halprin, like J. B. Jackson before, asserted the importance of a driver's embodied skills, and their kinaesthetic experiences of both the freeway and the landscape.[2] Freeways are seen to be practised and experienced as 'places', as distinctive systemic environments which are bound up with people's everyday experiences and actions: 'The freeways create a new geography and a new sense of place' (Brodsly 1981: 46). While Banham was clearly fascinated with the distinctive, exoticized spaces of LA and its freeways, he expressed a similar appreciation for Europe's largest multi-level junction (known as 'Spaghetti Junction'), situated on the M6 at Gravelly Hill, Birmingham. When it was opened in 1972, he wrote a review of this 'complex-*looking* intersection' for *New Society*, preparing an itinerary for 'kinaesthetes' wishing to tour 'the inner complexities of this agreeable little suburban megastructure' by car (Banham 1972b: 84, 85). The article was just one of many commentaries Banham wrote during the 1950s, 1960s and 1970s about the vernacular landscapes and pop-modern architecture of post-war Britain. Banham repeatedly encountered and wrote about distinctive, though often quite ordinary, structures and environments, tracing the ecologies of particular landscapes, spaces and places.

In contrast to Banham, Lawrence Halprin did more than simply write about freeway design and landscaping. In the 1960s he was commissioned to prepare the San Francisco Freeways Report (1962–4) and Panhandle Freeway Plan and Report (1963) for the California Department of Highways, and in 1965–8 he served as one of eight urban advisers to the Federal

Highway Administrator of the US Department of Transportation (Halprin 1986). In his writings, Halprin drew parallels between driving, highway design and such creative and dynamic artistic practices as sculpting, painting, calligraphy, choreography and dancing. He collaborated with his wife, the avant-garde dancer Anna Halprin, drawing upon theories of kinetic art, choreography and embodied movement, and developing a form of movement notation ('motation') designed to enable 'generalized notation of any motion through space', whether choreographing dance or 'visualizing the highway experience' (Halprin 1966: 87). In his book *Freeways* Lawrence Halprin included a series of 18 photographs of Anna engaged in a 'dance sequence under the freeway', reflecting his thoughts on the aesthetic and kinaesthetic relationship between human movement, architecture and the landscape (Halprin 1966: 20–1).

What the writings of Banham, Halprin and many others indicates is that there is a rich history of writings on driving in the landscape, as well as work by cultural commentators, artists, landscape architects, engineers and others who have attempted to comprehend, choreograph, and at times represent and notate, the embodied, kinaesthetic skills, habits and experiences of driving in the landscape. In the past decade or two, anthropologists, art historians and geographers have increasingly argued that landscape be turned 'from a noun to a verb', being approached as 'a dynamic medium' (W. J. T. Mitchell 1994: 1) which is worked (D. Mitchell 1996, 2001), practised (M. Rose 2002; Cresswell 2003), inhabited (Hinchliffe 2003), dwelt in (Ingold 1993; Cloke and Jones 2001), and moved through (see also Wylie 2002, 2005; Cresswell, 2003). Landscape is 'tensioned, always in movement, always in making' (Bender 2001: 3). Following these different engagements with landscape and movement, the writings of Halprin and Banham may be seen to form one strand in a much broader genealogy of sensibilities to movement in the landscape.

In this book I examine different moments and movements in the production and consumption of the landscapes of a modern British motorway: the first sections of the London to Yorkshire Motorway or M1. I show how lobby groups, politicians, preservationists, wealthy aristocrats and a range of professions invented and envisioned future British motorways in the early twentieth century, before examining how the landscapes of the M1 were planned, designed, constructed, landscaped and used in the 1950s and 1960s. The landscapes of the M1 have always been in a state of becoming, being actively worked through the movements and actions of surveyors, migrant labourers, construction machines, soil, concrete, rainwater, maintenance workers, drivers and passengers. Vegetation grows on the motorway verges. New technologies for governing the movements of drivers have been incorporated into the motorway's structures. Individual motorway journeys, media stories and the products of children's writers, pop bands and toy manufacturers have worked the landscapes and practices of the M1 into the national, as well as regional and local, imaginations. The landscapes of the motorway may be seen to be 'both a work and an erasure of work' (D. Mitchell 1996: 6), as particular movements and events, along with the effort involved in the design, construction, maintenance and use of these landscapes, are rarely evident or visible to motorway travellers.

In this introductory chapter I trace the theoretical background to my explorations of the geographies of the M1. In section one, I discuss the recent resurgence of work on mobilities in the social sciences and humanities, cautioning against suggestions that a 'new mobilities paradigm' is emerging. In section two, I discuss literatures on motor vehicles and driving, examining how the materialities of vehicles and practices of driving become bound up with distinctive subjectivities, ontologies, identities and mobilities, inculcating particular kinds of embodied skills and sensory engagements with the world. I argue against suggestions that driving is *asocial* and that roads are *placeless* spaces or 'non-places', tracing the distinctive ways in which drivers engage with their surroundings and communicate with other drivers. In section three, I examine how cultural commentators and scholars have tended to approach motoring as a purely visual experience, despite showing an awareness that motoring provides drivers and passengers with multi-sensory, kinaesthetic engagements with the landscape. I provide a detailed discussion of academic accounts of the visualities of motoring, before showing how a range of artists have explored the representational and non-representational dimensions of driving in the landscape. In section four, I examine both popular and academic writings on the histories and geographies of the modern road, highlighting the quite different status of 'the road' in British and American cultural imaginations. Finally, in section five, I outline the contents and principal arguments of the remainder of the book.

Mobilities

Movement, flow, fluidity and mobility are subjects of investigation across the natural, physical and social sciences. The collection *Patterned Ground* reveals how the flows and rhythms associated with such diverse phenomena as cities, glaciers, airports and lakes entwine and refract 'the natural' and 'the cultural' (S. Harrison et al. 2004). Human geographers have held a fairly long-standing interest in mobility, drawing upon a wide range of philosophical approaches – including positivism, phenomenology, Marxism and post-structuralism – to examine such things as the geographies of migration, cultural diffusion, transport, tourism and trade (Cresswell 2001, 2006). Mobility was frequently interpreted as an incidental, rational, universal or dysfunctional by-product of processes occurring in particular places, but in the past decade there has emerged a more extensive and critical academic literature which identifies 'mobility' as an important dimension in the shaping and practising of societies and cultures, spaces, places and landscapes (Urry 2000; Cresswell 2001, 2006). Mimi Sheller and John Urry (2006a, 2006b; see also Urry 2003a; Hannam et al. 2006) have referred to a 'mobilities turn' and the emergence of a 'new mobilities paradigm' in the social sciences, reflecting an increasingly post-disciplinary or inter-disciplinary intellectual agenda, the ascendance of particular strands of non-essentialist post-structuralist and feminist thought, and a focus on issues of identity, embodiment, performance, subjectivity, transnational migration, travel writing, globalization, tourism, mobile communications, the internet and the spaces of the airport, car and road. There is a danger that a language of 'turns' and 'paradigms' may lead academics to overstate the impact of this work, and overlook more firmly established lines of research (such as transport geography), as well as the diversity of these new agendas. Indeed, despite their talk of 'turns' and 'paradigms', Sheller and Urry are careful to argue that they are not 'insist[ing] on a new "grand narrative" of mobility, fluidity, or liquidity' that would repeat the mistakes of the wave of theorists who openly advanced 'nomadic theories' – celebrating, generalizing and frequently romanticizing the transgressive mobilities of the nomad, migrant and traveller – in an attempt to move away from sedentarist theories rooted in ideas of fixity (Sheller and Urry 2006b: 210; see also Kaplan 1996; Cresswell 1997, 2001).[3] Sheller and Urry (2006b: 211) state that they are more concerned with 'tracking the power of discourses and practices of mobility in creating both movement and stasis', echoing Tim Cresswell's long-standing concern to move away from both a 'sedentarist metaphysics' and a 'nomadic metaphysics', to focus instead on the 'politics of mobility' (Cresswell 2002: 11): 'Mobility, like social space and place, is produced. . . . any politics of mobility and any account of mobilities in general has to recognise the diversity of mobilities and the material conditions that produce and are produced by them' (Cresswell 2001: 20, 24).[4]

Modern western societies appear to *function* and *gain life* through the movements of all kinds of material and immaterial things, but they are heavily punctuated by sedentary assumptions and beliefs – for example, that citizens will have fixed dwellings, addresses, nationalities, and own or lease property (Cresswell 2006). Movement must be seen to occur for a (legitimate) 'purpose', and mobilities which are deemed unnecessary, subversive or pointless are frequently criticized and controlled by a range of authorities and commentators (Sibley 1994, 1995). Thus, while the movements of the business traveller, commuter, tourist, quarantined animal and air mail letter may be facilitated by politicians, businesses and planners, the movements of gypsies, refugees and migrant workers are commonly criticized and closely regulated (Sibley 1994, 1995; Cresswell 2001, 2006).

Mobilities and materialities are intricately entwined. In *The Railway Journey* Wolfgang Schivelbusch explored how travellers experience the world through the 'machine ensemble' of the railway or vehicle/highway (Schivelbusch 1986: 24), but the materialities of passports, border fences and such seemingly mundane things as shoes and walking boots are also caught up with, and inseparable from, particular mobilities, subjectivities and ontologies (Michael 2000; Divall and Revill 2005; Sheller and Urry 2006b). What's more, mobile subjects/objects do not simply float across spaces, places and landscapes; rather, their very mobilities continually rework and shape these places and landscapes (Massey 1991, 2000, 2005; Cresswell 2002, 2003; cf. Morse 1998). Cities, for example, are 'spatially open and cross-cut by many different kinds of mobilities' (Amin and Thrift 2002: 3; see also S. Graham and Marvin 2001; Sheller and Urry 2006a), but in this book and throughout the remainder of this chapter, I am concerned with the mobilities, materialities and practices associated with driving, and the spaces of the car, road and motorway.

Driving, Space, Social Relations

> Automobility is: 1. the quintessential *manufactured object* produced by the leading industrial sectors and the iconic firms within 20th-century capitalism . . . , and the industry from which the definitive social science concepts of Fordism and post-Fordism have emerged; 2. the major item of *individual consumption* after housing which provides status to its owner/user through its sign-values . . . 3. an extraordinarily powerful *complex* constituted through technical and social interlinkages with other industries . . . 4. the predominant global form of 'quasi-private' *mobility* that subordinates other mobilities . . . 5. the dominant *culture* that sustains major discourses of what constitutes the good life, what is necessary for an appropriate citizenship of mobility and which provides potent literary and artistic images and symbols . . . 6. the single most important cause of *environmental resource-use*. (Urry 2004b: 25–6)

John Urry has drawn upon theories of complexity to approach 'automobility' as a 'self-organizing autopoietic, non-linear system that spreads worldwide' and 'generates the preconditions for its own self-expansion' (Urry 2004b: 27). While I have reservations about the use of theories of complexity and systemic metaphors,[5] Urry very succinctly and effectively summarizes the ways in which the car industry, cars and driving (not to mention other motor vehicles) are socially, culturally, economically, politically, ethically and environmentally embroiled in our daily lives, whether we own a car, drive or not (Sheller and Urry 2000; D. Miller 2001; Wollen and Kerr 2002; Urry 2004b; Böhm et al. 2006). While 'immensely flexible', automobility may be seen to be 'wholly coercive', underpinning dominant assumptions about how people conduct and manage their lives across time and space (Sheller and Urry 2000: 743; cf. Morse 1998). Social scientists have recognized the importance of automobility for decades, but their discussions of the car and driving have been very specific and limited until fairly recently. Geographers have provided extensive studies of the modes of production associated with car manufacturing (Hudson and Schamp 1995), and the environmental implications and 'external costs' of car travel (Whitelegg 1997; cf. D. Miller 2001). Sociologists have focused on the working practices, unionization and affluence of the car worker (Goldthorpe et al. 1968–9; Beynon 1973; cf. G. Turner 1964), and the life and work of the lorry driver (Hollowell 1968), but it is only in the past decade or two that sociologists, anthropologists and geographers have begun exploring the more far-reaching ways in which cars (and occasionally other vehicles) are driven, consumed and shape our lives.[6]

Advertisers, car manufacturers, motoring journalists and drivers frequently suggest that it matters what vehicle or car we drive (on car advertising, see Wernick 1991; Dery 2006). Motor vehicles and their movements have long been enmeshed in gendered, racialized, sexualized, nationalized, localized and globalized processes of inclusion and exclusion, identity formation and stereotyping (Scharff 1991; Sachs 1992; O'Connell 1998; Katz 1999; R. Law 1999; Gilroy 2001; D. Miller 2001; Sanger 2001; Edensor 2004; Böhm et al. 2006; see chapter 5). Particular vehicles may be labelled as expensive, cheap, cool, youthful, boring, unreliable, or as masculine or feminine.[7] Hot-rod enthusiasts, 'boy-racers' and other customisers attempt to rebuild, restore and restyle their cars in an attempt to differentiate, individualize and 'improve' their appearance and performance (Relph 1976; Moorhouse 1991). The materiality of cars and vehicles is intimately entwined with the spaces, embodied actions, identities and subjectivities of *driving* (as well as simply *owning* a car), and it is important to recognize that there are clear differences between the experiences and embodied actions of drivers and passengers:

> In contrast to the passenger, the driver, in order to drive, must embody and be embodied by the car. The sensual vehicle of the driver's action is fundamentally different from that of the passenger's, because the driver, as part of the praxis of driving, dwells in the car, feeling the bumps on the road as contacts with his or her body not as assaults on the tires, swaying around curves as if the shifting of his or her weight will make a difference in the car's trajectory, loosening and tightening the grip on the steering wheel as a way

of interacting with other cars. . . . we must appreciate how driving requires and occasions a metaphysical merger, an intertwining of the identities of driver and car that generates a distinctive ontology in the form of a person-thing, a humanized car or alternatively, an automobilized person. (Katz 1999: 32, 33)

If politicians, environmentalists, economists and indeed geographers want to understand why people have an enduring attachment to their cars – to persuade them to move to more environmentally sustainable alternatives – it is vital that they understand the social relations, embodied practices and ontologies associated with driving. As Jeffrey Schnapp has shown, motor vehicles were the first, and remain the only, popular motorized '*driver*-centred' mode of transportation, combining the sense of freedom, control, independence and privacy that had previously been experienced with the bicycle and horse-drawn chariots and phaetons, with the sublime aura, mystery and seemingly effortless power, range and speed of mechanically powered modes of transport (Schnapp 1999: 3; see also Kern 1983; Thrift 1990; Sachs 1992). Motor vehicles shape our being, reconfiguring our sense of self and personal mobility. Through the act of driving, 'people' *become* 'vehicle drivers' – hybrid, collective or cyborg figures whose subjectivity and objectivity are (re)configured through the contingent, partial and momentary practice of dwelling in a vehicle and driving along the road (see Ross 1995; Graves-Brown 1997; Lupton 1999; Katz 1999; Michael 2000, 2001; Sheller and Urry 2000; Urry 2000, 2004b; Beckmann 2001; D. Miller 2001; Dant 2004; Edensor 2004; Featherstone 2004; Böhm et al. 2006; Merriman 2006b). As drivers gain experience of, and familiarity with, their vehicles, so the embodied skills, dispositions and actions of driving appear to be performed and practised in relatively unconscious, 'automatic', non-cognitive and unreflexive ways: '. . . the experience of driving is sinking in to our "technological unconscious" and producing a phenomenology that we increasingly take for granted but which in fact is historically novel' (Thrift 2004: 41). As an embodied, habitual, unreflexive and, some might say, non-cognitive activity, driving can appear fairly effortless, ordinary and mundane (Dant 2004; Featherstone 2004). Driving seems to entail a process of perpetual day-dreaming or forgetting, 'a spectacular form of amnesia' (Baudrillard 1988: 9), and a 'detached involvement' with one's surroundings (Brodsly 1981: 41; Featherstone 1998; Joyce 2003). Margaret Morse (1998) has described this as a process of perpetual 'distraction', while Jonathan Crary, in his work on attention and modern culture, describes the 'diffuse attentiveness and quasi-automatism' which is characteristic of both freeway driving and television watching (Crary 1999: 78). This is until something draws our attention, we become tired, lost or caught in a jam, and the 'orderliness of driving in traffic' breaks down (Lynch 1993: 155; Merriman 2006b; see also Latour 2005). As architectural critic Raymond Spurrier remarked in 1959: 'The driver should be as unconscious of the road itself as he is of what is going on beneath the bonnet. When either mechanism or road begins to obtrude, something has gone wrong' (Spurrier 1959: 245). As I discuss in chapters 3 and 5, a large number of experts and authorities – including landscape architects, engineers and civil servants – have stressed that carefully designed carriageways, signs and roadside planting can help maintain but must not distract the motorway driver's attention – forming a minor element in broader programmes to shape the experiences and conduct of motorway drivers (Merriman 2005b, 2006b).

Governing drivers and other road users has always been a contentious, politically sensitive activity. In early twentieth-century Britain, motoring offences brought many otherwise respectable, law-abiding, upper- and upper-middle class citizens into the nation's courts for the first time (Emsley 1993). As car ownership expanded, successive governments worked hard to introduce new motoring legislation and taxation without losing the support of the motoring public (in the past three decades this has been in the face of increasing global oil prices and pressure from environmental groups). Civil servants have long realized that to effectively govern the conduct of motorists across an extensive road network they need to supplement regulatory, disciplinary and juridical frameworks with liberal and educative programmes and technical devices – such as the Highway Code and Motorway Code (see chapter 5) – which facilitate and encourage motorists to learn new techniques for governing their own conduct and the movements of their vehicles (Merriman 2005b; cf. Joyce 2003).[8] Drivers may interpret and resist the formal rules of the road in a variety of ways, but the limited extent to which they can drive 'differently' or 'creatively', coupled with the severe consequences of rule-breaking and inattentiveness, lead them to perform in more or less socially acceptable ways. A misjudged turn of the steering wheel, press on the accelerator pedal or failure to glance in a mirror may result in a fatal accident, and yet such events are so frequent and 'normal' that it is only when they involve ourselves, friends, family or famous individuals that we look beyond the de-personalized and sanitized accident statistics and register the embodied effects and affects of such incidents (see chapter 5).

Driving is an important social, cultural, spatial practice, but it is not uncommon for academics to approach driving as a solitary, desensitizing, dislocated or asocial activity which generates experiences of placelessness (see Freund and Martin 1993).[9] In *Place and Placelessness* the humanistic geographer Edward Relph acknowledged that cars and other 'personal machines . . . offer us new options, comforts and experiences' (Relph 1976: 129), but they also desensitize us to, and separate us from, our surroundings, with modern twentieth-century roads representing both a feature and symptom of a flat, placeless geography:

> Roads, railways, airports, cutting across or imposed on the landscape rather than developing with it, are not only features of placelessness in their own right, but, by making possible the mass movement of people with all their fashions and habits, have encouraged the spread of placelessness well beyond their immediate impacts. (Relph 1976: 90; see also Kunstler 1994)

French anthropologist Marc Augé has made similar claims about the spread of 'non-places' such as motorways and airports in the late twentieth century, an era of 'supermodernity' (Augé 1995). On one level, Augé's descriptions appear to reflect the feelings of blankness, forgetting, indifference and ubiquity which *some*, perhaps many, travellers and consumers experience in fairly standardized, mundane and familiar environments (cf. Morris 1988). As Margaret Morse argues, 'practices and skills' such as driving and shopping constitute 'the barely acknowledged ground of everyday experience', and 'can be performed semiautomatically in a distracted state' (Morse 1998: 102; see also Crary 1999). Drivers experience 'a partial loss of touch with the here-and-now', all of which works to constitute 'an ontology of everyday distraction' (Morse 1998: 99; see also Crary 1999). Morse (1998: 103) goes on to describe the isolation and 'derealization' of freeway driving, but like Augé she overstates and over-generalizes the difference, novelty and dislocation of the experiences and environments associated with motorway driving. Firstly, as I have argued elsewhere, it is unnecessary to delineate a new species of place – i.e. 'non-place' or 'non-space' – to account for the detachment, solitariness, boredom and distraction which *some* drivers or passengers *may* experience on motorways; feelings which are just as likely to surface when one is at home or work (Merriman 2004b, 2006b). Secondly, Augé and others tend to overlook the history of such 'barometers of modernity' or supermodernity, for commentators have, in previous decades and centuries, associated feelings of boredom, dislocation, illegibility, excitement and shock with other previously new transportation and communication technologies, such as the railway in the nineteenth century (Thrift 1995: 19; see also Schivelbusch 1986; Crary 1999; Merriman 2004b). Thirdly, while Augé (1995: 79) argues that places and non-places are 'like palimpsests', being 'ceaselessly rewritten', it is only in his later writings and interviews that he supplements his auto-ethnographic reflections on travelling through spaces of supermodernity with an acknowledgment of the different experiences and degrees of access that individuals/groups – whether commuters, workers, or refugees – may have in/to such spaces as the airport (e.g. Augé 1999, 2004; cf. Cresswell 2001; M. Crang 2002a). Fourthly, Augé and other commentators often fail to examine the quite different embodied engagements and experiences of the driver and passenger. Fifthly, and finally – as Bruno Latour (1993) points out – Augé fails to register the many mediated or distanciated social-material relations and entanglements which emerge in these spaces, for he is attached to the idea that sociality is a function of face-to-face, unmediated communication. Non-places, Augé states, 'are the spaces of circulation, communication and consumption, where solitudes coexist without creating any social bond or even a social emotion' (Augé 1996: 178).

At first glance, motorway driving may appear to produce such experiences of detachment and solitude, but as a large number of sociologists, psychologists, anthropologists, geographers and ethnomethodologists have stressed, driving is a complex *social* practice and activity, and drivers do communicate and interact with people and all manner of things, inhabiting and consuming the spaces of the car and road in a myriad of distinctive ways (see, e.g., Goffman 1971; Dannefer 1977; Lynch 1993; Katz 1999; Amin and Thrift 2002; Featherstone 2004; Laurier 2004). In the contemporary West we tend to take our relations with cars and roads for granted, 'think[ing] our world through a sense of the self in which driving, roads, and traffic are simply integral to who we are and what we presume to do each day' (D. Miller 2001: 3). But cars and the spaces of the road have been, and are, shaped, consumed and inhabited in quite distinctive ways in different societies at different times, whether in early twentieth-century Britain (O'Connell 1998), contemporary India (Sardar 2002; Edensor 2004), Trinidad (D. Miller 2001), Cuba (Narotzky 2002) or amongst the Aborigines of the Pitjantjatjara lands in South Australia (Young 2001), to name but a few examples (see also Wollen and Kerr 2002). Drivers and passengers inhabit different vehicles in different ways, playing car games with other passengers, doing office work while on the move (Laurier 2004), talking to family members, listening to recorded music or the radio, singing (Brodsly 1981; Bull 2001, 2004; Schwarzer 2004), having sex in lay-bys, or day-dreaming and contemplating (Edensor 2003). Despite assertions to the contrary by some transport economists, driving-time, like all travel time, can be productive (Dant and Martin 2001; Lyons and Urry 2005). Driving entails intense 'affective and embodied relations', giving rise to a range of emotions and feelings, from fear and anger, to excitement (Sheller 2004: 221; see also Katz 1999). Drivers get frustrated, all too aware of the 'expressive limitations of their vehicles', and the fact that other drivers appear 'deaf to one's own concerns' (Katz 1999: 28; see also Lynch 1993). In extreme cases, frustrated and angry vehicle drivers may engage in acts of violent 'road rage' (see Katz 1999; Lupton 1999; Michael 2000, 2001). In other situations, drivers flash headlights, use indicator lights, sound their horn, shout, or make polite or rude gestures in an attempt to communicate their feelings or intended movements (see Katz 1999). Drivers continually predict and assess the actions and performance and movements of other 'vehicle drivers', judging motorists by their movements and stereotyping them according to their gender, ethnicity, age, class, nationality, or the appearance of their car (O'Connell 1998; Katz 1999).

Driving, Landscape, Visuality

Drivers inhabit, navigate and move through the spaces and landscapes of the road in distinctive, embodied ways facilitated by a range of technologies (Thrift 2004). Driving is not solely a visual experience, where 'the sights, sounds, tastes, temperatures and smells of the city and countryside are reduced to the two-dimensional view through the car windscreen' (Urry 2000: 63; see also Sheller and Urry 2000). An array of 'privatizing technologies' have affected the driver's and passenger's sensations of movement and their surroundings (Urry 2000: 63), but even today's air-conditioned, heated, suspensioned cars with ABS brakes, modern glazing, power steering, satellite navigation and sophisticated stereos afford multi-sensory affects and engagements with the landscape. Nevertheless, while a few academics and commentators have focused on drivers' multi-sensory and kinaesthetic inhabitations of the world, the vast majority of writings and interventions focus on the driver's or passenger's visual experiences of the landscape, and in these accounts the motorist's *vision* is rarely seen to be *embodied with*, and inseparable from, other sensory and kinaesthetic apprehensions of their surroundings (see Bull 2004, on the driver's audio-visual experiences).

In *America* Jean Baudrillard (1988), perhaps unsurprisingly, focused his attention on the visual spectacle of driving along the Los Angeles freeways and desert highways. Driving at speed through the desert was seen to create an 'invisibility, transparency or transversality in things', and what Baudrillard, citing Paul Virilio, termed an 'aesthetics of disappearance' (Baudrillard 1988: 7, 5). Baudrillard incorporates such effects and experiences into his own geography of LA and the American desert, but when Virilio addressed similar themes in his writings he abstracted the practices and experiences of driving from specific landscapes, spaces and times, tracing a dystopian, almost apocalyptic, futuristic, transhistorical and dislocated geography which effaced the multi-sensory nature of driving. In *Polar Inertia* Virilio looked to a future where 'the audiovisual feats of the electronic dashboard will prevail over the optical qualities of the field beyond the windscreen', and 'the temporal depth of the electronic image prevails over the spatial depth of the motorway network' (Virilio 2000: 15). Human physical movement would cease to be important, as 'dynamic automotive vehicle[s]' such as cars are replaced by 'the static *audiovisual vehicle*', marking 'the definitive triumph of sedentariness' (Virilio 2000: 18). Here, and elsewhere, Virilio's future predictions appear to fly in the face of contemporary sociological studies which demonstrate the continuing importance of corporeal travel and physical co-presence, alongside the increasing use and importance of virtual communication technologies (Urry 2002; 2003a). Virilio's predictions emerge from a fascination with the geographies of the screen, and comparisons between the experiences of driving and the visualities of the cinema, television and computer screen (see also Morse 1998). In *The Aesthetics of Disappearance*, Virilio sees the 'voyeur-voyager in his car' as analogous to 'the moviegoer', who 'knows in advance what he's going to see, the script' (Virilio 1991: 67–8). Windscreen and cinema screen are conceptualized as comparable framing devices or surfaces of projection, and here Virilio's thoughts parallel the writings of a long line of film theorists, social scientists and cultural commentators who have drawn parallels between a cinematic gaze and automotive visuality (see Friedberg 1993, 2002; Dimendberg 1995; Morse 1998; Larsen 2001; M. Crang 2002b; Schwarzer 2004; also Ross 1995). As French writer Octave Mirbeau remarked in 1908 on the impact of the motor car on 'modern man': 'Everywhere life is rushing insanely like a cavalry charge, and it vanishes cinematographically like trees and silhouettes along a road' (cited in Kern 1983: 113).[10] Cars, cinema and trains may be seen to act as 'vision machines' and virtual or physical transportation machines, but their users have had, and do have, very different motivations and embodied engagements with the spaces and landscapes which are 'projected' or 'refracted'.

It is useful, here, to compare and contrast the visualities enabled by railway travel with the visual engagements of the car driver and passenger. Schivelbusch famously argued that railway travel led to the emergence of a new mode of perception, a way of seeing that was 'panoramic', as the relative comfort, smoothness and speed of the ride led to a sense of detachment from the landscape, and the cultivation of distinctly different embodied practices and experiences from those associated with stage coaches (Schivelbusch 1986). Landscapes appeared to lose their depth, becoming 'evanescent' glimpsed scenes, with passengers trained in the arts of landscape painting and viewing finding the view disorienting and disagreeable: 'Those who were conditioned to looking at a landscape as a landscape painting, with a detailed foreground directing the eye to middle ground and distance, discovered the view from a speeding train could not be contained within this structure and that attempts to do so were unnerving, sometimes sickening' (Daniels 1985: 16). Observers compared the evanescence and apparent scrolling of the landscape past the train window with the popular panoramic and dioramic shows of the early nineteenth century, and Daniels remarked that 'those who enjoyed going to the panorama shows relished a spectacle more refined passengers found difficult to stomach' (Daniels 1985: 16).

A few academics have suggested that motorists experience the landscape in a panoramic manner similar to that of railway travel (Dimendberg 1995; Ross 1995), but while back-seat passengers may observe similar visual effects when staring out of the side windows of a car, a broad range of academics, cultural commentators and artists have emphasized that the visualities and embodied engagements of the driver and front-seat passenger are quite different to that of the railway passenger (Schwarzer 2004; also Liniado 1996; Featherstone 1998). Reyner Banham commented on the difference in *New Society* in 1972:

> The railway view presents a passive observer with a continuous panorama that unrolls from left to right, or other way about. One is very detached from it as its contents slide past according to the laws of parallax. Richard Hamilton once did a series of paintings about this very effect; he also did one only through a car windscreen and then gave up.[11]
>
> For the car-borne view is neither detached nor in parallax. The observer plunges continuously ahead into a perspective that is potentially dangerous and demands his active attention (nor is the passenger passive: watch his feet and hands, listen to his comments and warnings). (Banham 1972a: 243)

Mitchell Schwarzer associates this plunging perspective with what he terms 'dromoscopic perception', with 'a headlong immersion into a free space of movement around which buildings recede' (Schwarzer 2004: 99). This complex, plunging visual perspective is inseparable from the differentiated, performative, embodied actions of both driving *and* passengering, but throughout the twentieth century critical commentators, architects, artists and others differed in their opinions on the extent to which the dynamic, embodied experiences and visualities of car travel could be apprehended or represented.

On the one hand, engineers, architects, urban designers, landscape architects and psychologists have attempted to comprehend, codify and at times model the visual experiences of drivers *and* passengers. Since at least the 1940s, highway engineers have developed mathematical formulae to ensure that the aesthetic appearance of the alignment and curvature of roads is not irritating or confusing for drivers (Merriman 2001). In the early 1960s, as part of their study of Boston's urban expressways, the urban designers Donald Appleyard, Kevin Lynch and John Myer developed a system for 'recording, analyzing and communicating' the visual sequences presented to the car traveller (Appleyard et al. 1964: 19). The team were motivated by 'a desire to find a visual means for pulling together large urban areas', and they argued that high-speed expressways could provide 'a new means for making the structure of our vast cities comprehensible to the eye' (Appleyard et al. 1964: 63, 16).

On the other hand, a range of artists, architects and others have high-lighted the performative, non-representational nature of the visualities and broader sensibilities of vehicle driving and passengering. As the modernist American sculptor and painter Tony Smith remarked of a night-time drive along 'the unfinished New Jersey Turnpike' with three students in the early 1950s:

> This drive was a revealing experience . . . it did something for me that art had never done before. At first I didn't know what it was, but its effect was to liberate me from many of the views I had had about art. It seemed that there had been a reality there which had not had any expression in art. . . . I thought to myself, it ought to be clear that's the end of art. Most painting looks pretty pictorial after that. There is no way you can frame it, you just have to experience it. (quoted in Wagstaff 1966: 19)

Freeway travelling is approached as a kinetic, non-representational, performative engagement, invoking a visual aesthetic far in advance of contemporary forms of artistic representation and expression. Smith's personal revelation can be seen to resonate with both earlier and later artistic engagements with experiences of landscapes of mobility, but rather than follow Smith and express the futility of attempting to (re)present or express particular practices and movements, artists have more commonly chosen to experiment with aesthetic techniques that can articulate, refract or play with the dynamic, more-than-representational nature of driving and the view through the windscreen. The distinctive visualities of both *motoring* and *viewing passing vehicles* have been explored by an array of (largely 'modern') artists, from Henri Matisse in *The Windshield, on the Villacoubly Road* (1917), Giacomo Balla in his Futurist *Abstract Speed – the Car has Passed* (1913) and László Moholy-Nagy's experimental colour photograph *Pink Traffic Abstraction* (1937–40), to David Hockney's photographic montage *Pearlblossom Highway* (1986), pop art works by Roy Lichtenstein, Ed Ruscha and Richard Hamilton, and video art works by Rachel Lowe and Julian Opie (Millar and Schwartz 1998; British Council 2000; P. D. Osborne 2000; Wollen 2002; Horlock 2004).[12]

One of the most notable, often overlooked, creative attempts to engage with the visualities and phenomenologies of driving and passengering is the published diary of Brutalist architect Alison Smithson, *AS in DS: An Eye on the Road* (1983). In the early 1970s, Smithson – a member of the Independent Group with Richard Hamilton, Eduardo Paolozzi, Reyner Banham, her husband Peter Smithson and others – kept 'a diary of a passenger's view of movement in a car' during trips in the family's Citroën D.S.

(Smithson 1983: 15). After the text of the diary was complete, Smithson made sketches while on the move, and these multiple engagements with the passing landscape helped her to identify and engage with the ways in which 'the car-moved-seeing' produced a 'new sensibility', a new way of seeing, being in and moving through the landscape (Smithson 1983: 15–16). In a description of one of her many journeys between London and the family's home at Fonthill in Wiltshire, Smithson described the scene:

> . . . headlights are refracted by the mist into tiny, [sic] globules that change each oncoming aura into a Seurat-in-transit . . . this unreal fracturing of light, the gentle movement of the well-cushioned ride, somehow eats up the distance . . . the pointillist lights manoeuvre in the darkness . . . such sideways movements the more noticeable because of the otherwise uninterrupted steady forward movement of all the cars . . . now this car is holding its distance behind a constellation of ruby lights. (Smithson 1983: 97, ellipses in original)

Smithson's descriptions of her journeys along more or less familiar routes reveals an embodied kinaesthetic and visual sensibility to *particular* landscapes and driving environments which is all too lacking in the majority of academic writings about driving, passengering and the spaces of the road. Indeed, while drivers and passengers may develop particular dispositions and sensibilities to their surroundings which are embodied and practised on a fairly routine basis, different kinds of road and motorway facilitate or afford quite different experiences and styles of driving, and motorists frequently have quite specific reasons for driving along *particular* routes to reach their destinations.[13] Driving environments have quite specific geographies and afford quite specific mobilities.

Geographies of the Modern Road

Streets and roads have developed over centuries, but with the growth of motor car ownership during the twentieth century, many roads have undergone a significant transformation, as highway engineers, planners and politicians have adopted and implemented strategies and technologies aimed at shaping the movements and conduct of pedestrians, motorists and motor vehicles.[14] Critics have argued that increasing levels of motor vehicle ownership have led to the privatization of public roads, and the destruction of neighbourhoods, as the *social spaces* of *streets* are transformed into *asocial roads* dominated by unidirectional vehicular movement (see RTS 1997). As the anti-car, anti-road, anti-capitalist, pro-streets group Reclaim the

Streets argued in a propaganda poster distributed at their third London street party on the M41 West Cross Route on 13 July 1996:

> We are basically about taking back public space from the enclosed private arena. At it's simplest it is an attack on cars as a principle [*sic*] agent of enclosure. It's about reclaiming the streets as public inclusive space from the private exclusive use of the car. But we believe in this as a broader principle, taking back those things that have been enclosed within capitalist circulation and returning them to collective use as a commons.[15]

It is not just radical environmental groups who criticize the effects of cars on public space and busy roads on urban communities. In 1961 Jane Jacobs described '[t]raffic arteries, along with parking lots, filling stations, and drive-in movies' as 'powerful and insistent instruments of city destruction' (J. Jacobs 1961: 352). In 1974 geographer Ronald Horvath explained and mapped how 'automobile territory' – land devoted to the driving, parking and servicing of cars – had come to occupy a significant proportion of the surface area of American cities such as Detroit (Horvath 1974; see also Thrift and French 2002). As traffic levels increase, the ecologies and aesthetics of streets, roads and urban areas change (N. Taylor 2003), and yet few geographers have chosen to broaden their attention beyond the spaces of the bustling city street to examine the distinctive geographies of roads and motorways, as well as practices of driving. In this section I introduce the literatures on the histories and geographies of the street, road and motorway, suggesting why there are very few critical academic studies of Britain's roads and motorways.

There is an extensive literature on roads. Descriptive tracts and historical accounts have been published about Britain's roads for centuries – from the Roman *Itinerarium Antonini* (c. AD 200), through to descriptions in Daniel Defoe's *A Tour Through England and Wales* (1724–6) and more recent histories of the Great North Road and modern motorway (see the compendium Scott-Giles 1946). Engineers and planners have outlined the history of Britain's motorway-building programme (see Drake et al. 1969; Starkie 1982; Charlesworth 1984; Bridle and Porter 2002; P. Baldwin and Baldwin 2004),[16] but it is only in recent years that there has emerged a more critical (if diverse) historical and sociological literature on the spaces of the street, road and motorway. Sociologists and anthropologists have argued that modern roads and motorways are non-places or somewhat placeless spaces (Relph 1976; Augé 1995; cf. Merriman 2004b). There is a significant body of literature on cultures of street-walking and the spaces of the urban street, ranging from work on the gendering of street-walking and the activities of the *flâneur* and *flâneuse* in the nineteenth century, to writings on the ordering, lighting, regulation and surveillance of streets (e.g. Schivelbusch 1988; Tester 1994; Fyfe 1998; Nead 2000; Joyce 2003).

There is an expanding academic literature on the social, political, architectural, landscape and environmental histories of the spaces of the road and motorway, but aside from an array of social scientific writings on the British road protest movement (e.g. McKay 1996; Routledge 1997; Wall 1999) and a few specialized histories of motorway service areas, petrol stations and roads like London's Westway (e.g. McCreery 1996; Jones 1998; Croft 1999; D. Lawrence 1999; Robertson 2007), very little of this focuses on the spaces of Britain's roads. The roads of two countries tend to dominate the literature. Firstly, there are a large number of studies of the German *Autobahnen* which examine the extent to which their design, landscaping and promotion refracted Nazi political, social and aesthetic ideologies (e.g. Shand 1984; Gröning 1992; Boyd Whyte 1995; Dimendberg 1995; Rollins 1995; Zeller 1999). Secondly, there is an extensive academic literature on the highways, parkways and freeways of the USA, which, although diverse, appears to reflect the prominent position of 'the road' and the automobile in the American national imagination (e.g. Brodsly 1981; Berman 1983; Wilson 1992; Raitz 1996; Lackey 1997; Gandy 2002; Wollen and Kerr 2002; Krim 2005). Accounts of highways and being 'on the road' have assumed a notable place in both mainstream and countercultural imaginaries of the American nation, whether in books such as Jack Kerouac's 1957 *On the Road* (see Cresswell 1993), road movies such as *Easy Rider* (1969) and *Thelma and Louise* (1991) (see Cohan and Hark 1997; Eyerman and Löfgren 1995; Wollen and Kerr 2002), or in diverse representations of Route 66 (see Krim 2005). Roads and the vernacular modern landscapes of the roadside strip have also assumed an important position in the writings of American landscape historians, architectural historians, geographers and commercial archaeologists such as J. B. Jackson, John Jakle, Keith Sculle, Grady Clay and Arthur Krim (Jakle and Sculle 1994, 1999, 2004; Jakle et al. 1996; Raitz 1996; Jackson 1997; Krim 2005; also Venturi et al. 1972).[17] In contrast, British landscape historians and industrial archaeologists have appeared less keen to study Britain's roads, commercial architecture and modern vernacular landscapes.[18] What's more, as cultural commentators such as Will Self, Michael Bracewell and Stuart Jeffries have argued, Britain's roads are relatively short and congested, 'lacking the mystique' and expansiveness of America's highways (Picken 1999: 222):

> . . . Britain seemed so notably deficient in motorway culture compared with other countries, particularly the United States. The idea of a proper British road movie was laughable – there wasn't enough track. (Self 1993: 1)

Now we are a rain-soaked dime of a country, shrunk by roads into an awayday island where everywhere is near everywhere else, nowhere is worth going to and the journey in between is a misery. . . . In other countries, roads are the carriers of romance and they spawn genre movies and books. . . . Britain has a long way to go if it is to emulate America or any attractive car culture. (Jeffries 1998: 14)

In Britain, the cultural status of the motorway remains ambiguous, to say the least. . . . this country's experience of the motorway is comparatively young . . . and rooted, unwaveringly, in the very opposite of America's road-movie romance with the highway. (Bracewell 2002a: 285)

British roads, motorways and car journeys may not have been embroiled in the powerful discourses of automobility, freedom and romance which appear so pervasive in the USA, nor have they attracted the attention of many British academics, but they have, at various times, been romanticized and celebrated for their modernity, excitement, beauty, kitsch qualities and ethereality, as well as criticized for (or at least labelled as) being boring, dull, ubiquitous, dangerous, alienating, destructive, dystopian landscapes.

As I show in chapters 5 and 6, Britain's motorways were celebrated as exciting, experimental, modern landmarks and sites of travel in the 1950s and 1960s. The M1 caught the attention of board-game manufacturers, pop musicians, children's writers and a whole host of cultural commentators, but despite appearing in the occasional film (e.g. Albert Finney's 1968 *Charlie Bubbles*) and soap opera (e.g. *EastEnders*) there are no celebratory road movies set on the M1. During the 1970s, 1980s and 1990s the spaces of the motorway have featured in a series of more dystopian (sometimes critical, sometimes dark, sometimes nostalgic) narratives which remark on the marginal or anonymous nature of these landscapes – from J. G. Ballard's *Crash* (1973) and *Concrete Island* (1974), Peter Nichols' play *The Freeway* (1975), and Chris Petit's film *Radio On* (1979), to Trevor Hoyle's novel *The Man who Travelled on Motorways* (1979), Will Self's short story 'Scale' (1994), Michael Winterbottom's film *Butterfly Kiss* (1995), St Etienne's song 'Like a Motorway' (1994) and Black Box Recorder's 'The English Motorway System' (2000). In *London Orbital: a walk around the M25*, the writer, poet and film-maker Iain Sinclair describes a series of explorations of the landscapes, or 'acoustic footprints', surrounding London's infamous M25 orbital motorway (Sinclair 2002).[19] Sinclair decided to undertake the walk in 'the belief that this nowhere, this edge, is the place that will offer fresh narratives' on contemporary society (Sinclair 2002: 14). The M25 was Britain's first orbital motorway, which 'goes nowhere; it's self-referential, postmodern, ironic' (Sinclair 2002: 443), and in his characteristic style Sinclair interweaves descriptions of particular sites and landscapes he encounters – from modern shopping centres to old factories – producing a poetic topography of both the spectacular and the mundane, new and historic landscapes, sites of surveillance and seemingly lawless wastelands. Sinclair celebrates and reveres particular landscapes and sites of memory, as well as describing landscapes he detests. *London Orbital*, although somewhat unorthodox, can be placed in a growing list of popular 'biographies' of British roads, including Edward Platt's *Leadville: A Biography of the A40* (2000), Peter Boogaart's *A272 – An Ode to a Road* (2000), as well as photographic studies of the A1 by Paul Graham (1983) and Jon Nicholson (2000) (one could also include Martin Parr's (1999) *Boring Postcards*).

Academics, on the other hand, have been quite focused – and, at times, conservative – in their attention to the British motorway, and there are very few in-depth studies of the geographies of *particular* driving environments. Geographers have examined the effects of motorways on patterns of economic activity and regional development (e.g. R. H. Osborne 1960; Massey 1984; *The Geographical Journal* 1986; Hebbert 2000). In the early 1960s Jay Appleton examined the new geography of motorway construction as part of *The Geography of Communications in Great Britain* (Appleton 1962; see also 1960). More recently, Doreen Massey has described her memories of the A34, tracing the geographies of a road that 'is both local and global' and worked into people's lives in multiple ways: 'There have been many A34s in our lives' (Massey 2004; cf. Massey 2000, on the M1). There is a small, but significant, literature which has focused on the sociologies and geographies of driving along *particular* roads and motorways, from the Los Angeles freeways (Brodsly 1981; Katz 1999) and specific American parkways (Wilson 1992), to Malaysia's national expressway (Williamson 2003), India's roads (Edensor 2004) and various British motorways (Pearce 2000; Edensor 2003; Merriman 2003, 2004b, 2006b). What geographers, historians, sociologists and anthropologists have rarely done, however, is to examine how specific spaces of driving – particular roads and motorways – have been envisioned, planned, designed, constructed, landscaped *and* used (a notable exception here is Brodsly 1981).

Contents of the Book

In this book I examine how the first sections of England's London to Yorkshire Motorway (sometimes referred to as the London to Birmingham Motorway) were envisioned, designed, constructed, landscaped and used in the 1950s and 1960s. This was Britain's first *major* stretch of motorway, opened on 2 November 1959, eleven months after the opening of the 8¼-mile-long Preston Bypass section of the M6 (Britain's first motorway) in Lancashire on 5 December 1958. Prime Minister Harold Macmillan and the press had rightly celebrated the opening of the Preston Bypass as the launch of Britain's first motorway, but the opening of the first 72 miles of the London to Yorkshire Motorway (M1) attracted more widespread publicity and attention than the Lancashire motorway.[20] Here was *Motorway One*, which was far longer (and, indeed, wider) than the Preston Bypass. What's more, the M1 was a southern English motorway, located close to London and the offices and homes of the majority of national journalists, civil servants, politicians and influential cultural commentators. The M1 was easy to visit, observe, conduct experiments on and write about. The M1 was an exotic, distinctive, somewhat experimental space, and it emerged as a significant landscape/site of British modernity and post-war reconstruction, alongside the nation's new towns, expanding suburbs, tower blocks, new universities, schools and hospitals, and such prominent spaces and structures as Coventry Cathedral and the South Bank site of the Festival of Britain (on these and other sites of British modernity, see Saint 1987; Glendinning and Muthesius 1994; L. Campbell 1996; Matless 1998; Conekin et al. 1999; Bullock 2002; Conekin 2003; Gilbert et al. 2003). In recent years architectural historians, social and cultural historians and historical geographers (amongst others) have become increasingly interested in the histories and geographies of 1950s and 1960s Britain, and in this book I examine how the geographies of a linear motorway landscape refracted prominent attitudes and debates from the period, whether about modern design and architecture, scientific expertise and authority, the racism experienced by post-war immigrant labourers, or teenage consumption – to name just four examples.

As Britain's first motorways were constructed amidst the reconstruction programmes and consumer boom of the late 1950s and early 1960s it might appear judicious to conclude that they were a by-product of a war-time and post-war drive for economic, social and physical reconstruction. In chapter 2 I unsettle such a straightforward account, examining the attempts of politicians, influential motorists, industrialists, preservationists, engineers, landscape architects and road safety experts to promote motorway construction in Britain between 1900 and 1945. I examine the impact of the German National Socialist Party's *Autobahnen* on British attitudes to motorway construction. I show how planners, architects, engineers, motoring organizations and landscape architects argued that motorways would form an important component of Britain's post-war reconstruction, and I examine the debates which emerged between groups such as the Council for the Preservation of Rural England, the Roads Beautifying Association and the Institute of Landscape Architects about the design, landscaping and planting of Britain's roads and motorways in the 1920s, 1930s and 1940s.

In chapters 3 to 5 I examine how the landscapes of the M1 motorway were planned, designed, constructed, studied and used in the 1950s and 1960s. In the first of these chapters I examine debates surrounding the location, design, landscaping and planting of Britain's roads and motorways in the 1950s, focusing on the attitudes of landscape architects, engineers, architectural critics and government committees towards the design, landscaping and planting of the M1 and its service areas.

In chapter 4, I focus on the construction of the M1 in the late 1950s. I provide a critical examination of the narratives of construction that the contracting engineering firm John Laing and Son Limited presented to the public, future clients, company employees and local residents. I then examine a rather different representation of the construction of the M1: Ewan MacColl, Charles Parker and Peggy Seeger's one-hour folk Radio Ballad *Song of a Road*, which focused on the lives, work and oral traditions of the largely working-class migrant labourers and tradesmen whose biographies and geographies were largely overlooked in official accounts of the motorway.

In chapter 5, I examine how the M1 was used and consumed by motorists, commercial organizations, experts and the public in late 1959 and the early 1960s. I examine the attempts of politicians, journalists, the police and motoring organizations to predict and govern the conduct and movements of motorway vehicle drivers before and after the opening of the M1. I show how motorway driving was seen to produce distinctively new experiences, sensations, subjectivities and ways of being. I examine how the M1 was constructed and experienced as a space of modern consumption, catching the public's imagination, and becoming an important cultural reference point. I reveal how the motorway's service areas emerged as spaces of regulated consumption, and I discuss how the government's Road Research Laboratory approached the M1 as an experimental space of scientific inquiry, statistical calculation, accidents and death.

By the mid- to late 1960s the M1 was no longer seen to be new or unique, and motorways had become fairly familiar features of the English/British landscape. In the final chapter, chapter 6, I examine how attitudes to motorways and motoring changed in the late 1960s and 1970s, with the emergence of new environmental and conservation discourses, and as motorways were styled by writers and artists as dystopian and placeless landscapes. Finally, I argue that motorways are continually 'placed' through the practices and movements of millions of motorists, as well as motorway workers and local residents.

Chapter Two

Envisioning British Motorways

As Britain's first motorways were constructed amidst the reconstruction programmes and consumer boom of the late 1950s and early 1960s, it might seem appropriate to conclude that Britain's motorways are a by-product of this war-time and post-war drive for economic, social and physical reconstruction. In this chapter I seek to unsettle such a straightforward account by examining how politicians, motorists, industrialists, preservationists, engineers, landscape architects and road safety experts had been proposing the construction of motorways in Britain since the beginning of the twentieth century. In the first section I discuss attitudes to driving and Britain's road system in the first three decades of the century. The poor state of the nation's roads led a number of wealthy and influential motorists and business syndicates to promote the construction of private motorways in Britain. Motorways were identified as routes which could reinvigorate Britain's depressed agriculture and heavy industries, but critics started to question the financial viability of these private schemes and highlight the detrimental impact new motorways would have on the rural landscapes of counties such as Buckinghamshire, Surrey and Sussex. In the late 1920s and 1930s the roads lobby looked abroad for positive precedents, and in section two I show how the German National Socialist Party's *Autobahnen* had a profound impact on British attitudes to motorway construction in the inter-war years. Motorways became associated with speed, safety, beauty, national defence and, conversely, extravagance, militarization and dictatorships, and the politics and aesthetics of the German *Autobahnen* were frequently projected onto British proposals for motorway construction. In section three I examine the motorway proposals which emerged in late-1930s Britain, and in section four I trace the emergence of a professional consensus in war-time Britain which asserted that motorways should form a part of post-war reconstruction. Planners, architects, engineers, geographers, motoring organizations and landscape architects argued that motorways should form an important component of local, regional and national plans, but although senior civil servants saw their benefits, the government refused to commit to a definite programme of works. In the final section I examine debates over the appearance of Britain's roads and the conduct of motorists in the countryside, which surfaced in the inter-war years. The formation of groups such as the Council for the Preservation of Rural England in 1926, the Roads Beautifying Association in 1928 and the Institute of Landscape Architects in 1929 led to an increasing concern with the design, landscaping and planting of Britain's roads and motorways. Disagreements emerged between different organizations about the style of planting on Britain's roads, and these debates prefigured later discussions about the landscaping and planting of the M1 motorway in the 1950s, which I discuss in chapter 3.

Motoring and the Motor-Car Way, 1896–1930

The 1896 Locomotives on Highways Act – which allowed motorists to travel at speeds in excess of 4 mph, without being preceded by a flag bearer – helped to fuel a growing interest in motor cars and motoring amongst an array of wealthy, largely male late-Victorian and Edwardian aristocrats and entrepreneurs (Plowden 1973; O'Connell 1998). Motorists began to explore roads which had fallen into varying states of disrepair following the development of the railways in the 1830s, the decline of the stage-coach industry during the 1840s and the closure of the last turnpike trust in 1895. Road users reacted to the arrival and speedy progress of motor cars with a mixture of shock, bemusement, fear and excitement, while medical experts expressed concern, and local residents and farmers annoyance, as bodies, crops and homes were regularly engulfed in clouds of dust. Local police officers and magistrates treated motorists – particularly 'scorchers' (those speeding) – with varying degrees of severity (Emsley 1993). The authorities in Surrey, in particular, gained a reputation for their harsh treatment of offending motorists (Emsley 1993), while the growing number of speed traps led to the formation of the Automobile Association in 1905 – their first patrols (or scouts) warning motorists of the location of police patrols on the London to Brighton and London to Portsmouth roads (Barty-King 1980). In the 1890s, politicians, some of whom were themselves keen motorists, began to pay increasing attention to what was commonly termed 'the motor problem', and in 1903 parliament passed legislation outlining the law on dangerous driving, and requiring the licensing of drivers and the registration and numbering of vehicles (Plowden 1973; Emsley 1993). The growing popularity and impact of motoring also led a broad array of cultural commentators to celebrate, criticize or ambivalently remark upon the visual spectacle, experiences, etiquettes, dangers, speeds, economic potential and nuisances of motoring – the best known fictional, Edwardian commentaries on motoring being in E. M. Forster's *Howards End* (1910) and Kenneth Grahame's *The Wind in the Willows* (1908) (see O'Connell 1998; Thacker 2000).

With increases in car ownership and growing complaints about the presence of private motorists on poorly surfaced public roads, discussions began to focus upon the possibility of constructing new roads reserved for motor vehicles. Speaking in the House of Commons on 17 May 1900, Arthur Balfour – the future Prime Minister, Conservative Leader of the House, keen motorist and well-known 'scorcher' – stated that he often dreamt that 'we may see great highways constructed for rapid motor traffic, and confined to motor traffic, which would have the immense advantage, if it could be practicable, of taking the workman from door to door' (*Parliamentary Debates* 1900: 519). In January 1902, *The Automotor and Horseless Vehicle Journal* argued that the capabilities of the automobile could only be 'fully appreciated' after 'the construction of experimental motor roads' (*The Automotor* 1902: 138). Later that same year, the future-gazing socialist essayist and novelist H. G. Wells suggested that Britain's future road network should include 'specialized ways restricted to swift traffic' (Wells 1902: 9), while in August 1902 the first routes were actually proposed for the construction of motorways. Writing in *The Nineteenth Century*, a British civil engineer, B. H. Thwaite, advanced a call for the construction of a special toll motor-car and cycle way 'from London, through the centre of England', to Carlisle and Scotland (Thwaite 1902a: 306). Thwaite argued that a specially constructed roadway, with woodblock surface, could increase vehicle speeds and avoid the dust problems of conventional roads. At a time when British manufacturing firms were suffering from international competition, and rural areas were blighted by depression, Thwaite promoted his motor-car way as a catalyst to 'the agricultural and manufacturing interests of the country', particularly the English motor industry (Thwaite 1902a: 307). The proposals received widespread coverage in the motoring press, and John Montagu of Beaulieu – the motoring enthusiast, Conservative MP and prominent member of the Automobile Club[1] – published Thwaite's more detailed plans for a London to Birmingham motor-car way in his upmarket, weekly, glossy magazine *The Car (Illustrated): A Journal of Land, Sea and Air* (Thwaite 1902b, 1903; Beaumont 1903). Thwaite described how this arrow-straight four-lane 'carway' would stretch from Edgware (north of London) to Bickenhill (south-east of Birmingham); drawing in commercial vehicles and private cars from both cities and the numerous agricultural towns along the route. In an editorial accompanying one of the articles, Montagu expressed doubts about the viability of Thwaite's proposals (Montagu of Beaulieu 1902). As a politician, he was aware that any legislation might be blocked by powerful railway interests, while, as a motorist, he believed that a London to Brighton motor road would attract higher levels of traffic than a London to Birmingham motorway (Montagu of Beaulieu 1902, 1903). Indeed, Brighton was Britain's largest seaside resort at the turn of the century, and a large number of motorists were already setting out along existing roads to experience the delights of this popular town (Shields 1991; Walton 2000).

In November 1905, *The Motor Car Magazine*, edited by John Montagu, published lengthy extracts from a pamphlet *promoting* plans for a London to Brighton Motorway, and *advocating* the construction of 'a great system of motor roads . . . linking all the principal centres of the United Kingdom' (quoted in Montagu of Beaulieu 1905a: 402). The motor road, stretching from Norwood to Preston Park, was intended to be the first in a national network of motorways: providing swift travel for motorists; improving the conditions on ordinary roads for non-motorists; and having 'great strategic value' for the military (quoted in Montagu of Beaulieu 1905a: 402). A past president of the Institution of Civil Engineers, Sir Douglas Fox, prepared designs for the motorway, and the promoters aimed to introduce a parliamentary bill in 1906 that would enable construction (Montagu of Beaulieu 1905a, 1905b). On 10 January 1906, *The Car (Illustrated)* reported concerns about the ability of the promoters to raise the £3.5 million needed (Montagu of Beaulieu 1906a), while later that same month Montagu suggested that the Bill was unlikely to get through parliament, concluding that 'schemes to the north and west of England from London have, as far as I can judge, a much better chance' (Montagu of Beaulieu 1906b: 357). The Bill was eventually withdrawn.

In 1909 the Liberal government acknowledged the importance of the nation's roads by introducing a series of vehicle and petrol taxes which were intended to lift the financial burden of road improvements from local authorities and the Treasury (Plowden 1973). As part of these administrative changes, and amidst the furore caused by the increase in land taxes and death duties announced in his annual budget, Chancellor of the Exchequer David Lloyd-George introduced the Development and Road Improvement Funds Bill: proposing the establishment of a central Road Board that would be empowered to maintain roads and construct new highways limited to motor traffic (*Parliamentary Debates* 1909). Members of Parliament expressed widespread opposition to the Board's proposed powers to construct new motor roads, and what is surprising is that motoring groups such as the Royal Automobile Club supported this view – as they feared that the construction of special motorways might lead to the exclusion of motor cars from ordinary roads (Plowden 1973). This provision was eventually dropped from the Bill (see Development and Road Improvement

Funds Act 1909), but despite building no new roads, the Road Board brought considerable improvements to the nation's highways throughout the 1910s, tarring many existing routes.

The sale of cheaper, mass-produced models from the early 1910s, coupled with falling prices, the increasing popularity of hire-purchase agreements and an expanding second-hand market during the 1920s, brought car ownership within reach of the middle classes and, by the late 1930s, well-paid working class families (O'Connell 1998). The Ministry of Transport was formed in 1919, incorporating the Road Board, and by August 1923 the number of registered vehicles had topped one million (at 1.1 million, including 383,525 cars) (Plowden 1973). Early that same year John (now Lord) Montagu of Beaulieu threw his backing behind proposals to construct a private motor road between London and Birmingham (see Northern and Western Motorway 1923a). The promoters of the Northern and Western Motorway included the Earl of Radnor, the Chairman of Eagle Star Insurance (Sir Edward Mountain), the aircraft manufacturers Armstrong Whitworth, and the Shell Oil Company, and by July 1923 the syndicate had broadened their horizons, publishing plans for a 226-mile motorway, built in four stages, between Uxbridge (west of London, at the end of the projected Western Avenue) and Birmingham, Salford and Liverpool (Northern and Western Motorway 1923a). The syndicate's brochure described the poor state of the country's roads, before emphasizing the economic benefits which would accrue to traders, private motorists and local authorities following construction (Northern and Western Motorway 1923a; Montagu of Beaulieu 1923a). The motorway would raise the value of local land, encourage frontage development, help to reduce the cost of repairing ordinary roads, increase speeds, and save road hauliers up to ½d. per ton per mile (after a toll of ½d./ton/mile). Agriculture and other industries were expected to benefit from the improved transport links, while the motorway's construction was promoted as an important job-creation programme at a time of high unemployment – a strategy which was also being employed by the Ministry of Transport (Northern and Western Motorway 1923a; Jeffreys 1927a). The syndicate estimated that 14,000 men would be directly employed during construction, and another 26,000 indirectly employed (Northern and Western Motorway 1923a). The consulting engineers, Whitley and Carkeet-James, prepared cross-sectional drawings of the motorway and its junctions, and Montagu commissioned an artist to depict cars and goods vehicles travelling along a closely fenced, four-lane, single-carriageway motorway (Figure 2.1). The syndicate's proposals received mixed reactions amongst journalists, politicians and the public, and railway supporters, in particular, challenged the need for such a road. Montagu, as Chairman of the Northern and Western Motorway, was careful to stress that the motorway would complement rather than replace the nation's

Figure 2.1   An artist's impression of the Northern and Western Motorway, commissioned by Lord Montagu of Beaulieu, c.1923. Reproduced by courtesy of the National Motor Museum, Beaulieu.

railways (Montagu of Beaulieu 1923a, 1923d), while he used the railways as an example of how privately funded transport schemes could revive trade and halt depression (Montagu of Beaulieu 1923b). In the summer of 1923 the syndicate made a strategic decision to survey the sections between Coventry and Salford, which were deemed to pass through the areas with the highest levels of unemployment and 'distress' (Montagu 1923c: 44).[2] Local authorities were highly supportive, as this private road would not cost them a penny, and the potential to transfer goods between the Manchester Ship Canal at Salford and the West Midlands using the motorway was seen to provide a strong justification for the massive levels of public funding requested (up to 75% of construction costs)(Northern and Western Motorway 1923b). Local industrialists also supported the motorway, and the route was moved closer to Newcastle-under-Lyme after the syndicate received vocal support from Josiah Wedgwood and other influential figures in the Potteries (Tritton 1985).

Writing on 'the future' in his 1923 book *The Road*, the writer, historian and poet Hilaire Belloc made reference to the syndicate's proposal, which at the time of writing had 'very nearly materialized and . . . was on the point of becoming law' (Belloc 1923: 196). Belloc knew that there were 'considerable social and political obstacles' to the construction of such motor roads, but he insisted that they were vital to ensure the provision of 'communications corresponding to modern needs' (Belloc 1923: 198, 194). The syndicate's Motorways Bill received its first reading on 15 November 1923, but the dissolution of parliament for the General Election in December 1923 led to its swift demise (*Parliamentary Debates* 1923). In March 1924 the government's Trade Facilities Act Committee announced that a failure by the motorway syndicate to raise adequate funds meant that the Committee were 'not prepared . . . to give any [financial] assistance' (*Parliamentary Debates* 1924a: 2138), and although the Bill was reintroduced with some success in April 1924 (*Parliamentary Debates* 1924b) it became a victim of another General Election in October 1924 (Blizard 1926; Tritton 1985).

During 1924 Lord Montagu turned his attention to other transport schemes, including a proposed Thames Express boat service (Tritton 1985), but the parliamentary agent for the Northern and Western Motorway syndicate, George Pearce Blizard, was still campaigning for a motorway from 'the Midlands to the Mersey' in October 1929 (*Manchester Guardian* 1929; see also Blizard 1926). In 1925 Blizard had become chartered secretary to London and South Coast Motorways Limited, but while earlier proposals for a London to Brighton Motorway had been criticized for their financial viability, this scheme was *also* criticized by prominent individuals concerned with preservation and amenity issues. In a letter to *The Times* in April 1925, Lawrence Chubb, Secretary of the Commons and Footpaths Preservation Society, expressed concern that this motorway would bisect a number of rights of way and 'famous Surrey commons' (Chubb 1925: 15), while S. D. Adshead, Professor of Town Planning at University College London, saw the motorway as an unwelcome urbanizing force which would 'penetrate the most secluded parts of Surrey and Sussex' to the dismay of rural residents and travellers alike (Adshead 1925: 15). This was a time when planners and preservationists were becoming increasingly concerned about the effects of unplanned growth, and Adshead would go on to serve on the Executive Committee of the Council for the Preservation or Rural England, which was set up by Patrick Abercrombie and like-minded planners in 1926 to promote the planning *and* preservation of the English countryside. In the opening of his letter to *The Times* in 1925, Adshead stressed the irony of the motorway proposals emerging just days after 'the excellent letter [to *The Times*] by Professor Abercrombie bewailing the encroachment of town on country' (Adshead 1925: 15; cf. Matless 1998).[3] Abercrombie and Adshead agreed that town and country should be kept separate and distinct, but while Abercrombie and other planner-preservationists believed that modern roads and housing could, if well planned, fit into rural surroundings (Matless 1998), Adshead's letter reveals a general hostility to development of *all* kinds in rural Sussex and Surrey – which would see a conversion of 'agricultural land into building land'

(Adshead 1925: 15). This attitude was most clearly revealed in January 1929, when Motor Roads Development Syndicate Limited introduced the Southern Motor Road Bill into parliament. Adshead penned another letter to *The Times*:

> . . . there may be occasions when such speed ways will one day take a legiti-
> mate part in connexion with the transport system of the country; but there
> are certain areas through which they ought never to be allowed, and it is
> doubtful if the time is ripe for their introduction anywhere. . . . It would mean
> the development of a corridor of suburbanism extending down every side
> road. . . . At a time when regional planning is aiming at the preservation of
> wide areas of primitive scenery, . . . ought we not to hesitate and consider
> very seriously whether or not adjoining South Coast pleasure resorts would
> not in the long run be seriously damaged by connecting themselves to London
> with speedways, which would only accelerate the already rapidly diminishing
> seclusion of the rural scenery with which they are now surrounded and behind
> which they now repose? (Adshead 1929: 8)

Town and country should be kept distinct. Distinct towns should be cushioned by preserved countryside. Suburbanization must be avoided (Adshead 1929). Other objectors highlighted the errors of helping private companies to develop and despoliate the countryside. *Laissez-faire* emerges as the enemy of the preservationist, planner and local campaigner alike (Matless 1998). The prominent horticulturist and Surrey resident Sir William Lawrence expressed his faith in public authorities, but not private companies, to 'carry out their works with as little interference with the amenities of the countryside as possible' (Lawrence 1929: 12). Surrey and Sussex councils also opposed the scheme, expressing concern that they might have no powers to control these spaces: '. . . if no power of public control is given the fences may become lines of advertisement hoardings' (*The Times* 1929a: 6; cf. Matless 1998). The Sussex Agricultural Association, MPs, local authorities and two highly organized protest groups opposed the scheme on amenity, planning and financial grounds, and, as with earlier proposals, the Bill was withdrawn amidst rumours of financial difficulties and in view of a forthcoming General Election (Buxton 1929; Carkeet James and Douglas Cooper 1929; *The Times* 1929a, 1929b). The motorway promoters appeared very quiet during the depression years of the early 1930s; a time when there was little hope of gaining parliamentary support for the construction of private motorways at the public's expense. The British government did instigate the construction of a number of arterial roads and bypasses during the 1920s and 1930s – including the Great West Road, North Circular Road, Croydon Bypass and Kingston Bypass – but only 127 miles of new road had been built by 1927 (Jeffreys 1927a). The construction programme was cut back following the financial crisis of 1931, and those roads which were completed were not the motorways demanded by private promoters (Jeffreys 1949).

The German *Autobahnen*: The Politics and Aesthetics of a Nation's Roads

During the 1920s and 1930s, attention began to turn to the motorway construction programmes being pursued abroad. The extensively landscaped American parkways caught the attention of British lobbyists and landscape architects during the 1920s and 1930s (*Transport Management* 1927), but it was the construction programmes and promotional exercises of two of Europe's Fascist regimes which were to have the greatest impact on thinkers in Britain (Matless 1990). In 1926 Mussolini's government hosted the fifth International Road Congress in Milan and Rome, and prominent British lobbyists, civil servants and journalists – including William Rees Jeffreys[4] and George Blizard (Roads Improvement Association) and Colonel Charles Bressey (Chief Engineer of the Ministry of Transport) – attended the Congress and toured the *autostrade* (Blizard 1926, 1928; Jeffreys 1927b, 1949). In 1929 Jeffreys met Italian government officials and inspected further sections of *autostrade*, but just four years later he and many others turned their attention to the extensive network of motorways being constructed by the National Socialist Party in Germany.

The German National Socialist Party commenced the construction of a national network of 4300 miles of *Autobahnen* in 1933 (Shand 1984; Dimendberg 1995). *Autobahn* construction was promoted as the key to economic stability and job creation at a time of high unemployment, while the comprehensive reach of this future national network was seen to 'represent the most vivid expression of the unshakeable unity of the Reich' (Todt 1938: 274; see also Boyd Whyte 1995). The *Autobahnen* served as highly visible, concrete and far-reaching symbols of the 'New Germany' and Nazi ideals, and these new motorways – along with the 1936 Olympic Games and annual Nuremberg rallies – became an important focus for German propaganda abroad, as well as at home (Griffiths 1983; Dimendberg 1995). Writing in the 1938 English-language propaganda book *Germany Speaks* – in which '21 leading members of party and state' discussed a range of foreign and domestic issues – Hitler's General Inspector of German Roads, Fritz Todt, described how the 'great importance of the Reich motor roads has long since been recognised in foreign countries' (Todt 1938: 274).[5] The Third Reich had hosted the International Automobile Show in Berlin in 1933, the seventh International Road Congress in 1934 and the second International Congress for Bridges and Overground

Structures in 1936 (Todt 1938; Sachs 1992). Documentary films and technical booklets were produced in English and German, foreign dignitaries and inspection parties were invited to view the *Autobahnen*, celebratory articles appeared in the National Socialist journal *Die Straße* (The Road), and inauguration ceremonies by Adolf Hitler and senior party figures were widely publicized (Sachs 1992; Dimendberg 1995). The government issued postage stamps of the *Autobahnen* in 1936 (Sachs 1992), and all of these events and media worked to create a spectacle of the *Autobahnen*, promoting National Socialist ideals and achievements at home and abroad (Shand 1984; Dimendberg 1995; Boyd Whyte 1995; Rollins 1995).[6]

In 1937 the German government invited Britain's two largest motoring organizations, the Automobile Association and Royal Automobile Club, to lead a delegation of experts to inspect Germany's roads. The AA and RAC created an organizing committee with representatives of the British Road Federation and the Parliamentary Road Group, and they formed a 224-strong party – the 'German Roads Delegation (1937)' – who toured Germany for ten days between 24 September and 3 October 1937 (*The Times* 1937a). The Delegation included journalists, planners, landscape architects, representatives of motoring organizations and the motor trade, 57 members of both houses of parliament, and 88 county councillors and county surveyors (*The Times* 1937a, 1937b). The German government used the visit as an opportunity to promote the design, landscaping and construction of the *Autobahnen*, but they also used it to promote the broader 'engineering' of the German nation – architecturally, socially and politically. Alongside tours of 545 miles of *Autobahnen* and 280 miles of ordinary highways, the Delegation were entertained by German politicians, local authorities and motoring organizations, who took them to the Munich Oktoberfest, Leipzig Opera House, Heidelberg Castle and the Olympic Stadium in Berlin (*The Times* 1937c). The report of one delegate, Malcolm Heddle (County Surveyor of Aberdeen), describes the numerous tours and functions they attended, including a 'canteen lunch in an Autobahn workers' camp', a 'special steamer sailing down the Rhine' and a rally on Tuesday 28 September: 'Left Nuremberg by special train for Berlin at 8.45a.m. to witness demonstration where Signor Mussolini and Herr Hitler were both present. Left Berlin at 10.45p.m. by special train for Munich.'[7] The German government embroiled the visitors in a propagandist performance of the nation. Germany was presented as a politically, aesthetically and technically modern state that respected long-established traditions. Modern motorways, modern stadia, modern engineering techniques, modern Fascist political ideologies, and traditional values, festivals and historic sites become juxtaposed and aligned.

At the time of the Delegation's visit, British politicians were becoming increasingly alarmed at the rate of Germany's rearmament and at its repeated calls for the acquisition of an empire (Griffiths 1983). With these concerns, members of the German Roads Delegation attempted to defend their visit to the press. Upon his departure Sir Stenson Cooke, Secretary of the AA, explained that 'the first and only object of the tour was to study the technical achievements exemplified in the new German roads' (*The Times* 1937b: 8). Cooke was quite clear that the observation and appreciation of Germany's modern motorways could be separated from an appreciation of German national politics. Other commentators and members of the Delegation appear to have been less able to make this distinction – in some cases this was due to their personal politics and admiration for Nazi ideals, in other cases this was due to a conviction and concern that the *Autobahnen* were military achievements. The Delegation included three politicians who served on the council of the pro-German (but not wholly pro-Nazi) Anglo-German Fellowship: Lord Eltisley (Chairman of the County Councils' Association and a former Conservative MP), Norman Hulbert (Conservative MP and member of the RAC) and Sir Assheton Pownall (Unionist MP). Two other delegates – Conservative MPs Edward Keeling and Alfred Denville – openly supported General Franco's actions in the Spanish Civil War, while one of the journalist delegates, L. A. Mohan of the magazine *Modern Transport*, had stood as a candidate for the British Union of Fascists in 1936 (Griffiths 1983). In December 1937, another delegate, Frank Clarke MP, penned an article on the Delegation's trip for the pro-German, and frequently pro-Nazi, *Anglo-German Review* (F. Clarke 1937; Griffiths 1983). Clarke stressed that he was 'impressed not so much by the National Socialism as by the national discipline', particularly the 'personal adoration' of the working classes towards their leader, the absence of 'lounging and lipsticks', and the dignity of the *Autobahn* workers (F. Clarke 1937: 9). Clarke's praise extended to the design, construction and landscaping of the *Autobahnen*, but he made it clear that the 'ruthlessness of war-time necessities', which Hitler had brought to *Autobahn* construction, was 'incapable of reproduction in Britain' (F. Clarke 1937: 9). Whether he regretted this lack of British 'ruthlessness' is unclear.

Writing on 'The new German motor-roads' in *The Geographical Magazine* in January 1938 – two months prior to the *Anschluss* with Austria – Alan Brodrick was quite clear that nationalist politics and 'military considerations had played an important part' in the design of the *Autobahnen*:

> The roads which follow the frontiers lie well back from them so as to be beyond the range of artillery fire. . . . The Autobahn network may possibly be completed by the end of 1940, for more and more men will be transferred to road-work as the rearmament of Germany progresses towards its completion. These great roads are, indeed, as much an essential part of the

rearmament programme as is the motorization of the army; for they are absolutely necessary if the new army is to be the instrument for rapid and decisive action which the present-day rulers of Germany wish to make it. (Brodrick 1938: 209)

Brodrick's fears were confirmed in March when German troops crossed the border with Austria, and a special correspondent of *The Times* observed 'a long train of lorries and armoured cars and soldiers on motor-cycles' congesting the *Autobahn* between Munich and the border near Salzburg (*The Times* 1938f: 12). Twenty-seven days later Hitler inaugurated 'the construction of the Austrian network of Reich motor roads' (*The Times* 1938g: 15). In November 1938, following the installation of a 'new Czechoslovak government with fascist tendencies', the German government commenced the planning and construction of an *Autobahn* across Czechoslovakia, which was intended to unite the Austrian capital Vienna with eastern Germany and Berlin (Pospisil 1982: 396). These *Autobahnen* were important devices to help enable and consolidate German imperial and military power, providing a concrete symbol of Germany's colonial ambitions and expansion.[8]

With hindsight, historians have emphasised that the *Autobahnen* were not designed largely for military traffic, but foreign observers frequently associated *Autobahn* construction with the rearmament and expansion plans of an undemocratic nation (Shand 1984; Boyd Whyte 1995; cf. Rollins 1995). As the editor's summary preceding Brodrick's January 1938 article explained:

> Dictatorship expresses itself naturally in grandiose public works, of which the German *Autobahnen* afford a striking example. Democratic governments, subject to critical opposition and accountable for every item of expenditure, cannot afford to burden the exchequer from motives of self-advertisement, still less for the sake of unavowed aims. . . . Fascinating, therefore, as the new German motor roads undoubtedly are to a people so motor-conscious as our own, the extent to which they offer a model to be imitated in Great Britain can only be assessed after full consideration of the circumstances in which they were built; and for this purpose Mr Brodrick's article supplies valuable material. (*The Geographical Magazine* 1938: 193)

Motorways become associated with extravagance, self-advertisement and the military manoeuvrings of a dictatorship. Britain, as a democracy committed to economic prudence and austerity, simply could not afford them (Brodrick 1938).

Other commentators took a different approach. If the *Autobahnen* were a prominent architecture in Germany's preparations for war and expansion, a network of British motorways and a British national roads authority could be established for purposes of national defence. During parliamentary debates on the Trunk Roads Bill in December 1936, Lord Parsonby of Shulbrede expressed concern that the Bill – which would bring 4500 miles of through-roads under government control – was simply part of Britain's plans to rearm in response to events abroad:

> Their intention may be to increase the flow of commercial traffic so that these [trunk] roads may be used for the great commercial vans and trolleys which are increasing greatly on the roads. But I believe most of all that this [the 1936 Trunk Roads Bill] is really part of the policy of what is euphemistically and very inaccurately called 'defence', driving big roads from all centres to the ports, owing to the enormous importance of a free flow of traffic from munitions factories, from aeroplane factories and from all the various military centres to the coast should the war break out. (*Parliamentary Debates* 1936: 666)

Lord Parsonby felt that military strategies were being disguised by a language of economic planning and progress, but as Britain became increasingly aware of Germany's military intentions, commentators spoke with some urgency about the role of good roads in the government's '*emergency* defence programme' (*Roads and Road Construction* 1938: 355). Road improvements and road construction programmes became inserted into a hierarchy of national defence initiatives. Some planners saw roads as relatively unimportant compared with the provision of 'adequate air-raid shelters' (*Roads and Road Construction* 1938: 355). Other commentators argued that good roads would be vital for efficient evacuation and the transporting of domestic and military goods. As Geoffrey Boumphrey (1939: 158) put it, 'it would be foolish to argue that our unprecedented expenditure on armaments should deter us from embarking on this equally necessary and truly creative work'. Politicians and lobbyists continued to enfold new roads and motorways into debates about national and international geopolitical strategies throughout World War II and into the post-war, cold-war years (see BRF 1951a). Nevertheless, the majority of professional commentators continued to admire foreign motorways in isolation from their political context.

The German Roads Delegation forwarded their final report to the Minister of Transport in January 1938, expressing their 'firm opinion that a national scheme should be framed without delay for a series of motorways, assigned for execution in an ordered sequence, as supplemental to the existing highway system' (GRD 1938: 12; *The Times* 1938a). The Delegation's report appears to have drawn extensively upon information provided by Fritz Todt and the *Reichsautobahnen*, using German data to highlight the low accident rates and individual benefits associated with the motorways.

Motorway construction was seen to reduce congestion, decrease vehicle running costs, reduce accidents, and be cheaper than modifying existing roads (GRD 1938). German propaganda helped support the Delegation's case for the construction of British motorways, but the Delegation's visit and final report was also used in German propaganda promoting the *Autobahnen*. In *Germany Speaks*, Fritz Todt stated that he was 'proud of the complimentary remarks' in the German Roads Delegation's final report, and was pleased that further opportunities to 'strengthen the relations so happily inaugurated between German and British road experts' had emerged (Todt 1938: 275–6). Todt led a return delegation to study 'London's traffic problems' in November 1937, while one of the British delegates, Professor R. G. H. Clements of Imperial College London, made plans to take his students to Germany in late 1938 (Todt 1938: 276).

R. G. H. Clements was a key member of the German Roads Delegation, as well as a respected academic and member of the Ministry of Transport's 'Experimental Works on Highways (Technical) Committee'. He was one of four men representing the RAC in the Delegation, and Charles Bressey later described him as occupying 'the only Chair of Highway Engineering in the British Empire' (Bressey 1944: 42). On his return from the tour Clements wrote *The System of Motor Roads in Germany: A Record of Facts and of Technical Details*, which the Delegation published in October 1937. The booklet was styled as a record of 'the actual facts touching specific phases of the problem and its solution' to enable individual members of the Delegation to draw their own conclusions (Clements 1937: 5). The booklet contained sections on 'concrete', 'bituminous surfaces' and 'constants of design', but it was the sections on 'safety features', 'æsthetics' and 'labour' which were felt to outline many of the primary advantages of the *Autobahnen*. Clements argued that the *Autobahnen* provided high levels of employment, prevented ribbon development, encapsulated principles of good design and landscaping, and reduced the number of accidents by 83% (Clements 1937). The accident statistics were provided by the German government and proved to be of great interest to members of the Delegation, which included representatives of the Pedestrians' Association and National Safety-First Association, both of whom had advocated motorway construction to improve road safety in the mid-1930s (*Roads and Road Construction* 1935). Safety, efficiency and beauty become entwined in a rhetoric of good design based on 'simple and logical' principles, with dual carriageways, 'clean and bold' alignment, clear and minimal signage, super elevated curves, minimization of access and crossing points, and maintenance of a constant traffic flow, combining to create a safe, efficient, economical and well-designed motorway (Clements 1937: 10, 12).

Foreign visitors frequently marvelled at the architecture, engineering and landscaping of these modern *Autobahnen*. Clements remarked that,

'By general consent no feature of these roads aroused greater interest than the sheer clean beauty of the work accomplished' (Clements 1937: 24). He celebrating 'the vigorous sweeping curves of the alignment', the incorporation of local landscape features into landscape designs, the careful moulding of cuttings and embankments, and the absence of roadside advertising and telegraph poles (Clements 1937: 24). The Delegation's final report provided a much firmer outlining of the benefits the German *Autobahnen* and future British motorways could bring to the landscape: 'Works of this character, designed to definite standards and with appropriate treatment, sited on lines away from centres of population, would best conserve the amenities of the countryside' (GRD 1938: 15). The Delegation explained how the German government had employed landscape experts to design aesthetically pleasing, gently curving, modern high-speed motorways, but as a senior member of the Delegation pointed out in *The Times*, serpentine roads could have strategic as well as aesthetic advantages:

> ... a properly constructed motor road follows the configuration of the country; . . . and 'marries' (to use a term of golf-course architecture) with the scenery. It is important from the point of view of military and of civic society that the road should not be unduly straight. A perfectly straight road is much more vulnerable to aerial attack than one which winds a serpentine course, and a road varied by gentle curves is easier and more restful to drive along than a dead straight line, which becomes dangerous on account of the soporific effect of monotony. (Wolmer 1937: 15)

While military strategies may not have been the prime motivation for the construction of the *Autobahnen*, Lord Wolmer and other observers were quick to draw conclusions. Planning and design become associated with military intentions and advantage, but the politics of the German National Socialists also became entwined with the aesthetics of the *Autobahnen* in less obvious and less spectacular ways. Fritz Todt explained how the intention was that the *Autobahnen* would be 'absolutely adapted to the German landscape', encapsulating an 'artistic feeling and a love of Nature' and reflecting the 'deeper and spiritual movement of the National Socialist revolution' (Todt 1938: 272–3). In early 1934, Todt approved the employment of a team of *Landschaftsanwälte* (landscape advocates) to coordinate the design, landscaping and planting of the *Autobahnen*. The *Landschaftsanwälte* were heavily influenced by the ecological, anthropocentricist and regionalist ideas of the early twentieth century *Heimatschutz* ('homeland protection') movement (Rollins 1995), but the close association of the *Landschaftsanwälte* with the National Socialist Party makes it futile to attempt to dissociate the landscaping policies employed on the *Autobahnen* from Nazi ecological, racial and nationalist ideals (Gröning 1992; Dimendberg 1995; Matless 1996, 1998).

It is important to point out that the *Landschaftsanwälte* advocated certain planting policies which were being widely practised in Britain at this time. R. G. H. Clements praised the Germans for planting the *Autobahnen* with 'trees, herbs and grasses native to the district' (Clements 1937: 25), and a focus on vegetation that was native to a local area, region or nation was commonly justified in Britain using ecological and aesthetic criteria that emphasized that only 'native' species would survive or look right in particular surroundings. The difference was that a number of the German *Landschaftsanwälte*, including the most senior landscape advocate, Alwin Seifert, were members of the Nazi Party, and in several plans they espoused an exclusionary, nationalist and racialized planting and landscaping policy (Gröning 1992; Rollins 1995; Dimendberg 1995). Seifert is said to have 'freely used the jargon of the racist Right', and although his ecological and aesthetic idea of *bodenständige* ('rooted to the soil') cannot simply be equated with the racist ideology of *Blut und Boden* ('blood and soil'), he drew frequent links between the health of landscapes and peoples, and he called for the 'Germanizing' of the Polish landscape following its accession to Germany (Rollins 1995: 503, 509, 511). As British commentators tended to consume the spectacle of the *Autobahnen* through Nazi Party propaganda, these more sinister policies were frequently hidden from view. It is only more recently that the writings and planting policies of the *Landschaftsanwälte* have come under extensive criticism, and they were easily overlooked by foreign observers commenting on the design and landscaping of the *Autobahnen* in the 1930s.

Motorways for Britain? National Plans, National Defence

The German Roads Delegation visit had a significant impact on the views of many of its influential delegates. Before the tour, motorway promoters were still advocating the construction of individual sections of motorway, which it was suggested would improve the economic geographies of specific regions. After the tour, the Delegation's organizers and an increasing number of delegates asserted that a *national network* of motorways must be planned. The Delegation's conclusions differed from the prevailing attitudes of a number of its member institutions, most notably the Automobile Association. During the late 1920s and 1930s, the AA had campaigned for the improvement of Britain's main arterial roads, supporting the centralized control of 4500 miles of through-routes under the 1937 Trunk Roads Act, but they were less keen about the idea of constructing an entirely new system of motorways (Barty-King 1980). In contrast, the British Road

Federation had paid very little attention to the condition of Britain's roads at all. The Federation were formed in 1932 to campaign against government restrictions on commercial vehicles – which were designed to limit their competition with the railways (see BRF 1933). In 1937 the Federation broadened their remit, highlighting the poor state of the nation's roads. In March 1937 they published *Get a Move On*, urging the government to develop 'an active and progressive road construction policy' and build roads 'fit for the traffic of the future' (BRF 1937c). Photographs of the German *Autobahnen* made it clear what the Federation desired, and in August 1937 they outlined their case in the booklet *National Motor Roads*: 'In every modern state there is a growing realisation that patchwork remedies will not do and that instead there must be a bold replanning of the national road systems' (BRF 1937a: 8). Modern motor roads must lie at the heart of every modern state, whatever its politics, improving the convenience of road transport, increasing safety and aiding industrial efficiency. What's more, 'developments in the science of warfare' make motorways 'imperative as part of a general scheme of National Defence' (BRF 1937a: 4). War would create increased traffic flows and render junctions vulnerable to aerial attack, and Britain must follow 'those continental countries' that were clearly aware of the strategic value of motorways:

> ... in addition to the movement of troops, military vehicles and munitions, the extent to which this country relies on imported foodstuffs, makes it essential that the ports, and especially the West Coast ports, should be linked up with the great centres of population by adequate roads running straight from the docks where the cargoes are discharged. (BRF 1937a: 5)

The British Road Federation engaged with a range of debates and fears in an attempt to muster support for motorway construction, and in mid- to late 1937 they publicized a scheme to construct 2108 miles of new motorways linking Britain's main ports – including Glasgow, Holyhead, Cardiff, Southampton, Bristol, London, Hull and Plymouth – with other key centres of industry and population – such as Birmingham, Manchester and Sheffield (see BRF 1937b; Wolmer 1937). Motorway construction would be focused on Britain's strategically important and less vulnerable central and western regions, and no motorways would be built in the nation's attack-prone eastern and south-eastern extremities, such as Kent, Sussex and East Anglia. The geographies of motorway planning and construction would have to be adapted to war-time strategic needs.

During 1937 organizations intensified their calls for the construction of British motorways, and Minister of Transport Dr Leslie Burgin warmed to their proposals following his own exposure to the spectacle and propaganda associated with the German *Autobahnen*. Burgin had met Fritz Todt during his tour of London's roads in November 1937 (Todt 1938). In January 1938 Burgin's family spent five days touring Germany and viewing the *Autobahnen* as 'guests of the German government' on their return from a Swiss skiing holiday. Burgin's notes on the trip reveal that he journeyed on an *Autobahn* 'from Munich to the Austrian frontier and back' in 'a high power Mercedes Benz, sitting at the side of Dr Todt', and his account of this 'delightful' experience is somewhat poignant given the German Army's more sinister advance along the same *Autobahn* to effect the *Anschluss* with Austria two months later.[9] Burgin and his officials concluded that Britain did not require an extensive network of newly constructed motorways, but they were open to proposals to construct single sections of experimental motorway, where it could be shown that this was cheaper than upgrading existing roads. Experimental motorways might in turn provide engineers and civil servants with the experience necessary to judge the merits of these new roads. What's more, the government already had preliminary plans for a number of schemes. In July 1937 the Highways and Bridges Committee of Lancashire County Council had submitted plans to the Ministry to construct a 62-mile North–South motorway between Carnforth and Warrington, while that same year preliminary surveys were undertaken for the construction of a new motorway between London and Birmingham, close to the route of the Northern and Western Motorway's 1923 scheme.[10] Interest in both schemes prompted the Ministry's Chief Engineer, Frederick Cook, to instruct his staff to survey the route between London and Preston to determine 'the relative advantages of (a) its improvement to the required standard width, (b) a new route for mixed traffic, and (c) a motorway'.[11] In early 1938 Burgin recommended that Cabinet approve the construction of an experimental motorway in Lancashire, but Chancellor of the Exchequer Sir John Simon pressured Burgin to withdraw the proposal due to 'the national financial position'.[12]

Motoring and engineering associations continued to lobby the government. In May 1938 the County Surveyors' Society published a scheme for a national network of 1000 miles of motorway that had been inspired by the experiences of their members who had 'formed part of the German Roads Delegation'.[13] The Society's scheme (Figure 2.2) was more modest than the British Road Federation's extensive and defensive plan, and the network was designed to complement Britain's existing trunk roads and 'meet the ordinary commercial needs of traffic' rather than be planned to aid 'national defence'.[14] In November 1938 prominent members of the German Roads Delegation, including Lord Eltisley and Roger Gresham-Cooke, decided to continue the Delegation's work and promote their conclusions through a new organization entitled the Modern Roads Movement (*The Times* 1938c).[15] The German Roads Delegation and the Modern Roads Movement were fairly successful in presenting their views to the

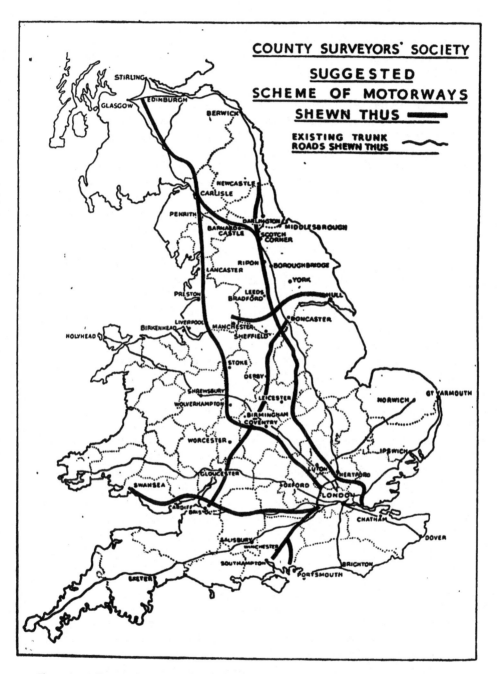

Figure 2.2 'Suggested scheme of motorways' by the County Surveyors' Society, May 1938. Map published in the British Road Federation's *Roads and Road Transport*, 1944. Reproduced by kind permission of the County Surveyors' Society and the British Road Federation.

government, and perhaps their biggest coup was in ensuring that no less than ten of their influential representatives presented evidence on the benefits of motorways to the Select Committee of the House of Lords on the Prevention of Road Accidents (see Select Committee 1938, 1939). In April and May 1938 and February 1939 representatives of the Royal Automobile Club (R. G. H. Clements and C. W. Evans), the British Road Federation (Roger Gresham-Cooke and Earl Howe), the Society of Motor Manufacturers and Traders (Colonel C. D. McLagan), the National Safety-First Association (F. G. Bristow), the County Surveyors' Society (D. H. Brown) and the House of Lords Road Group (Lords Eltisley, Sandhurst and Teynham) submitted evidence to the Committee on their impressions of the German *Autobahnen* and their effects on German road safety. The Select Committee considered the Delegation's publications and data provided by Fritz Todt's department, but while the Delegation had advocated the construction of a national network of motorways, their witnesses were mindful of shifts in the national mood and economic situation. The witnesses suggested that an experimental motorway, preferably between London and Birmingham, rather than in Lancashire, should be constructed before a national network was started. As Lord Eltisley conceded on 20 February 1939, the Delegation's proposals had been 'somewhat modified perhaps by the fact of the national emergency', and they did not 'feel justified in recommending a thing until it is really tried out' (Select Committee 1939: 116). In their final report, the Select Committee stated that the Delegation had provided 'striking proof' of the important effects of the *Autobahnen* on accident statistics, but they did criticize their 'whole-hearted admiration' for Germany's solutions, which failed to address 'the fundamental differences between Germany and Great Britain' (Select Committee 1939: 42–3). The Committee concluded that 'they would like to see an experimental motorway constructed', possibly between London and Birmingham, and that 'careful statistical and scientific examination' should be used to decide whether future motorways be built (Select Committee 1939: 43). The Report was broadly in line with government policy. Two months later the Ministry of Transport responded with a statement that constructed an equally particularist geography of Britain – asserting that events in Germany must not dictate whether Britain constructed motorways:[16]

> It would be hazardous in the extreme to assume, however, that what may be suitable in Germany will necessarily be the best method of achieving that object in this Country. The geographic and economic differences between the two Countries are wide; there is a vast difference in the distribution of population, in the distances and character of country between main centres of population and in the standard and effectiveness of existing

communications. For a wide variety of reasons, geographical as well as social, the cost per mile of constructing a motor road in this Country would be considerably greater than in Germany; more bridges would be necessary, land would be dearer, labour conditions are different. Even the effect on the accident toll is far from being so certain as enthusiasts have assumed; two-thirds of our accidents occur in built-up areas and the traffic in these areas would not be substantially reduced by the withdrawal of the long-distance traffic which would be able to use motor roads. On certain of our main roads, for which figures have been obtained, less than twenty-five per cent of the vehicles travel more than thirty miles. (Ministry of Transport, quoted in Drake et al. 1969: 39)

The geographies and economies of Germany and Britain are seen to be very different, with Britain's geography, economy, and travel patterns being suited to multi-purpose roads rather than brand-new limited-access motorways. Britain's transport problems are seen to require British solutions, and while British officials had previously expressed a great interest in motorway construction abroad, the situation had now changed.

Lobby groups continued to argue their case. In July 1939 the British Road Federation underlined the 'chaotic conditions' on Britain's roads in one of its last peacetime publications, *From Here to There* (BRF 1939). The Federation stressed that bold plans already existed. Official reports such as Sir Charles Bressey and Sir Edwin Lutyens' (1938) *Highway Development Survey 1937 (Greater London)* and the final report of the Select Committee of the House of Lords on the Prevention of Road Accidents advocated the construction of motorways – whether as single experiments, arterial motorways radiating from London, or, in the case of the County Surveyors' Society's proposals, as a national network.[17] The British Road Federation insisted that 'THERE MUST BE A NATIONAL PLAN AND TIME SCHEDULE', and it was the business of all of their member organizations, and of every citizen, 'TO SEE THAT THE PLANS ARE NOT PUT IN A PIGEON HOLE' (BRF 1939: 20–1). The Federation were being optimistic. The government had already effectively shelved these plans, and the outbreak of war in September led to dramatic changes to the working of Britain's transport infrastructure.

Motorways, War and Reconstruction

With the outbreak of war, Britain's transport infrastructure was tested in unprecedented ways. In the first three days of the war, 1,474,000 mothers, children, disabled citizens, teachers and helpers were evacuated by train and bus from many of Britain's main industrial cities and ports (Calder 1969: 38). The government's Railway Executive Committee coordinated the activities of the 'Big Four' railway companies and the London Passenger Transport Board. Britain's 60,000 independent road hauliers were coordinated by Regional Transport Commissioners, and the 2000 miles of canal which had not fallen into complete disrepair were used to transport bulk goods such as coal, tar, oil, granite and grain (Ministry of Information 1942). Petrol rationing was introduced and enforced black-outs brought chaos to the roads. Kerbs, bollards, trees and vehicle bumpers were painted white to help motorists drive in the dark, but this was to little avail (Brendon 1997). Official statistics for pedestrian deaths rose from 3100 in 1938 to 4500 in 1939 and 4800 in 1941. It was only with the ter-mination of the private petrol ration in July 1942 that figures dropped to 3000 in 1943 (Sillitoe 1973; Luckin 1997). Petrol rationing led an increas-ing number of travellers to brave the already overcrowded railways and buses, which suffered extensive bomb damage during the blitz (Ministry of Information 1942). As roads campaigners had predicted in the 1930s, it was not only Britain's urban transport infrastructure, but also the nation's western communications which came under increased pressure; and Britain's western ports suffered from heavy congestion and crippling labour shortages until the government intervened in March 1941 (Calder 1992). When the German Army reached the English Channel in May 1940, fears of invasion brought yet more measures. On 30 May signposts containing geographical information such as place or street names were removed, and barricades, tank traps and pillboxes were erected along roads (Calder 1992). The importance of transport to the war effort was emphasized when the Ministry of Information published a propaganda booklet for the Ministry of War Transport in 1942. The booklet, *Transport Goes to War*, conveyed a message common to other pamphlets of the time, positioning transport as central to the health and actions of the national body at war: '. . . it is the blood circulating from the body into . . . [the] fighting arm and fist' (Ministry of Information 1942: 6).

During World War II many engineers, planners and architects under-took official tasks, whether designing emergency housing schemes, sitting on government committees or completing military service (Gold 1997). Design professionals working in Britain started to engage in discussions with each other at conferences and on committees and radio programmes, and their largely progressive visions of a reconstructed nation were gradu-ally presented to the public through a diverse range of media: Penguin books, exhibitions, films, radio programmes, official plans, novels and advertisements (Matless 1998). Motorways frequently appeared in these visions and plans. During 1942, 1943 and 1944, the Institution of Civil Engineers' Post-war Development Panel on Roads, the Institution of Municipal and County Engineers, the London Regional Reconstruction Committee of the Royal Institute of British Architects, the newly formed

Standing Joint Committee of the RAC, AA and Royal Scottish Automobile Club, the Institution of Highway Engineers and the British Road Federation all independently published reports or plans advocating the construction of national networks of motorways (ICE 1942; IMCE 1942; LRRC 1943; SJC 1944; IHE 1943; BRF 1944).[18] Motorways also formed an important component in officially sanctioned reconstruction plans published by the government. In his *Greater London Plan 1944*, Patrick Abercrombie envisaged 'ten express arterial radial routes', radiating from one express ring road, to convey motor traffic between London and other important regional and national centres (Abercrombie 1945: 71). J. H. Forshaw and Patrick Abercrombie's earlier *County of London Plan* (1943) conveyed a similar vision for the reconstruction of central London, with eleven new arterial roads radiating from the city. In both plans motorways and dual carriageway roads appear as one of a number of important strategies to 'hold country and city in regional balance' (Matless 1998: 205).

Reconstruction became a key focus of debate within many government departments. In August 1942 Frederick Cook, Chief Engineer in the Ministry of War Transport, wrote an influential 15-page memorandum on post-war motorways, in which he underlined the importance of developing a road network for the 'social, industrial and economic needs of the post-war world'.[19] Motorways might, he argued, prove to be more costly than improvements to existing roads, but they could improve vehicle efficiency and road safety, help national defence, preserve amenities, improve the delivery of fresh produce from Britain's prime agricultural areas, and support Britain's industries (particularly the newer light industries).[20] Cook's memorandum referred to the findings of the Royal Commission on the Distribution of the Industrial Population (the Barlow Commission), whose final report had been published in January 1940 (Royal Commission 1940). At the heart of the Commission's report lay observations made by distinguished geographer Professor Eva Taylor (University of London) on behalf of the Royal Geographical Society, in which she noted the increasing concentration of industry and population into a coffin-shaped area encompassing London and the Midlands, Lancashire and Yorkshire conurbations (E. G. R. Taylor et al. 1938; Royal Commission 1940). Taylor's industrial coffin lay at the centre of Frederick Cook's 'Map no. 5', onto which were shaded the principal locations of 'market gardening & cash crop farming', towards the edges of the coffin, and the principal areas of 'predominantly beef & dairy farming', many of which lay inside the coffin.[21] Cook used the map to suggest that motorways built to serve Britain's densest areas of population would also help farmers transport fresh produce more swiftly over longer distances, and one Ministry civil servant was excited by the health benefits which might result:

I am also struck by the fact that the motor ways which he [Cook] regards as most important will help the market gardening industry. The increase in the consumption of fresh vegetables by working classes is a point of great importance in the nutrition policy which I hope and believe H. M. G. will carry through after the war.[22]

Motorways might improve the health of the urban masses – addressing concerns which war-time conscription and evacuation, health surveys and organicist tracts had been highlighting for some years (Matless 1998) – but economic arguments still prevailed. Cook's memorandum was significant. It was the first extensive government appraisal of the pros and cons of motorways, and it persuaded a number of senior civil servants that the time had come for a public statement of policy. No official statement appeared. Speaking at the launch of a 'road-safety campaign at Bournemouth' in May 1943, the Parliamentary Secretary to the Ministry of War Transport, Philip Noel-Baker, stated that the government were 'far advanced with . . . plans for beginning the work of providing the means for motor roads and the segregation of fast motor traffic' (paraphrased in *The Times* 1943: 2). Noel-Baker's remarks were rather premature, and he was less committal in subsequent speeches and parliamentary debates, refusing to detail future government policy, expenditure, and indeed whether a *national* network of motorways would be needed (*Parliamentary Debates* 1943, 1944). Motorways remained at the edge of the political map until mid-1945. I return to the government's post-war plans in chapter 3.

Motorways and the British Landscape

Motorways, planning, preservationism

In this final section I discuss the debates about the design, landscaping and planting of Britain's roads which emerged in the inter-war period, and came to inform government policy on the design and landscaping of roads and motorways in the post-war years. During the 1920s, 1930s and early 1940s engineers, politicians, wealthy motorists, planners and motoring organizations envisioned Britain as an imperial power that could be improved and strengthened by the construction of a national network of motorways. Motorways would have economic, road safety and strategic benefits, but it was stressed that they should also be 'fitted' to the landscape, bringing motorists into new relationships with their surroundings. In letters to *The Times* in April 1925 and January 1929, prominent planner-preservationist S. D. Adshead warned of the ribbon-development and suburbanism that might spring up if the government agreed to the construction of a London to Brighton motorway. The Northern and Western Motorway syndicate aimed to encourage frontage development and they suggested that garden cities, factories and airfields might be constructed close to motorway junctions. During the 1920s, motorists – in private cars and charabancs – ventured into the countryside in increasing numbers, attempting to seek out beauty-spots they had read about in guide books, which were being written and published to cater for and cultivate specific tastes, mobilities and sensibilities (Lowerson 1980; Matless 1998; Liniado 1996; O'Connell 1998; Gruffudd et al. 2000). Planner-preservationists and modern architects associated with groups such as the Council for the Preservation of Rural England (CPRE), the Design and Industries Association (DIA) and the Society for Checking the Abuses of Public Advertising (SCAPA) apportioned a considerable degree of blame for the despoliation of rural England and Britain to the disordering presences of motor vehicles and motorists (Matless 1990, 1998; O'Connell 1998). In his 1934 book *A Charter for Ramblers* the philosopher, preservationist, broadcaster and walker Cyril (C. E. M.) Joad complained that areas such as the Lake District had become 'vulgarized' by a flood of 'crude, untutored people' transported by 'motor and charabanc' (Joad 1934: 174). The poor conduct, philistinism and ignorance of these people was seen to render them devoid of aesthetic insight and appreciation, as they followed other motorists to crowded, commodified beauty-spots and the 'intolerable' modern inn, 'the A. A. hotel, complete with motorists who, all liver and no legs, are drinking whisky in the lounge' (Joad 1934: 28). Joad felt that vehicles and motorists were degrading the countryside that he enjoyed to walk, but many other planner-preservationists and commentators were less sweeping in their judgements and criticisms (Matless 1990, 1998). In his 1935 introduction to the topographic book *The Beauty of Britain*, writer, broadcaster and preservationist J. B. Priestley accepted that although high-speed driving was not to his taste, motorists might have discovered a new and potentially valuable way of seeing and experiencing the landscape:

I shall be told that the newer generations care nothing for the beauty of the countryside, that all they want is to go rushing about on motor-cycles or in fast cars. Speed is not one of my gods; rather one of my devils; but we must give this devil its due. I believe that swift motion across a countryside does not necessarily take away all appreciation of its charm. It depends on the nature of the country. With some types of landscape there is a definite gain simply because you are moving so swiftly across the face of the country. There is a certain kind of pleasant but dullish, rolling country, not very attractive to the walker or slow traveller, that becomes alive if you go quickly across it, for it is turned into a kind of sculptured landscape. As your car rushes along the rolling roads, it is as if you were passing a hand over a relief map. Here, obviously, there had been a gain, not a loss, and this is worth remembering. The newer generations, with their passion for speed, are probably far more sensitive than they are thought to be. Probably they are all enjoying aesthetic experiences that so far they have been unable to communicate to the rest of us. We must not be too pessimistic about young people if they prefer driving and gulping to walking and tasting. (Priestley 1935: 2–3)

In this description – which prefigures later observations, from different angles, by J. B. Jackson, Reyner Banham and landscape architects such as Brenda Colvin, Sylvia Crowe and Lawrence Halprin – Priestley suggests that younger drivers may have developed very different sensibilities to landscape and the experience of speed. Movement animates a landscape, creating sculptural qualities and generating new experiences. Priestley is able to tolerate these different aesthetic sensibilities, and in his 1934 book *English Journey* his open-mindedness and ambivalence reveal themselves in an account of a bus journey he had taken from Leicester to Nottingham, in which two boys express excitement at the passing of petrol stations:

> They are lucky, these lads, for if their taste does not improve, they will be able to travel on all the main roads of England in an ecstasy of aesthetic appreciation. Perhaps many of their elders think the petrol stations pretty. Perhaps the confounded things *are* pretty, and we are all wrong about them. (Priestley 1934, cited in Matless 1998: 57)

Priestley places petrol stations, bypasses, bungalows and other modern developments in a modern, post-World War I, Americanized 'Third England' that is distinct from a first, 'old England' of cathedral cities and traditional country pursuits, and a second England, associated with the nineteenth-century urban, industrial Midlands and North (Priestley 1934; Baxendale 2001). Priestley may have been ambivalent about the geographies of this new 'Third England', but other planner-preservationists were more forthright in their views. Motor vehicles, motorists and the unplanned ribboning, *laissez-faire* development and urban clutter which followed them along the roads were seen to bring chaos, ugliness, indefiniteness and disorderliness to rural England (Matless 1990, 1998). Here were the tentacles of unplanned growth, spreading out from the city, which Clough Williams-Ellis lambasted in his landmark 1928 preservationist text *England and the Octopus*. Planner-preservationists did not simply oppose all modern development or look back to a 'golden age'; rather they called for planned order and the application of modern design principles to create structures which were true to the present age (Matless 1998). The campaigning of the CPRE, Design and Industries Association and preservationists such as Clough Williams-Ellis and Patrick Abercrombie led to some successes. The

Restriction of Ribbon Development Act of 1935 provided one partial, if rather late, solution to uncontrolled growth, while pressure on private companies renowned for advertising in the countryside led one of them, Shell-Mex and BP Limited, to started removing outdoor advertising hoardings from fields and enamel signs from garages and petrol stations in the late 1920s. In the 1930 Design and Industries Association pamphlet *The Village Pump: A Guide to Better Garages*, Shell-Mex were marked out 'as an honourable exception' amongst petrol companies (DIA 1930: 1), and in an advertisement at the end of the Design and Industries Association's 1930 yearbook *The Face of the Land*, Shell announced that their 'ways are different': 'Shell began removing its advertisement signs from the countryside as long ago as 1926' (Shell, in Peach and Carrington 1930; see also Hewitt 1992).

In his introduction to *The Face of the Land*, Clough Williams-Ellis suggested that the Ministry of Transport and local authorities should replace the 'concrete track lined with bungalows and bill-boards', which passes as 'the typical English road', with motor roads based on 'the more gracious American "Park-way"' (Williams-Ellis 1930: 22). The CPRE's General Secretary, H. G. Griffin, corresponded with New York parks commissioner Robert Moses, the Department of Metropolitan Parkways in Boston and German ecologist Walther Schoenichen to gather information on the design and landscaping of the parkways and *Autobahnen* (Matless 1990, 1998).[23] Planner-preservationists praised the planning and order of these modern roads, but by the mid-1930s they and other commentators shifted their attention from the Italian *autostrade* and American parkways to the German *Autobahnen*. In *The Shape of Things*, published in 1939, the design critic and preservationist Noel Carrington contrasted the 'German autobahn' and 'English arterial road' as examples of 'plan and afterthought' (Carrington 1939: 133; see also Matless 1998: 60). Illustrations contrasted the two types of road: a purpose-built, functional *Autobahn* – replete with modern structures, dual carriageways and appropriate landscaping; and a single-carriageway, chaotic English road – which had suffered from an absence of planning and the erection of advertising hoardings, a poorly designed petrol station, a sham mock-Tudor inn and a roundabout ringed with municipal railings (Carrington 1939). Carrington blamed the state of England's roads on 'the present English form of government', which 'does not favour good design', but this did not mean 'that an authoritarian regime is superior because the new German roads are superior to ours' (Carrington 1939: 134). If a democratic country such as the USA could have good roads, then so too could Britain, and the solution was to vest more power in 'the hands of men professionally trained and elected ultimately by citizens who are intelligently interested in the design of their surroundings' (Carrington 1939: 134).

Modern designs for modern roads

Architectural journalists and organizations drew similar contrasts between British and foreign roads. A special 1937 issue of *The Architectural Review* on 'Roads' contrasted a backward-looking, English obsession with 'preservation' with the 'boldness and comprehensiveness of thought' of German engineers who had employed modern, functional designs 'to create amenity' (*The Architectural Review* 1937: 164). Ribbon development must be avoided, junctions and landscaping must be carefully thought out, and architects must realize that road furniture is only a functional tool to choreograph and govern the performances of drivers: 'It must be obvious that a complicated system of studs, traffic lights, roundabouts and beacons is simply a complicated system of stage props designed . . . on the principle perhaps that the stage must be set before the actor can begin to do his business' (*The Architectural Review* 1937: 156). Local authorities must cooperate with one another to simplify and coordinate road design, and *The Architectural Review* suggested that British planners and designers seeking to develop a distinctively British, functional, 'modern idiom for road accessories' should look no further than the 'black-and-white posts, traffic-lights and beacons' associated with Britain's nautical design tradition (*The Architectural Review* 1937: 169). Simple, functional, modern roads should reflect positive aspects of British architectural and landscape traditions, and the design principles which lay behind these and other architectural visions were reiterated in the handbook to the Royal Institute of British Architects' exhibition 'Road architecture: the need for a plan', which was opened on 1 March 1939 by Labour MP, former Minister of Transport and Leader of London County Council Herbert Morrison: 'It follows, if our social and economic life is to run smoothly, that the roads must serve us adequately and efficiently, with an absolute "fitness for purpose"' (RIBA 1939: 53). 'Fitness for purpose', one of the favourite maxims of early- and mid-twentieth century British modernists, must inform any future road aesthetic. What's more, national planning would be required to create a road network and shape a national geography suited to war as well as peace: 'Defence may reorganise the map, and the roads are inextricably part of the map. . . . Evacuation, military needs, food supply, industrial dispersion, and again – evacuation. These are the words that may mould our plan, and a plan we must have' (RIBA 1939: 58).

RIBA's exhibition included photographs and diagrams showing the current state of Britain's roads, positive precedents at home and abroad, and future plans for the nation.[24] The British Road Federation – who had helped design the exhibition – provided a mechanical model for the exhibition that included model cars moving along dual-carriageway roads and across clover-leaf junctions (see RIBA 1939; Shipp 1939). Paul Rotha's Realist Film Unit were commissioned to produce a 17-minute film, *Roads*

Across Britain, which 'put simply but dramatically the case for new roads in Britain', combining footage of Britain's existing roads with a sequence shot on New York parkways (Rotha 1939: 133).[25] The exhibition catalogue and displays emphasized that it was imperative for Britain's roads to be designed to fit into the landscape. Architecture, planning and design are located as vital disciplines in this respect, but planner-preservationists, landscape architects, local authorities and horticulturists also pushed for vegetation to be planted along the nation's highways.

The Roads Beautifying Association and roadside planting

The Roads Improvement Act of 1925 had empowered local authorities to purchase roadside land for planting purposes, but the ongoing neglect of many verges, and a lack of horticultural knowledge amongst county council staff, led Dr Wilfrid Fox – a leading dermatologist, businessman and amateur horticulturist – to write a letter to *The Daily Express* calling for a national programme of highway planting (Spitta 1952).[26] Fox was a friend of the *Express*'s editor, Ralph D. Blumenfeld, who arranged for him to meet Minister of Transport Colonel Wilfrid Ashley. After discussions between Fox and the Ministry's Chief Engineer, Charles Bressey, Ashley invited Fox 'to assemble a voluntary organization of horticultural experts, for the purpose of advising local authorities on the planting and preservation of roadside trees' (Spitta 1952: 5). The inaugural meeting of the Roads Beautifying Association, held at the Ministry of Transport's headquarters on 25 July 1928, was chaired by Wilfrid Ashley and attended by prominent horticulturalists, arboriculturists, county councillors, transport experts and politicians – many of whom joined the Association's Executive Committee (*Daily Express* 1928).[27] Wilfrid Ashley became President, Wilfrid Fox became Secretary, and William J. Bean (retired Curator of the Royal Botanic Gardens, Kew) acted as horticultural adviser – a position which was financed by the Ministry of Transport between November 1929 and November 1931, when the national economic crisis forced the Treasury to withdraw the funds (RBA 1932).[28] The Association's Executive Committee included Charles Bressey, Lord Montagu of Beaulieu, several county council leaders and such horticultural luminaries as Dr Arthur Hill (Director, Royal Botanic Gardens), Sir William Lawrence (Chairman, Royal Horticultural Society), F. R. S. Balfour, Winchester nurseryman Edwin Hillier, Gerald Loder and the wealthy banker, horticulturist and rhododendron hybridizer Lionel de Rothschild.[29]

The Association's horticultural work, fund-raising and promotion activities were fairly broad. Funds were raised from private subscriptions, donations by county councils and organizations such as the AA, and profits from the sale of publications and calendars. From 1938 the Association were paid by the Ministry of Transport to act as their official advisers on roadside planting (RBA 1939).[30] A Children's Section was established which had 100 members by 1938, including whole schools, Girl Guides groups, Princesses Elizabeth and Margaret Rose, and many children of adult members (notably Christopher Robin Milne, son of A. A. Milne) (RBA 1938: 29). The Association undertook plantings to mark the Silver Jubilee of George V in 1935, and they became a component body of the Coronation Planting Committee formed to encourage plantings commemorating George VI's coronation in 1936 (RBA 1935, 1937). The first roadside plantings were undertaken on the new Kingston Bypass in Surrey in October 1928, and work on the Liverpool–East Lancashire Road, London's North Orbital Road and bypasses across the Home Counties soon followed (RBA 1930a). By 1936, the Association had planted over 300 miles of road with 50,000 standard trees (RBA 1936), and a list of 'what we have done' published in 1939 revealed how county councils outside of South-East England were increasingly receiving advice – the counties with the most schemes designed by the RBA up until 1939 being: Surrey (10), Staffordshire (9), Leicestershire (8), Hampshire (8), Lancashire (6) and Yorkshire (5) (RBA 1939: 19). The Association undertook private work on non-road schemes to boost their funds. Planting schemes were prepared for Senate House at the University of London, Guildford Cathedral, Bournville Village Trust in Birmingham, Southampton Docks, a BBC transmitter in Droitwich, several housing estates and playgrounds, and a 'Home for Jewish Incurables' (RBA 1935: 10). During 1935 the Roads Beautifying Association were approached by the government's Special Areas Commission, who were tasked with reinvigorating the economies of Britain's 'depressed' industrial regions. The Association felt that there was 'no more worthy or deserving work' than the 'improvement of the distressed areas' (RBA 1937: 8). The planting of a new park in Jarrow (RBA 1935) was followed by a more ambitious programme of works for North Eastern Trading Estates Limited, who were developing a modern, landscaped industrial estate on fields near Gateshead (RBA 1937; Linehan 2003). The RBA's scheme reflected contemporary discourses which presented a 'troubled assessment of [the] community, landscape, body and soul' of these 'depressed' regions (Linehan 2003: 133), and the Association firmly believed that their planting programmes could lift the economic and psychological depression of these areas:

> The Commissioners for Distressed Areas also felt that the ugly and bleak conditions in the neighbourhood of Gateshead were in some measure responsible for the mental depression of the inhabitants, and we were asked to advise on the planting of the slag heaps and other unsightly remains of a past industry. (RBA 1937: 8–9)

Modern planting schemes could not only beautify roads for the appreciation of motorists, horticulturists and local residents. Planting schemes could also be used to improve modern industrial landscapes, ensuring that the 'factory sites of to-day are no longer necessarily ugly, nor merely unobtrusive', but are 'being made a positive element for beauty in the landscape' (RBA 1938: 6). The Roads Beautifying Association, like the CPRE, felt that functional planting and landscaping, coupled with careful planning and design, *could* ensure that modern structures and activities blend into and actually improve the English landscape. However, the two organizations did differ in their opinions on the types of vegetation which should be planted alongside roads.

The RBA's planting schedules for the Kingston Bypass included a range of flowering plants and shrubs (including primroses, bluebells, daffodils, crocuses and roses) and 43 different species of tree (including Lombardy Poplars at crossroads, birch, chestnut, maple, scarlet oak and avenues of flowering cherries) (*The Times* 1928). Critics immediately began to attack the proposals from a range of angles. Horticultural writer Edwin Campbell felt that the very idea of planting and beautifying modern roads contradicted their function, as they were planned and constructed 'for utilitarian purposes' (E. Campbell 1928). More extensive disagreements surrounded the species of vegetation the Association advocated, and this sparked a series of debates which rumbled on throughout the 1930s, 1940s and 1950s. In the *Daily News* on 29 December 1928, it was suggested that expert horticulturists were criticizing the species of tree planted on the Kingston Bypass, as they were 'not suitable for the soil', were 'too exotic', and there were 'not enough characteristically British trees . . . selected' (*Daily News* 1928). Wilfrid Fox associated these criticisms with a stultifying preservationism that ignored the positive impact of successive waves of progressive landscapers and agriculturalists who had introduced species and improved the English landscape:

> The truth is that the scenery of England has changed throughout the ages and is always in a state of transition, never static, . . . and what we have got to do is to hand on to posterity something which is worthy of our present knowledge. . . . It is impossible to stand still in either science or art, and landscape horticulture is surely a combination of both. . . . Why should the architects be the only revolutionaries? For good or ill they have in this era definitely left their stamp on our cities, but they deny the horticulturists the right to do the same thing on our roads. (Fox 1944: 233, 235)

Fox presents the Roads Beautifying Association as the only truly progressive organization concerned with modernizing and improving the English landscape, and he and other influential members of the RBA were keen to contrast their planting policies and landscaping philosophy with what they felt to be the more conservative, preservationist tendencies of the National Trust and CPRE (see RBA 1936, 1939). The distinction is interesting. Key figures in the CPRE were themselves pushing for the adoption of positive planning principles and a modernist design aesthetic to bring a sense of order to the English landscape, but the RBA overlooked the CPRE's entwining of planning, modern design and preservationism, projecting themselves as the only truly radical, progressive body seeking to modernize and revolutionize the English landscape.

Tensions had first emerged between the Roads Beautifying Association and the CPRE at the time of the RBA's formation in July 1928. The CPRE felt that the creation of a new Association was unnecessary, as they 'already had the whole question under consideration'.[31] The two organizations worked independently, but with the drafting of the government's Trunk Roads Bill in 1936 the CPRE saw an opportunity to influence their approach to the landscape treatment of trunk roads. The CPRE established a Trunk Roads Joint Committee in November 1936, and although Wilfrid Fox represented the RBA on this committee, Fox felt they would simply duplicate *his* Association's work.[32] H. G. Griffin responded, emphasizing that the CPRE had been considering these issues since before the RBA were formed.[33] The disagreements quickly escalated after Lionel de Rothschild (Chairman of the RBA's Technical Sub-Committee) wrote a letter to *The Times*, in which he criticized the creation of a new committee and suggested that Britain's competing and overlapping amenity organizations should be merged into one National Landscape Trust (L. N. de Rothschild 1936a: 17; 1936b). A reply was despatched by the Earl of Crawford and Balcarres (President of the CPRE and also, notably, a Vice-President of the RBA) (Crawford and Balcarres 1936), and the 'incident' appeared to quieten down after the CPRE agreed to change the Committee's title to the Joint Committee of the Council for the Preservation of Rural England and the Roads Beautifying Association.[34] The Committee published their final report in August 1937, in which they emphasized that Britain's trunk roads must harmonize with local and regional surroundings, advertising and ribbon development must be controlled, and 'exotic species' should be confined to 'urban and sophisticated areas' (Trunk Roads Joint Committee 1937: 17). Botanist Professor Edward J. Salisbury (University College London) had a significant influence on the Report's planting proposals, which appear to have struck a compromise between the landscape and horticultural tastes of the Roads Beautifying Association and CPRE.[35] In 1938 the two organizations agreed to cooperate on matters of roadside planting – sending representatives to each other's meetings and holding joint discussions on themes such as 'Planting the highway of to-day' (RBA

1939: 13). Nevertheless, the CPRE and RBA continued to disagree on many issues, and by this time other architecture, planning, horticulture and preservationist organizations were advancing equally strong opinions on the landscaping and planting of roads.

Landscape architects and the modern road

In the late 1930s and 1940s, landscape architects entered the debate. In an article published in the Institute of Landscape Architects' journal *Landscape and Garden* in 1939, Brenda Colvin – one of the founding members of the Institute, and their first female President (1951–3) – criticized the prevailing British obsession with trying to ' "beautify" the road', arguing that planting must be used 'to knit the highway into the landscape':

> . . . unfortunately most of the planting that is being done still shows a misunderstanding of the principles involved, and an almost pathetic lack of vision. The logic of much of it seems to be based on the assumption that since flowering trees and shrubs are pretty and excite our admiration, the more of these and the greater the number of varieties we plant along the roads the more the roads will be 'beautified'. (Colvin 1939: 86)

Colvin doesn't name those who are doing the misplanting, but contemporary readers would have known that she was referring to the work of the Roads Beautifying Association and the local councils who followed their advice. At one level Brenda Colvin's views do not appear that different from those of the RBA, for while she stressed that exotic trees and flowering shrubs *may* be appropriate for urban areas – where speeds are inevitably slower and the scale more domestic (see Colvin 1939, 1948) – the Roads Beautifying Association stated their commitment to ensuring that only 'wild species' would be planted along country roads, and that 'garden hybrids and varieties' would be limited to urban or semi-urban areas (RBA 1935: 15). The difference was that while Fox justified the RBA's planting policy in horticultural and ecological terms, he overlooked the criticisms of landscape architects such as Colvin who argued that they took no account of the speed of vehicles, the experience of moving through the landscape, and the aesthetics and overall landscape design of the road:

> Travelling at anything over thirty miles an hour, the details of flower and leaf count for very little; form and mass, light and shadow are the materials we

must make use of, and these are also the requirements from the point of view of the more distant observer in the countryside. (Colvin 1939: 88)

As Colvin stated in her 1948 book *Land and Landscape*, beautiful modern roads would only result from a 'more fundamental' approach to the *landscape* than the rather superficial and largely *horticultural* approach adopted by the Roads Beautifying Association (Colvin 1948: 244).

The timing of Colvin's criticisms is significant. During World War II, prominent figures in the Institute of Landscape Architects attempted to broaden their profession's sphere of influence, shedding the Institute's image as a 'domestic garden society' (Jellicoe 1985: 9).[36] This was a time when a small, but influential, number of landscape architects (particularly those qualified as architects) were sitting on government committees and engaging in debates on reconstruction, while a large number of architects and planners were elected as members of the Institute of Landscape Architects, decreasing the proportion of members who had trained as horticulturists and gardeners (Fricker 1969; Jellicoe 1985).[37] At a meeting to discuss ILA policy in November 1942 – which was attended by representatives of planning, architecture, horticulture and amenity groups – the Vice-President of the Institute, Lady Allen of Hurtwood, stressed that future emphasis must be placed on 'the social value of our profession in a democratic age', as landscape architecture had previously 'been too closely identified with designing and making private gardens and estates' (Allen 1943: 5). Landscape architecture must be conceived as 'a *new* national service', and central and local government and other organizations must recognize the role that the profession could play in shaping the nation's public spaces (Allen 1943: 5). During the war and post-war years, new kinds of design work started to become available to landscape architects, and this was particularly noticeable in the commissions received by the President of the Institute of Landscape Architects, Geoffrey Jellicoe, as early as 1943.

Jellicoe was a qualified architect who had become Principal of the Architectural Association School of Architecture in 1939. During the war he became engaged in designing emergency housing projects for the Ministry of Supply and coordinating bomb repairs in the Islington area (Jellicoe 1987; Spens 1992, 1994). In 1943 Jellicoe undertook commissions to prepare landscape plans for cement works in Buckinghamshire and the Hope Valley of the Derbyshire Peak District (Jellicoe 1980; Spens 1994). Jellicoe was a well-respected design professional who consistently argued that large-scale, modern, industrial schemes such as power stations or cement works *could* be made to fit into sensitive landscapes. Other private commissions followed. In late 1943, the British Road Federation employed him to prepare their new exhibition, 'Motorways for Britain', which was launched at 22 Lower Regent Street on 9 December 1943. In an advert in *The Times*, the BRF pointed to the modern, landscaped vision of high-speed motorways, which Jellicoe projected in the exhibition:

> Models, plans and photographs show how British technique has designed a complete plan for Britain's post-war road programme – 1,000 miles of motorways and their connecting roads designed to harmonise with typical British scenery. The profusion of exhibits covers every feature of the new roads, and shows how they save time, money, and above all, lives. They are modern and practical, but the art of the landscape architect has invested them with charm and beauty. Here is a fascinating preview of post-war reconstruction for everybody. (BRF 1943: 3)

Jellicoe's main contribution was in preparing 12 'faked views' (Jellicoe 1945: 90) in which he montaged drawings of modern motorways onto photographs of well-known beauty-spots, to show how motorways 'not only need not spoil the landscape, but should enhance it' (*Wartime Journal* 1944: 18). In the post-war years landscape architects paid particular attention to the perspective of the driver moving along the road, but in these 12 views Jellicoe and the British Road Federation were more concerned to persuade the public and politicians that motorways could fit into 'different types of English scenery' (Jellicoe 1945: 90). Jellicoe's views showed how motorways could be successfully integrated into wild or cultivated landscapes, traversing meadows, scaling hillsides, passing through mountain ridges or alongside rivers. In one view, near West Meon in Hampshire, Jellicoe sunk the motorway in a ditch, providing bystanders with a visual effect that was similar to an 'eighteenth century ha ha' (*Wartime Journal* 1944: 18). The 'faked views' appeared in a number of wartime and post-war publications published by the British Road Federation and landscape architects (Curnock 1944; Jellicoe 1945; BRF 1946, 1948, 1949; Colvin 1948; Nockolds 1950). Five views were reprinted in George Curnock's *New Roads for Britain* and another five in Harold Nockold's *Roads: The New Way*, which the British Road Federation published in 1944 and 1950, respectively. The cover to Curnock's booklet depicted a book of plans of 'New roads for Britain', with the right-hand page serving as a window through which one could glimpse one of Jellicoe's projected motorway landscapes (Figure 2.3). If the reader opens the cover they are presented with the entire view of the Trough of Bowland in Lancashire (Figure 2.4), with Jellicoe's two motorway carriageways straddling a winding river to create 'an entirely new landscape and . . . a simple waterside parkway' (*Wartime Journal* 1944: 19). These landscaped roads were far removed from the Roads Beautifying

Figure 2.3 The front cover of George Curnock's *New Roads for Britain*, published by the British Road Federation, 1944. Reproduced by kind permission of the British Road Federation.

Figure 2.4 Photomontage by Geoffrey Jellicoe of two motorway carriageways passing alongside the banks of a river in the Trough of Bowland, Lancashire. Frontispiece to George Curnock's *New Roads for Britain*, published by the British Road Federation, 1944. Reproduced by kind permission of the British Road Federation.

Association's decorative planting schemes. Jellicoe considered the siting and alignment of the road, as well as proposing landscaping and planting which extended far beyond the immediate verge. As I show in chapter 3, his sketches appeared on the covers to two influential British Road Federation pamphlets aimed at politicians charged with preparing post-war motorway legislation. The pamphlets, Jellicoe's sketches and the broader lobbying activities of the British Road Federation had a notable impact, persuading politicians that motorways could fit into the English landscape, and that motorways would form an important part in the nation's reconstruction.

Chapter Three

Designing and Landscaping the M1

In this chapter I examine how the M1 was planned, designed and land-scaped as a distinctively modern, functional space in the early post-World War II period. In section one I show how politicians, lobby groups and landscape architects envisioned a carefully landscaped network of motor-ways which would reinvigorate the national economy and bring health and vitality to the 'national body'. I then examine how economic crises and subsequent cutbacks in capital investment programmes led to delays in the commencement of the motorway programme, prompting groups like the British Road Federation and Roads Campaign Council to step up their pressure on the government during the mid- to late 1950s. In the second section I examine the work of regional planners, geographers and engineers, who, following the announcement of the routing of the London to Yorkshire Motorway (M1) in September 1955, considered how the new motorway would bind together the regions. In section three, I examine how prominent landscape architects such as Brenda Colvin, Sylvia Crowe and Geoffrey Jellicoe pushed for the appointment of landscape architects who could ensure that Britain's future roads and motorways would be designed, landscaped and planted in a manner that would enhance the landscape and could be appreciated by motorists travelling at speed. In section four I examine how the first sections of the M1 were designed and landscaped during the late 1950s. The government's Advisory Committee on the Landscape Treatment of Trunk Roads expressed concern at the designs prepared by Sir Owen Williams and Partners for the bridges, embankments and vegetation on the motorway. Discussions focused on the need to main-tain a sense of flow in the landscape and the visual effects of different species of vegetation when seen at speed. In the fifth section I examine debates about the design of the first two service areas on the M1. Discus-sion focused on whether these were urban, rural or suburban spaces, how they should be designed and landscaped, and what an appropriate 'road style' should look like. In the final section I examine the controversies surrounding the design of the new motorway signs prepared by Jock Kinneir, Margaret Calvert and the government's Advisory Committee on Traffic Signs for Motorways; debates which centred on the appropriateness and effectiveness of particular styles of lettering and colours in the English landscape. Throughout this chapter I examine how all manner of inhuman materials – from trees and shrubs, to signs and bridges – became enmeshed in political and aesthetic debates about the design of modern British motorways.

Legislating and Campaigning: Towards a National Motorway Network

On 6 May 1946 Minister of Transport Alfred Barnes presented the House of Commons with the Labour Party's ten-year plan for the reconstruction of Britain's road network. Over the period of the plan the government would repair bomb-damaged roads, eliminate accident black-spots and bring about the 'comprehensive reconstruction of the principal national routes' (*Parliamentary Debates* 1946: 591). Motorways would form the framework to this national plan, binding together the nation and the regions to facilitate economic growth. A map of projected motorways was exhibited for the benefit of MPs in the House of Commons Tea Room (Figure 3.1). As the map reveals, new motorways would be built on similar alignments to what are now motorways M1, M4, M5, M6 and M62. A London Orbital Road would be constructed, while a number of existing trunk roads would be improved, including the A1, A9, A12, A20, A23, A30 and A74.

With the announcement of the programme, Ministry of Transport engineers started to investigate the proposed alignments of the motorways, and in 1948 the government introduced a Special Roads Bill to parliament which would provide the legal powers required to construct limited-access motorways. Discussion of the Bill focused on the effects of construction on the nation's economy, agriculture and scenery. One Conservative Member of Parliament, Christopher York, expressed concern at the replacement of 'good agricultural land' with 'unproductive slashes across the land' (*Parliamentary Debates* 1948: 1768, 1769), but the vast majority of MPs referred to the positive benefits motorways would bring to the country. Conservative MP and preservationist Edward Keeling – a member of the Georgian Group, Executive Committee of the National Trust and ally of the CPRE – pointed to the potential value of motorways in the English landscape: '. . . a motorway, if it makes no unnecessary gashes in the landscape, if it is designed to blend harmoniously with the country it pierces,

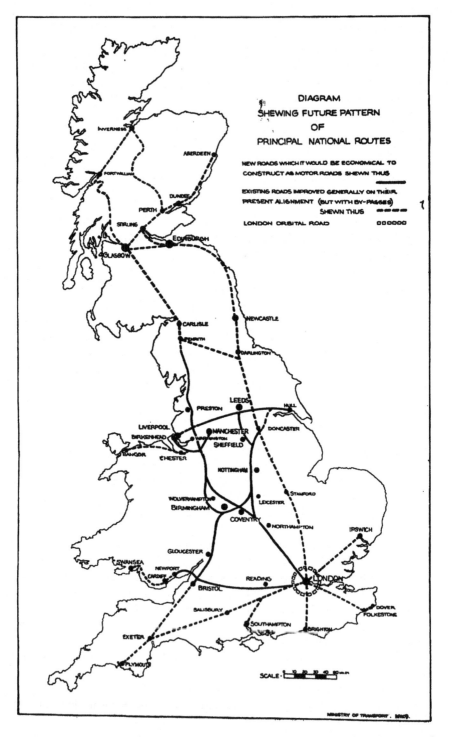

Figure 3.1 The Ministry of Transport's 1946 map of future national routes, including motorways and reconstructed trunk roads. Source: Author's collection.

if local stone and other local materials are used so far as possible, need not harm, and may even enhance, the natural beauty of the scene' (*Parliamentary Debates* 1948: 1758). The British Road Federation made similar points in their propaganda booklet *The Case for Motorways*, which was distributed to politicians during the passage of the Bill in 1948 and 1949 (BRF 1948, 1949). The Federation argued that: 'Aesthetically, and for the practical purpose of attracting more tourist traffic, motorways will open up new vistas for the road travellers, and provide opportunities for landscaping which will improve the countryside' (BRF 1948: 6). These positive landscaped motorways were illustrated on the cover to *The Case for Motorways*, which sported one of Geoffrey Jellicoe's 'faked views' of a motorway curving into the distance along the side of the Malvern Hills in Worcestershire (Figure 3.2). With this sketch Jellicoe had sought to prepare a 'logical . . . very simple, but very subtle design' in which the two carriageways were positioned 'either side of an existing hedge' – preserving an important landscape feature and, more importantly, reducing the parallelism and linearity of the motorway (Jellicoe 1945: 91).

Jellicoe's sketch fronted a booklet that was dominated by economic arguments and statistical evidence emphasizing the importance of these new motorways to the national economy. Many MPs were clearly impressed by the authority of their conclusions. Minister of Transport Alfred Barnes referred to the expertise and 'valuable publicity' provided by the British Road Federation on the economics of motorways (*Parliamentary Debates* 1948: 1741). Labour MP Arthur Symonds referred to the '*prima facie* case . . . made out for these motorways in the literature . . . we have received' (*Parliamentary Debates* 1948: 1761), while other MPs referred directly to a series of BRF publications that had been sent to politicians, including *The Case for Motorways*, *The Road Way to Recovery* and a report on the *Economics of Motorways* prepared by a Joint Committee of the British Road Federation, the Institution of Highway Engineers and the Society of Motor Manufacturers and Traders (BRF 1948, 1949; Brunner 1948; Joint Committee 1948).

The key force behind the first two BRF publications was economist Christopher Brunner, General Manager of the Shell-Mex Oil Company and Chairman of the BRF's Publicity Committee. Brunner was a respected statistician who had authored books on transport and oil economics, but in developing his arguments and translating his authority to a non-expert readership, Brunner was quick to draw upon analogies which had long been used to describe transport flows in Britain and abroad. Bodily and specifically health-related metaphors loom large in writings on transport and planning. In *The Road Way to Recovery* Brunner's first diagram featured a human hand overlain with a map of Britain, on which the main lines running across the palm of the hand correspond to the position of the

Figure 3.2   Cover of the British Road Federation's *The Case for Motorways*, 1948, featuring a photomontage by Geoffrey Jellicoe. Reproduced by kind permission of the British Road Federation.

motorways proposed in the 1938 County Surveyors' Society scheme (Figure 2.2). Brunner's text developed the analogy through a discussion of this feminized nation and hand: 'Roads are one of Britain's most valuable assets – her very life-line' (Brunner 1948: 4). Brunner continued, switching to the more common analogy of roads as the arteries or veins of a national body: 'As the blood-stream is to the body, carrying nutrition to all its parts, so is an adequate highway system to a modern community' (Brunner 1948: 7). Efficient flows through a modern transport infrastructure (particularly roads) are seen to be vital to the health of the nation, and planners and economists had been deploying similar bodily and health-related analogies as far back as the seventeenth century, when William Harvey discovered the circulation of blood (Sachs 1992; Sennett 1994).

In their 1938 booklet *The Crisis and the Roads* the British Road Federation described 'congestion on the roads' as 'congestion of the nation's arteries': 'A man cannot put up a successful fight if his blood flows sluggishly and the life-streams of the nation must flow freely if we are to defend ourselves successfully in the event of attack' (BRF 1938: 1).[1] The BRF's concerns reflected broader national debates about the population's health and fitness. In 1937 the government set up the National Fitness Council to initiate a National Fitness Campaign, aiming to address long-standing concerns about the poor health of military recruits at a time of rearmament and German military aggression, but also reflecting 'a wider ethos of fitness as a positive social force' that was evident in the popularity of rambling and cycling, and of organizations such as the Youth Hostels Association, the Ramblers' Association, the Women's League of Health and Beauty, and the Boy Scouts and Girl Guides (Matless 1998: 91). In their 1939 'Road Architecture' exhibition catalogue, the Royal Institute of British Architects stressed that roads 'are the veins and arteries of Britain' (RIBA 1939: 53), and that national planning was essential to provide adequate roads for 'the needs of industry, leisure and defence': 'We must analyse anew the map of Britain, for if the blood is not to run sluggishly in the veins, and if the arteries are not to harden, then the *roads must change* and that change must have as its basis nothing more or less than a National Plan' (RIBA 1939: 54). Writing in his book *British Roads* (1939) Geoffrey Boumphrey suggested that Britain, as an advanced industrial nation, simply must have better roads. A great nation could be compared with an athlete, just as weaker nations could be seen to have the relatively poor health of sedentary bodies: 'The athlete has need of a better circulation than the sedentary worker' (Boumphrey 1939: 156). In 1942 Mervyn O'Gorman – a prominent electrical and aeronautical engineer, and former vice-chairman of the Royal Automobile Club – referred to motorways as the necessary 'back-bone' of the nation and post-war reconstruction (O'Gorman 1942: 5), while in 1958 Sir Herbert Manzoni, City Engineer and Surveyor of

Birmingham, compared the 'congestion on our roads' with an aggressive cancer:

> It probably started in the organs of this country, which are the urban areas and, like cancer, it has spread; it has got into the blood-stream and is now evident in many places throughout the country. Like cancer, if it is unchecked, it will strangle the country. There can be no doubt about that; if we do nothing for a few years, traffic congestion will strangle Great Britain. Fortunately, unlike cancer, it is curable. (Manzoni 1958: 300)

The consequences of poor roads are compared with a variety of medical conditions, and it is inevitably the planner, engineer or architect who must step in and heal the national body, practising their expertise as would the physicians of the new National Health Service (Bartram and Shobrook 2001; Hughes 2002).

During parliamentary debate on the Special Roads Bill, Minister Alfred Barnes admitted that the continuing downturn of the economy and resulting cutbacks in capital investment programmes precluded the government from starting work on the motorway programme in the near future (*Parliamentary Debates* 1948). The British Road Federation continued to lobby parliament throughout the early 1950s. They highlighted the urgency of constructing motorways in Lancashire (BRF 1952) and between South Wales and England (BRF 1951b), as well as stressing the vital role that motorways could play in the government's rearmament programme (BRF 1951a). In February 1955 the Conservative government's Minister of Transport and Civil Aviation, John Boyd-Carpenter, finally announced the start of an 'Expanded Road Programme' that would include the construction of the first section of the London to Yorkshire Motorway (M1), as well as the Dartford Tunnel under the River Thames, and bypasses at St Albans, Doncaster, Preston and Lancaster (*Parliamentary Debates* 1955). The programme was a mere fraction of that proposed in 1946, and frustrations about the delays surfaced in a series of events during 1955. In March the BRF joined eleven other pro-roads organizations to found the Roads Campaign Council, a group who aimed 'to organise a short, sharp publicity campaign demanding increased expenditure on roads' (Plowden 1970: 70). In February, meanwhile, Deputy Engineer-in-Chief at the War Office Brigadier Thomas I. Lloyd had initiated a campaign for the conversion of Britain's unprofitable and inefficient railways into a network of 20,000 miles of motorway reserved for fast traffic (T. I. Lloyd 1955).[2] Two years later Lloyd detailed his arguments in *Twilight of the Railways* (1957), and in 1958 he helped to form the Railway Conversion League, who campaigned throughout the 1960s for the closure of unprofitable railway lines – building upon the British Railway Board's 1963 report *The Reshaping of*

British Railways (BRB 1963) – and arguing for their conversion into new roads and motorways.

Locating the M1: Regional Planning, Local Protests and the Authority of the Engineer

The exact line of the London to Yorkshire Motorway was not published until September 1955, but the projected route had been planned on a county basis from the late 1940s.[3] County councils lying along the route mapped the projected alignment of the motorway on the county development plans which they submitted to the Ministry of Housing and Local Government in the early 1950s (Doubleday 1951; Sterne 1952; Jeffery 1959). In 1949 geographer Eva Taylor had urged planners to consider the bigger picture when developing their county plans, planning future developments in relation to broader regional and national geographies (E. G. R. Taylor 1949). For regional planners and geographers such as Taylor, the routing of motorways only made sense as part of regional and national plans, and it is significant that the London to Yorkshire Motorway passed along the central axis of the coffin-shaped industrial belt Taylor had identified in her evidence on the distribution of the industrial population presented to the Barlow Commission in the late 1930s (E. G. R. Taylor et al. 1938; see chapter 2). Motorways are presented as key infrastructures to bind together regions and strengthen and rebuild the nation (cf. Gruffudd 1995a). Motorways become key to the forging of a modern, regional England (and Britain), and this latter point was confirmed in two articles about the M1 motorway which were published in geography journals in 1960.

The first article, 'The Communications of Watford Gap, Northamptonshire', was written by Jay H. Appleton, and published in the *Transactions and Papers* of the Institute of British Geographers (Appleton 1960). Appleton would go on to author *The Geography of Communications in Great Britain* (1962) and, notably, *The Experience of Landscape* (1975), but in this article he explained why generations of engineers had routed 'at least half a dozen lines of communication', including canals, railways, roads and a motorway, through the Watford Gap, 'all of which have been, are, or will be of major importance in the economic geography of Britain' (Appleton 1960: 215). The physical geography of the region and specifically Watford Gap – comprising of two valleys cutting through the Northamptonshire Uplands – was identified as a major force in making this a prime location for so many major lines of communication passing between London and the South-East, and the Midlands and North-West (Figure 3.3; see also Figure 3.4). Watford Gap is constructed as a key passage point in the economic,

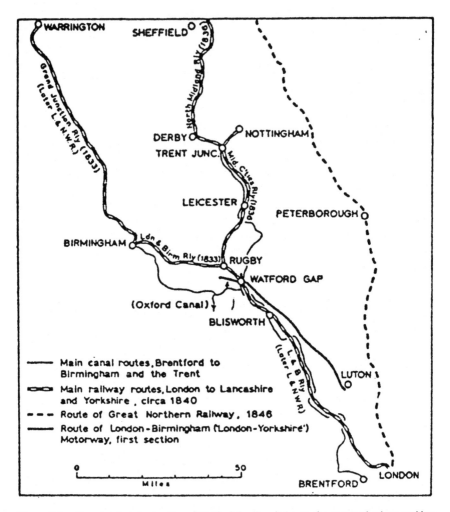

Figure 3.3   Map showing the location of Watford Gap in relation to the communications corridors linking London, the Midlands and the North-West. Reproduced from *The Institute of British Geographers Publication No.28: Transactions and Papers 1960*, p. 223, by kind permission of the Royal Geographical Society with the Institute of British Geographers.

communications and physical geographies of the Midlands and the nation, while more recently the Gap has been constructed as an important site and on a notable boundary in the cultural and political geographies of the nation (see chapter 6). The second article that I want to consider was published in *The East Midland Geographer* in June 1960. The article, by University of

Nottingham geographer Richard H. Osborne, examined the likely importance of the projected second, Crick to Doncaster, section of the M1 on
the economy, industry and population distribution of the East Midlands
(R. H. Osborne 1960). Osborne predicted that the motorway would
strengthen the North–South axis of the region: speeding up cross-regional
journeys, encouraging long-distance travel, influencing the location of
future industrial and urban growth, facilitating new commuter routes and
fostering increased demand for housing (R. H. Osborne 1960). The M1
would reshape the geographies of this newly conceived region and of a
modern regional England, and East Midlands planners were urged to
control the effects of the motorway on the region's villages, towns, cities
and countryside.

Siting . . . and objecting

In 1951 Sir Owen Williams and Partners were commissioned to carry out
preliminary surveys for the route of the London to Yorkshire Motorway
between London and Doncaster.[4] Sir (Evan) Owen Williams (1890–1969),
managing partner of the firm, had established his reputation as a consulting
engineer and architect during the 1920s and 1930s (Cottam 1986b; Stamp
1986; Yeomans and Cottam 2001; Saint 2004).[5] In 1921 he was appointed
consulting engineer to the 1924 British Empire Exhibition at Wembley
(which included the construction of Wembley Stadium), and his work with
architect Maxwell Ayrton on the Exhibition's structures was followed by
their collaboration on at least ten reinforced concrete bridges for the Ministry of Transport between 1924 and 1930 (Cottam 1986a, 1986b). During
the 1930s Williams strengthened his reputation as one of Britain's leading
experts in concrete reinforcement, acting as consulting engineer, and frequently architect, of a series of important and celebrated modernist buildings, including the Empire Pool in Wembley (1934), Pioneer Health Centre
in Peckham (1935), Daily Express buildings in London (1931), Glasgow
and Manchester (both 1939), the Dorchester Hotel in London (1930) and
the widely praised Boots 'Wets' Factory in Beeston, Nottingham (1932)
(Cottam 1986b). Sir Owen 'dazzled the [modern architectural] avant-
garde' with his pioneering, modernist, functional concrete and glass designs,
but unlike them he did not draw his influences from European modernism,
and he was an established, Establishment (rather than avant-garde) figure
(Saint 2004: 2; Stamp 1986).

On 16 September 1955 Sir Owen Williams and Partners were asked to
prepare *detailed* designs and oversee the construction of the 140 miles
of the London to Yorkshire Motorway connecting the St Albans and
Doncaster bypasses.[6] The line of the first section of the motorway (Luton

to Rugby) was advertised in the press, and the public were provided with three months to lodge formal objections.[7] Planning legislation required that the government and engineers consider local and national agricultural needs when locating the motorway,[8] and in papers delivered to the Royal Institute of British Architects in 1956 and the Royal Institution of Chartered Surveyors in 1958 Sir Owen explained how he had consulted land utilization maps, respected the rights of landowners and tenants, and attempted to preserve the amenities of the countryside (Sir E. O. Williams 1956, 1958a, 1958b).

Sir Owen's considerations did not prevent 142 parties lodging objections to the line of the motorway by 24 December 1955.[9] The majority were acting on the advice of land agents and the Country Landowners' Association, objecting to the lack of detailed information about the access roads, bridges and tunnels that would enable them to cross the motorway and reach their land. As these objections couldn't be answered until a later stage, the Conservative government's Minister of Transport and Civil Aviation Harold Watkinson (1910–95) decided to overrule the objections and finalise the scheme on 28 March 1956, on the condition that: '(a) A full letter of explanation should be sent to those objectors whose ground of objection was severance of their property or lack of information about compensation or accommodation works; and (b) interviews should be held with those objectors who proposed alternative lines of route'.[10]

The decision not to hold a public inquiry had been carefully considered. It was only later, in the 1970s, that campaigners such as John Tyme regularly and very publicly challenged the powers of government, and criticised the undemocratic, top-down nature of the public inquiry process (Tyme 1978; see also Powers 2004). Nevertheless, this was a period when there were widely publicized challenges to the continued occupation of land which had been requisitioned for military purposes before and during World War II – notably at Crichel Down in Dorset, where there was a highly publicized public inquiry in 1954 (Brown 1955; Wright 1995). As a Ministry of Transport and Civil Aviation official remarked in March 1956, objections to the line of the M1 would have to be carefully addressed, as 'the present climate of public opinion is very sensitive to any form of property dis-possession'.[11] The consulting engineers held meetings with 18 objectors who proposed realignments of the motorway in May and June 1956, but only three modifications were accepted: a realignment to avoid three cottages at Whilton Locks, Northamptonshire; a rerouting of the motorway near Kilsby for economic reasons; and a realignment away from the village of Tingrith in Bedfordshire following the objections of 31 villagers (*The Times* 1955).[12] The other proposed changes were rejected in similarly worded letters sent by the Ministry of Transport and Civil Aviation in May 1956:

Sir Owen Williams explained to you the difficulty in trying to move the pro-
visional line of the motorway further west as you had suggested. . . . He
showed you on the large scale map the points at which the existing draft line
was fixed by geographical and other features, and in particular showed how
it would be virtually impossible to provide adequate connection between
properties to the east and the west of the motorway when constructed if the
line were in fact marked as you suggested.[13]

In seeking to justify the Ministry's decisions, the civil servant referred to
three incontestable, authoritative presences which should persuade landowners of the necessity for the proposed route – namely the physical geography of the landscape, a respected engineer and a map. Sir Owen Williams
is presented as the mediator, performing the material evidence made mani-
fest on a map and on the ground. He is presented as an expert whose
experience, engineering knowledge and ability to read and revision the
landscape through a 1 in 2500 map makes his decision about the motorway's alignment incontestable. Sir Owen's personality was presented as
vital to the sensitive negotiations leading up to this decision, and in John
Laing and Son Limited's 1959 booklet on the construction of the motorway, engineering historian L. T. C. Rolt suggested that it was Sir Owen
Williams's personal approach and diplomacy which had resulted in the
withdrawal of all but two of the formal objections (Rolt 1959).

During the 1950s and 1960s opposition to the routing of new motorways
was rarely couched as opposition to the construction of motorways or to
top-down planning strategies *per se*, and protesters tended to focus on the
problems and merits of specific routes. The first sections of the M1 crossed
high-value agricultural land in counties such as Northamptonshire, but as
all of the possible routes were likely to affect productive farmland, amenity
groups did not challenge what they acknowledged to be an important
national scheme. The routing of the second section of the motorway
through Leicestershire in 1957 and 1958 was more controversial, and it
led to what was probably the first major campaign against the route of a
British motorway. The motorway was to pass through the popular amenity
area of Charnwood Forest, but this led to opposition from a large number
of local and national groups, including the CPRE, the Ramblers' Association, the Leicestershire Trust, Nature Conservancy, the National Parks
Commission and the Royal Fine Art Commission. An alternative route was
proposed that would avoid the Forest and pass through the Soar Valley,
but this line was heavily opposed by local farmers, the National Farmers'
Union and the Country Landowners' Association. Economic and amenity
value were counterbalanced and opposed by different organizations and
experts, and a final route was eventually chosen that would avoid the best
areas of both Charnwood Forest and the Soar Valley (Merriman 2001).[14]

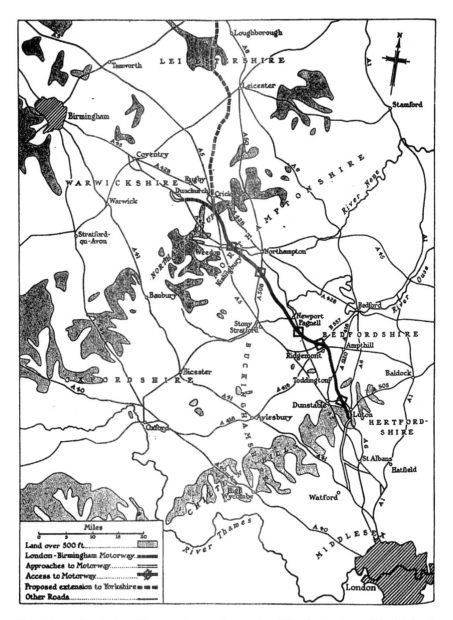

Figure 3.4 Map by A. J. Thornton showing the location of the first sections of the London to Birmingham/Yorkshire Motorway. *The Geographical Magazine*, volume 32, no. 5, October 1959, p. 242. © Geographical, the magazine of the Royal Geographical Society. Reproduced by permission.

The exact line of the first sections of the London to Yorkshire Motorway was finalised in mid-1957. The motorway would extend from Luton in Bedfordshire to Crick in Northamptonshire, with a Birmingham spur connecting with the A45 south of Rugby (Figure 3.4). At its southern end the motorway would join onto the St Albans Bypass Motorway, a 17-mile-long forked approach motorway which was opened as part of the London to Yorkshire Motorway in November 1959. Later in this chapter I examine the motorway's design and landscaping in detail, but in the next section I examine the broader attitudes of landscape architects and the government's Advisory Committee on the Landscape Treatment of Trunk Roads to the design and landscaping of Britain's roads and motorways in the 1940s and 1950s.

Landscape Architecture and the Post-war, Modern Road

During World War II, prominent members of the Institute of Landscape Architects were successful in aligning their profession with the prevailing 'planner-preservationist imagination' (Matless 1998: 223), and with the end of hostilities landscape architects started to gain commissions to work on major public projects, including the Festival of Britain sites and new towns such as Stevenage and Harlow (Clark 1951; Scott 1999). As building restrictions were lifted in the early 1950s, and construction work diversified throughout the late 1950s and 1960s, landscape architects won commissions to landscape new university campuses, schools, forestry plantations, reservoirs, factories, power stations and housing estates (see Mauger 1959). During the late 1940s and early 1950s, economic crises and a focus on 'essential' reconstruction work delayed all major road construction programmes, but the Institute of Landscape Architects worked hard to ensure that their professional members *would* be employed alongside engineers to lay out all future roads and motorways. Three prominent landscape architects – Brenda Colvin, Sylvia Crowe and Geoffrey Jellicoe (all past or future presidents of the ILA) – worked particularly hard to shape government policy and professional opinion on the design and landscaping of roads and motorways.

In her 1948 book *Land and Landscape*, Brenda Colvin argued that landscape architects and engineers must ensure that modern dual carriageway roads are 'fitted' to the contours and existing features of the landscape, so that the road 'will seem to belong happily to its surroundings' and the driver will be kept interested and enlivened (Colvin 1948: 244, 247). Landscaping and planting must be functional: breaking the 'mechanical monotony of engine sound and road surface', keeping drivers 'alert and vigilant', preventing dazzle, framing attractive views, screening eyesores and breaking up the parallelism of the road (Colvin 1948: 246). Colvin stressed that the danger was one of doing too much, in too much detail: the 'English have become too garden-city minded' (Colvin 1948: 249). Small-scale ornamental plants might be suited to urban gardens 'seen at a walking pace', but the 'dramatic variations' characteristic of the English countryside would 'too easily be blurred and lost to the motorist by a lavish use of trees and shrubs of exotic or garden type', which would also prove costly to maintain (Colvin 1948: 248). Local or regional vegetation could best highlight Britain's 'natural landscape variety', while the speeds and scale of modern motoring were ideal for the modern motorist to appreciate the beauty and regional variations of the nation's landscape (Colvin 1948: 248). Here we see that Colvin was clearly mindful of the kinds of practical, non-representational engagements with and experiences of architecture and landscape which geographers have more recently suggested academics and architectural commentators need to explore (see Lees 2001). Movement and speed are seen to be vitally important to the way we see, encounter and inhabit Britain's landscapes, and the role of the landscape architect and engineer must be to translate the speed, scale and function of a particular road into an appropriate landscape.

Colvin's friend and fellow landscape architect Sylvia Crowe provided a more extensive discussion of these themes in her 1960 book *The Landscape of Roads*, and both authors were keen to present solutions to landscape problems (Crowe 1960). While Ian Nairn's highly influential 1955 *Outrage* special issue of *The Architectural Review* had presented a rather gloomy account of the spread of a universal suburbia or 'subtopia' across the English landscape and along England's roads (Nairn 1955), Crowe's writings presented the architecture, planning and design community with positive examples of how modern industry, reservoirs, forestry plantations, power stations and new roads could be fitted into, and even enhance, the landscape (Crowe 1956, 1958, 1960, 1966; Collens and Powell 1999).[15] Her 1956 book *Tomorrow's Landscape* was presented by the architectural critic Eric de Maré as a 'practical guide to the proper adjustment of our landscape' and as the first constructive reply to Nairn's 'prophecy of doom' (de Maré 1956: 121). Crowe suggested that while modern structures were frequently built on a vast scale, divorced from our humanized landscape, 'we are faced with the alternatives of either linking them by siting and design with the existing scale or of creating around them a new landscape related to their own scale' (Crowe 1956: 15). In the case of roads it was the design-speed which would, above all else, affect the scale of the road and its place in the landscape:

> The faster the speed for which it is designed, the further it must depart from
> the old pattern of the humanized landscape. This conflict between machine

speed and human speed is part of the problem which confronts us throughout our mechanized civilization. (Crowe 1960: 13)

As driving speeds increase, the landscape of the road must become more expansive, coherent and free from excessive detail and distractions. A view designed to be appreciated for five seconds by a motorist travelling at 60 mph would require an opening, or frame, 418 feet larger than if the same view were to be framed for a pedestrian travelling at 3 mph (Figure 3.5). The challenge becomes one of composing a landscape which can be viewed or 'read' at speed, and Crowe suggested that landscape architects could learn a great deal from previous landscape and artistic traditions that developed 'principles of penetration and the moving viewpoint' (Crowe 1960: 34). In particular, two artistic traditions could be contrasted with the 'static' viewpoints of the 'classic conception of a landscape' (Crowe 1960: 33). Firstly,

modern painting and sculpture which exploits the strong directional line exploring the depths of a composition, . . . [and, secondly] the English landscape

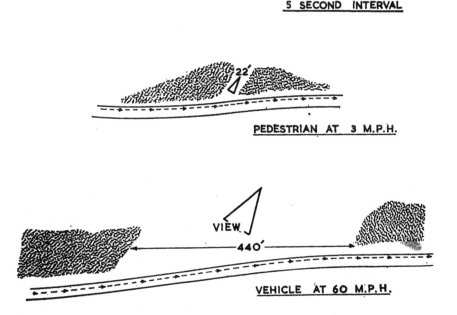

Figure 3.5   Landscapes composed for appreciation by pedestrians and motorists travelling at different speeds. Illustration by John Brookes, from Sylvia Crowe's *The Landscape of Roads*, 1960, p. 34. Reproduced by kind permission of Emap Construct, Elsevier and John Brookes.

school [which] developed the older Chinese conception of a landscape of
movement, to be enjoyed as an unfolding scroll. (Crowe 1960: 33–4)

By placing the modern motorway in this history of sensibilities to movement in landscape art and design, Crowe not only suggests that these traditions may provide 'a valid starting point' for landscape architects and
engineers designing the modern road, but that the landscape architect and
their finished landscapes are continuing a long-established artistic tradition
(Crowe 1960: 34).

Crowe's genealogy of landscape design appears to owe much to a paper
on the landscaping and design of motorways which her friend Geoffrey
Jellicoe had delivered to the Town Planning Institute in 1958 (Jellicoe
1958; see Crowe 1960). Jellicoe opened his paper with a critical discussion
of the landscaping and design of motor roads in Germany and the USA.
Like Colvin and Crowe, he argued that British engineers could draw important lessons from their design, but all three were adamant that the distinctiveness and diversity of Britain's landscapes necessitated the formation of
a British motorway modernism, combining an awareness of foreign precedents with a sensitivity to 'our own traditions and national characteristics',
which are 'nowhere better expressed in Landscape than in the great English
park' (Jellicoe 1958: 275, 276). Jellicoe, like Crowe, provided a somewhat
compressed history of English landscape design. He pointed to the important lessons of the 'art of the picturesque', before describing how the work
of Humphry Repton was instructive for landscape architects, for it was he
who taught us that 'a road that is agreeable to drive along, is also agreeable
as static scenery in the surrounding landscape' (Jellicoe 1958: 276). As
Stephen Daniels has shown, the mobilities of late eighteenth- and early
nineteenth-century polite society – when Repton was conducting his work
– became associated with new techniques for not only designing, but also
experiencing and conducting oneself in, the landscape (Daniels 1996,
1999), but Jellicoe stresses that while Repton sees the road as essentially
'subsidiary to the park, . . . in modern England it is the road that organizes
the landscape through which it passes' (Jellicoe 1958: 276).

Twentieth-century landscape architects argued that motorways must be
designed around the movements and embodied vision of the high-speed
motorist and composed from specific features on a site, but Jellicoe also
outlined a number of basic visual effects which appealed to the human eye
and could be adjusted to the scale of any road. These effects were seen to
be present in a nineteenth-century watercolour, *The Shadowed Road* (often
titled *Landscape with Cottages*) (Figure 3.6), by the Norwich School painter
John Crome (1768–1821), which reveals 'a complex of tree foliage, the
incident of a cottage, the glimpse of a distant view, and an overall play of
light and shade' (Jellicoe 1958: 276; on Crome, see Goldberg 1978;

Figure 3.6 *The Shadowed Road,* c.1808–10. Watercolour by John Crome (1768–1821). The painting is usually titled *Landscape with Cottages.* Reproduced by permission of V&A Images/ Victoria and Albert Museum.

Hemingway 1979). While Jellicoe acknowledged that the picture was composed on an inappropriate scale for a motorway, the landscape architect merely had to translate the scene to the dimensions, scale and speeds of a modern road. The architectural critic Raymond Spurrier wondered whether Britain's highway engineers would pay any attention to Jellicoe's suggestions, as 'the average landscape of the average motor road in Britain' exhibited none of the compositional elements present in *The Shadowed Road* (Figure 3.7) (Spurrier 1959: 242). Britain's modern roads were badly aligned, boring, adorned with poorly designed signs and vegetation, and there were few positive British examples to which landscape commentators could turn for inspiration. Crowe (1956, 1960) praised the siting and engineering – but not the planting carried out by the Roads Beautifying Association – on the Mickleham Bypass and Bix-Henley Road, while Jellicoe lauded the designers of the Oxford Bypass for their separation of dual carriageways, incorporation of existing trees and hedges, and creation of 'a scenery of the highest order' (Jellicoe 1958: 280). A photograph of the Oxford Bypass by Geoffrey Jellicoe's wife Susan was included in his article and Crowe's *The Landscape of Roads* with admiring captions, while a sketch which appears to be based upon this photograph was presented by Raymond Spurrier as 'the Shadowed Road – modern style' (Figure 3.8): the antidote to the average British road (Spurrier 1959: 244).

John Crome's watercolour of *The Shadowed Road*, right, shows us a pattern of road and landscape in the horse-drawn ages, adapted to the needs of its traffic, integrated with the surrounding scenery, interesting as an unfolding experience to the traveller. The average landscape of the average motor-road in Britain, below, is none of these things, except possibly the first, but even that is doubtful, since the boredom of its endless road-ribbon and its undesigned surroundings can adversely affect a driver's attentiveness. In the article which begins opposite, Raymond Spurrier emphasizes the need for motorways to be fully thought-out visual conceptions, comparable in quality to the horse-drawn road, but re-phrased on the scale of high-speed travel.

A 99
1 mile

Figure 3.7 'The average landscape of the average motor-road in Britain . . .'. P. 242 of *The Architectural Review*, volume 125, April 1959, prefacing an article entitled 'Caution – road works' by Raymond Spurrier. Reproduced by permission of The Architectural Review/Emap Construct.

7, the Shadowed Road—modern style. Divided carriage-ways welded together with the landscape by planting and landform, and by light, shade, and texture.

Figure 3.8   'The Shadowed Road – modern style'. Illustration from an article entitled 'Caution – road works' by Raymond Spurrier, *The Architectural Review*, volume 125, April 1959, p. 244. Reproduced by permission of The Architectural Review/Emap Construct.

The alignment of the views of Jellicoe, Crowe and *The Architectural Review*'s Raymond Spurrier in their optimistic hope that contemporary landscape architects, engineers and designers could create modern motorways inspired by English landscape traditions is not surprising. While Jellicoe's approach to landscape design drew upon such diverse influences as abstract modern art, sculpture and, from the early 1960s, Jungian psychology (see Jellicoe 1987; Spens 1992, 1994; S. Harvey 1998), the approaches that he, Crowe and Colvin proposed for landscaping Britain's roads and motorways had notable parallels with arguments about the importance of picturesque theories to the planning and design of Britain's townscapes and landscapes, which key figures associated with *The Architectural Review* had been developing from the early 1940s (see de Wolfe 1949; Pevsner 1956, 1968; Cullen 1961; Banham 1968; Daniels 1993; Matless 1998, 2002; Powers 2001, 2002; Bullock 2002). Nikolaus Pevsner, Hubert de Chronin Hastings, James M. Richards and Gordon Cullen argued that eighteenth-century picturesque principles could provide a useful precedent for contemporary town planners, architects and landscape architects; showing how they might compose informal and varied layouts, views and relational compositions by using the materials – and respecting the distinctive design aesthetics – 'found' on a particular site (de Wolfe

1949). Just as Jellicoe thought about how drivers would encounter and move through the recomposed landscape of *The Shadowed Road* (modern style), twentieth-century reformulations of the picturesque were framed as an opportunity to understand mobile encounters with 'the embodied, the differentiated, the phenomenal world' (de Wolfe 1949: 362; see also Cullen 1961). *The Architectural Review* described how the picturesque layout of the South Bank site of the Festival of Britain – which was celebrated by admirers of the picturesque, but highly criticized by a new generation of Brutalist architects and critics such as Reyner Banham – was 'contrived for the benefit of the moving, not the stationary, spectator' (*The Architectural Review* 1951: 78), while in 1956 the *Review*'s Art Editor, Gordon Cullen, pointed to the need to understand 'vision in motion' and establish a clear visual design code in order that roads may be considered as townscape or landscape (Cullen 1956: 243; see also Cullen 1961). In a discussion of Jellicoe's ideas and Britain's future motorways in *The Architectural Review*, Spurrier was quite clear that it was our 'national tradition – the technique of creating a series of unfolding views, so firmly established by the eighteenth century Picturesque Movement, that should so obviously be applied to motorway design' (Spurrier 1959: 244). Picturesque principles should be fused with modern design principles to create a distinctively English motorway aesthetic.

Drawing upon traditions of landscape design, modern art and highway engineering, Colvin, Crowe and Jellicoe provided persuasive written accounts of principles for designing and landscaping roads and motorways, but as high-profile landscape architects they were keen to influence local and national government policy. During World War II Brenda Colvin prepared *Trees for Town and Country*: a guide to aid post-war reconstruction which included sections on road and street planting, and was published for the Association of Planning and Regional Reconstruction in 1947 (Colvin 1947).[16] The previous year she had chaired an Institute of Landscape Architects committee and prepared their report on *Roads in the Landscape* (ILA 1946). Between 1949 and 1954 Colvin served as the Institute's representative on a Council for the Preservation of Rural England committee concerned with the landscaping of roads (Joint Committee 1954), while in 1955 she was appointed as the Institute's representative on the government's newly established Advisory Committee on the Landscape Treatment of Trunk Roads.[17] The Landscape Advisory Committee, as they were commonly known, included such key figures as Clough Williams-Ellis (Council for the Preservation of Rural Wales), George Langley-Taylor (CPRE), Lord Rosse, Wilfrid Fox (Roads Beautifying Association), Lord Bolton (Royal Forestry Society of England and Wales), Sir Eric Savill (Deputy Ranger, Windsor Great Park) and Dr

George Taylor (Keeper of Botany, British Museum). At the Committee's inaugural meeting in April 1956, their chairman, Sir David Bowes-Lyon – President of the Royal Horticultural Society, and brother of Queen Elizabeth the Queen Mother – expressed his hope that they would 'advocate the use of indigenous trees and discourage the use of foreign trees . . . which were uncharacteristic of the region'.[18] This approach echoed the principles of such organizations as the ILA and CPRE and was implicitly critical of the work of the Roads Beautifying Association, and after just four meetings, one site visit and an argument over central reservation planting, Wilfrid Fox resigned – citing his fundamentally different 'outlook' to 'the Chairman and other vocal members of the committee' as the reason.[19] The Committee's preference for indigenous, native species resonates with the push in Nazi Germany to plant native German species along the *Autobahnen* in Germany and Poland, but the discussions in 1950s Britain reflected ongoing debates amongst horticulturists, ecologists and landscape architects about the abilities of different plant species to survive and look right in the English landscape rather than an entwining of exclusionary nationalist political ideologies with debates about landscape, ecology and race (cf. Shand 1984; Dimendberg 1995; Rollins 1995; Matless 1996; Zeller 1999).[20]

The Landscape Advisory Committee's first major project was to approve the detailed architectural designs and planting programme for the initial sections of the London to Yorkshire Motorway (M1). The route and preliminary designs had been established by Sir Owen Williams and Partners before the formation of the Committee. In 1956 civil servants expressed concern that consultation with the new committee might seriously delay the motorway, and so they were instructed to focus their attention on the more superficial aspects of design and planting:

> In the case of the London–Yorkshire Road . . . it has taken . . . several years to find a satisfactory line. If a road of this description were then to be submitted to a Committee who would no doubt want to go over it from end to end and criticise it from an aesthetic point of view and the consultants had then to examine their alternative suggestions, I am afraid that the making of the scheme and the actual construction would have to be very considerably postponed.[21]

This decision spurred the Institute of Landscape Architects to pressure the government to appoint qualified landscape consultants to advise on the design of the M1 and all future motorways. Geoffrey Jellicoe used his role as a Royal Fine Art Commissioner to ensure that the Commission's secretary pressed the Ministry on the matter,[22] which led government officials and Sir Owen Williams and Partners to hold a meeting to discuss the appointment of suitable consultants with the President of the ILA (Richard Sudell) and Jellicoe in July 1956.[23] The Institute of Landscape Architects suggested five consultants who could be employed on the five sections of the motorway: Brenda Colvin, Sylvia Crowe, J. W. M. Dudding, Richard Sudell and L. Milner White.[24] After discovering the high fees recommended by the ILA, the Ministry decided that the consulting engineers should employ their own consultants,[25] and this decision – coupled with Sir Owen's appointment of two foresters, Archibald P. Long CBE and Mr A. J. M. Clay, rather than qualified landscape architects – dismayed leading members of the Institute of Landscape Architects, Royal Fine Art Commission and Royal Institute of British Architects.[26] During 1958 and 1959, when the shape of the M1 was starting to materialize on the ground, all three organizations made their views clear in letters and press releases forwarded to the government and newspapers (*The Architects' Journal* 1959; *JILA* 1959; *Parliamentary Debates* 1959a; RFAC 1959; *The Times* 1959a).[27] The Secretary of the Royal Fine Art Commission, Godfrey Samuel, expressed the view of his colleagues in a letter to the Ministry of Transport and Civil Aviation in December 1958:

> As far as the first phase of the London–Yorkshire Motorway is concerned, the work in progress was recently inspected by representatives of the Commission and confirmed the impression that had been gained from the original drawings and model that the scheme would have benefited from collaboration with a landscape consultant at the design stage. It is now clearly too late, however, for this omission to be made good.[28]

The consensus amongst these organizations was that landscape consultants must be appointed to work with engineers on motorways at an early phase. In a letter published in *The Times* in May 1959, the then President of the Institute of Landscape Architects, Sylvia Crowe, stated that the Landscape Advisory Committee was no 'substitute for built-in professional advice', and that 'those trained to assess the character of a landscape' must form an important part of the planning team 'from the reconnaissance stage onwards' (Crowe 1959a: 11). The Ministry of Transport eventually decided to appoint a landscape architect to their staff in 1961. Geoffrey Jellicoe was asked to submit the names of suitable candidates, and one of these, Michael Porter – who had 'worked under Miss Sylvia Crowe at Basildon New Town' – was appointed to the new post in July 1961.[29] This was too late for the first sections of the M1.

'A New Look at the English Landscape': Landscape Architecture, Movement and the Aesthetics of a Modern Motorway

A road is a flow channel; its virtues will be those of smoothness and easy flow – minimum changes of velocity in any direction. Its visual virtues will be similar; no abruptness, no interruption, no fussiness, until the road superimposes its own slow steady rhythm of turnout, service area and major destination on to the undertones of change of geology and land use. (Harris 1959: 162)

In his review of the M1 for *The Architects' Journal*, civil engineer Alan Harris captured the emphasis of a broad range of landscape architects on the importance of a sense of flow and a mobile viewpoint in designing the landscapes of roads and motorways. With almost no involvement by the Landscape Advisory Committee or qualified landscape architects in the detailed design of the M1, it is not surprising that commentators stressed that the first sections of the motorway lacked the characteristics of a good modern motorway landscape. In *The Landscape of Roads* Sylvia Crowe compared the poor design of the M1 with the positive landscaping on the Ulm to Baden-Baden *Autobahn* in Germany (Figure 3.9). While the German motor road is seen to have a 'fluid plasticity', 'smooth transition between road and countryside' and landforms which are shaped and related to the surrounding landscape, the M1 is held to be afflicted by harsh, angular lines and landforms that act as a 'jarring element', divorcing the road from the landscape (Crowe 1960: 94, 93; see also *Proceedings* 1961). Flow and movement emerge as positive aspects of the aesthetics of landscape, which Crowe and others contrast with the negative interruptions, disruptions and angular jarring effects of a poorly designed motorway. The routing of the M1 through 'The Midlands Plain' had made it likely that it would interrupt the 'intricate and flowing landscape' of the area, but Crowe stressed that many of the disruptive forms and features of the motorway *could* have been avoided with appropriate landscaping (Crowe 1960: 57).

Modern structures in the landscape

In 'The London–Birmingham Motorway: a new look at the English land-scape', which appeared in *The Geographical Magazine* in October 1959, Brenda Colvin criticized the 'hard sharp lines and clumsy angles' of the motorway embankments, before focusing on Sir Owen Williams and Partners' distinctive, standardized, concrete two-span over-bridges (Figure 3.10). Despite concerns that Sir Owen's standardized bridges might 'lead to monotony', the Royal Fine Art Commission had approved their design,

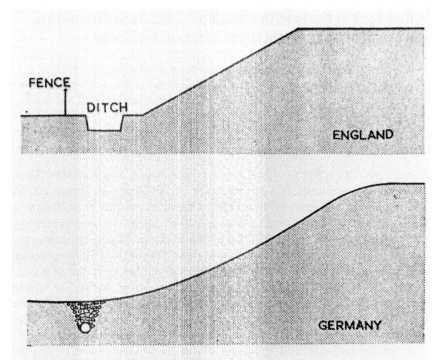

Fig. 64. The autobahn embankments flow smoothly between road and countryside, as opposed to the angular interruptions of the British motorway.

Figure 3.9   Motorway embankments in England and Germany. Illustration by John Brookes, from Sylvia Crowe's *The Landscape of Roads*, 1960, p. 95. Reproduced by kind permission of Emap Construct, Elsevier and John Brookes.

subject to 'improvements to detail', in January 1958.[30] Brenda Colvin was more critical, expressing concern that Sir Owen's over-bridges 'seem very heavy in design': 'The central supporting pillar spoils the flow of open view under the arch, and the solid concrete parapet increases the apparent depth of the arch and the sense of its weight to an extent which is all the more oppressive because so frequent' (Colvin 1959: 243). At a time when many engineers were designing light, unobtrusive, clean-lined pre-stressed concrete bridges – with open metal rails replacing a solid parapet – Sir Owen's reinforced concrete over-bridges were criticized by a broad array of architects, engineers and landscape architects (*Proceedings* 1961). The central columns, solid parapets and reinforced concrete design had been adopted for reasons of cost and speed of construction, and Sir Owen defended their

Figure 3.10   Two-span over-bridges spanning the M1 motorway. Reproduced by permission of Laing O'Rourke Plc.

design by stating that 'they have a shape that will always be remembered' (cited in Cottam 1986a: 139), and that they were characteristic of the modern era: '. . . in the design of the structures . . . regard had been paid to the spirit of the age, to the genius of the age. They were in a bold, massive manner' (paraphrased in *Team Spirit* 1958f: 5).[31] In his presidential address to the Royal Institute of British Architects in November 1959, Basil Spence – a member of the Bridges Committee of the Royal Fine Art Commission – explained how the 'breadth and strength' of the bridges reminded him 'of some of the great Roman works' (Spence 1959: 36). Spence looked forward to using the motorway, as he was overseeing the execution of his design for the new Coventry Cathedral, and its opening meant he could 'now get to Coventry without the awful frustrations of the A5' (Spence 1959: 36). Few architectural commentators shared his enthusiasm or repeated his praises.

Sylvia Crowe suggested that the bridges were rather 'static' when compared with the light *Autobahn* bridges, appearing as 'rough knots' and providing 'a visual check' on the sense of flow in the landscape (Crowe 1960: 94). This need for movement and continuity was seen to work in several directions, and Crowe argued that Sir Owen's bridges impeded the flow of the landscape both *along* and *across* the motorway: '. . . they divided the landscape between one side of the road and the other and gave the impression that they were impeding the passage of traffic' (Crowe 1962: 225). Raymond Spurrier likened the 'visual interruption' experienced by the motorist to a 'thump, thump, thump against the sensibility with painful monotony like the expansion joints along a bad concrete road' (Spurrier 1961: 230). Members of the Landscape Advisory Committee expressed concern 'at the heavy appearance of the bridges' after a visit to the motorway in May 1959 (*The Surveyor* 1959: 623),[32] while Brenda Colvin's replacement on the Committee, President of the ILA James Adams, criticized the 'brutal bridges in careless contexts' (J. W. R. Adams 1962: 4). Alan Harris criticized the 'surpassing ugliness' of the over-bridges (Harris 1959: 165), and Ian Nairn – author of *Outrage* and Assistant Editor of *The Architectural Review* – criticized the bulk and mass of the bridges in a reference to 'Sir Owen Williams's deplorable attempts to outdo Vanbrugh' (Nairn 1963: 426). An array of other commentators renowned for their quite different attitudes to modern architecture also voiced criticisms of the bridges. Reyner Banham, the architecture and design critic who detailed the rise of New Brutalist architecture in 1950s Britain (see Banham 1966), was especially hostile to the design of the motorway's 'coarse, cheap bridges' which announced 'the ugliest piece of motor road in the world' (Banham 1972a: 242).[33] Sir Owen Williams was accused of reviving a 'face-grinding Victorian "economy"' that was evident in 'practically everything one sees along M1' (Banham 1960: 786). He had designed some of Britain's most celebrated modernist structures of the inter-war years – including, as noted above, the Boots 'Wets' factory at Beeston, Nottingham – but while he was 'one of the white hopes' of the British Modern Movement in the 1930s, Banham felt that his post-war constructions had been a great disappointment to a new generation of British architects and critics (notably the Brutalists) (Banham 1960: 786).[34] His pre-war functional structures, 'unsullied by aesthetic intentions', had been displaced by a 'deliberately *anti*-aesthetic' approach in his post-war motorway work (Banham 1960: 784).

Writing from a very different critical position in his 'Men and buildings' column in the *Daily Telegraph*, the poet, conservationist and architectural commentator John Betjeman referred to the landscaping and bridges on the M1 as 'matters of lasting regret' (Betjeman 1960: 15). Betjeman's rival broadcaster, guide-book writer and architectural commentator Nikolaus Pevsner added to the barrage of critical writings and reviews in introductions to his guides to *Northamptonshire* and *Buckinghamshire* in *The Buildings of England* series:

> . . . more than 130 BRIDGES had to be built. . . . Especially surprising are the supports between the traffic lanes in the N and S directions: a kind of elementary columns [*sic*], without base and capital, but with an abacus – a curious period suggestion, not called for in this forward-looking job. Sir Owen Williams evidently wanted to impress permanence on us, and permanence is a doubtful quality in devices connected with vehicles and means of transport. Elegance, lightness, and resilience might have been preferable. . . . On the motorway elegance was arrived at only in the foot-bridges. Even retaining walls, revetments, etc., are of concrete blocks. (Pevsner 1961: 66; see also 1960)

As the two-year-old motorway was 'the most prominent structure' built in Northamptonshire in the twentieth century, Pevsner felt obliged to include it in his architectural history of the county, but he was careful to add that the motorway was not typical of the capitalized category/style 'MODERN ARCHITECTURE': 'So the Motorway is MODERN ARCHITECTURE only with reservations' (Pevsner 1961: 66). In reflecting on the entire Buildings of England series in 1974, Pevsner inferred that he had got rather caught up in the moment, including the motorway in *Northamptonshire* for its novelty and modernity rather than for its architectural quality and historical importance: '. . . the appearance of the motorways. One got used to them quickly, and it seems odd already now that only twelve years ago I sacrificed one of my one hundred illustrations to so rich a county as Northamptonshire for the purpose of showing the M1' (Pevsner 1974: 17). In 1959, 1960 and 1961, the M1 appeared new, modern and exciting to Dr Pevsner. By 1974 it had aged and was just another motorway – a theme I return to in chapter 6.

Architectural and design commentators inevitably focused their attention on the 59 two-span bridges passing over the motorway, but critics also cast their judgement on a range of other structures, many of which were invisible to the motorist's eye.[35] Alan Harris felt that the four-span farm access bridges passing over the motorway were 'more attractive visually' than the two-span road bridges (Harris 1959: 164). He was less complementary about the four pre-cast, cantilevered railway bridges which took railway lines under the motorway and were so complicated 'as to be positively comic' (Harris 1959: 164). Spurrier felt these bridges performed their function 'with the grunting for effect of a second-rate acrobat and still leaves an impression that it would be unsafe to venture beneath such a complicated structure' (Spurrier 1961: 234). Complicatedness is contrasted with functional design, but for Sir Owen these cantilevered bridges were simple, functional structures designed so that their construction would cause minimal interference to railway traffic. Other structures gained more favourable reviews. Where the motorway's embankments were high enough, mass-arch tunnels carried roads and access-ways under the motorway. In contrasting a mass-arch cattle creep with a square-topped under-pass (Figure 3.11), Sylvia Crowe reserved light praise for the arched design, which was 'not quite so gloomy' as the square under-bridge, and served its function without an 'ornamental parapet' (Crowe 1959b: 160). Brenda Colvin described the larger mass arch road tunnel at Newport Pagnell as having 'a curious gigantic beauty' (Colvin 1959: 246). Alan Harris felt that its design was 'very impressive' (Harris 1959: 165). Pevsner was less approving, complaining that such structures 'impress by a cyclopean rudeness rather than elegance' (Pevsner 1961: 66). Pevsner, like Banham, Crowe, Colvin, Jellicoe and Nairn, was quite clear that modern architecture

Fig. 8 (above). *A square-topped underpass, marked by a solid, ornamented parapet, and with stone-pitching at the sides. Fig. 9 (below). A higher, arched concrete cattle creep, used where head-room permits, which fits in more happily and is not quite so gloomy.*

Figure 3.11 Two contrasting underpasses designed by Sir Owen Williams and Partners, from an article by Sylvia Crowe in *The Architects' Journal*, volume 130, 10 September 1959, p. 160. Reproduced by permission of The Architects' Journal/ Emap Construct.

and modern bridges could be attractive and well designed, but the design of motorway structures must reflect an appropriate, contemporary, functional modernism, and their form and presence must be light, almost imperceptible, and not appear to impede movement and the visual flow of the landscape.

Landscaping and planting

Landscape architects could do little about the design of the bridges, but they did suggest techniques for integrating the motorway's structures into their surroundings and maintaining the visual flow of the landscape. Sylvia Crowe explained how 'the functional use of planting [could] produce the link between the landscape of speed and the landscape of nature', and how 'massed planting' could improve 'the bad shape of the banks and the appearance of the bridges' (Crowe 1960: 95, 116; cf. Crowe 1959b). Functional planting could help unify the motorway and the landscape, reinstating a sense of flow and guiding the driver's vision in an appropriate manner. Planting could screen unsightly views and help prevent monotony and boredom, but Brenda Colvin stressed that without the influence of the Landscape Advisory Committee and Royal Fine Art Commission the M1 may have 'had a ribbon of Forsythia and other garden shrubs on the central reserve . . . and subtopian decoration on side reserves and embankments' (Colvin 1959: 246). The planting proposals were prepared by Sir Owen Williams and Partners.

Archibald Long, the Consultant on Forestry and Landscape employed by Sir Owen Williams and Partners, was a well-respected forester – being President of the Society of Foresters of Great Britain, a retired Director of Forestry for Wales and former Assistant Commissioner for Forestry for England and Wales – but his first landscape report included species of tree and shrub which the Landscape Advisory Committee deemed to be too fussy, ornamental, colourful and urban for a modern rural motorway (Sir OWP 1957). At a meeting of the Committee in July 1957, Sir Eric Savill questioned proposals to plant forsythia and pyracantha, while Sir David Bowes-Lyon and the whole committee 'agreed that flowering plants of a semi-garden character were misplaced in real countryside'. Sir Ralph Clarke of the Royal Forestry Society for England and Wales voiced concerns over proposals to plant Austrian Pines, as 'less ugly conifers [are] available', while copper beech, purple sycamore and whitebeam were thought unsuitable as 'the Committee did not favour colour variations in foliage other than shades of green'.[36] Mr Long revised the planting schedules, which were rejected again in January 1958. The Committee urged the consultants to simplify their proposals, avoid ornamental species of tree and shrub, focus on indigenous trees, pay attention to the speed and experiences of motorists, and submit detailed illustrated planting plans rather than lists of species.[37] Visits to the construction site in June 1958 and May 1959 confirmed the Committee's view that simple, large-scale massed plantings would be essential on such a vast and fast motorway. As Clough Williams-Ellis stated after the visit in May 1959: 'Traversing the actual carriageway one realized more vividly than ever the immense size. . . . It so far transcends the hitherto generally accepted human scale as actually also to dwarf nature itself . . . there is a danger that any landscaping effects may merely produce a niggling and irritating triviality.'[38] The tour brought Clough into close contact with Sir Owen's over-bridges, and he expressed concern that they were 'a good deal heavier and more clumsy . . . than was suggested either by the drawings or models'.[39] Planting would be needed to mask the abutments of bridges and the sharp ends to parapets, soften the hard lines of embankments and cuttings, create screens and frame views.

The consultants' modified and simplified planting plans were finally accepted by the Landscape Advisory Committee in October 1959, and 72,050 trees and 4700 shrubs were planted by Turfsoils Limited in the winter of 1959–60. By this time, Sir William Ling Taylor was overseeing the planting programme for Sir Owen Williams and Partners, and twenty-five species of tree and ten of shrub were planted in the first winter, although 81% of the trees were of just five common, long established species: Alder (10,000), Ash (11,000), Common Oak (20,000), Scots Pine (10,000) and Spanish chestnut (7600) (Sir OWP 1960).[40] The Landscape Advisory Committee felt it had succeeded in preventing Sir Owen's landscape consultants from urbanizing or suburbanizing the motorway with detailed ornamental species that would interrupt the flow of the landscape and distract drivers, but disagreements soon emerged over another issue: the design and landscaping of the first two service areas at Newport Pagnell and Watford Gap.

Towards a Road Style: Service Areas in the Landscape

During 1955 and 1956 the Ministry of Transport and Civil Aviation consulted a wide range of organizations about the facilities they would like to see on Britain's future motorways.[41] Petroleum companies, road user groups, commercial motoring organizations, caterers and hoteliers were sent questionnaires and attended consultation meetings, and their views were used to shape government policy on the location, design and use of service areas. The service areas on M1 were sited by Sir Owen Williams and Partners, who submitted their proposals to the Ministry in February 1956.[42] Sir Owen's son and business partner Owen Tudor Williams visited the USA to look at the country's motor roads and rest areas in 1956, and this led the firm to propose that four service areas be sited on the motorway at intervals of approximately 12 miles, with every third site being a major service area and intervening sites being smaller, minor areas.[43] The two types of site would 'service' different kinds of motorist and contain different facilities. The two major areas – located south of the Birmingham Spur (M45) at Watford Gap, and close to Toddington village, north of Luton – would cater for all classes of road user, while the minor service areas at Newport Pagnell and Rothersthorpe were to be designed primarily for lorry drivers rather than 'the travelling public' (Sir OWP 1957: 6). Plans for a fifth service area close to Redbourn on the St Albans Bypass were dropped due to widespread opposition and legal difficulties in February 1959,[44] while in March 1958 the Ministry decided that only two of the four sites, at Newport Pagnell and Watford Gap, would initially be developed as 'prototype' service areas.[45] Companies were invited to tender for leases to design and operate the two sites, and the proposals were assessed on financial grounds and 'the experience and standing of the firms and the adequacy of fuelling and catering arrangements proposed'.[46] Blue Boar Company Limited – an operator of breakdown services and transport cafés in the Watford Gap area – won the lease to design and operate Watford Gap service area, and Motorway Services Limited – a partnership of Blue Star Garages and caterer Fortes – acquired the lease for Newport Pagnell service area.[47]

The designs for Newport Pagnell and Watford Gap service areas received widespread criticism from the Royal Fine Art Commission, Royal Institute of British Architects and the government's Landscape Advisory Committee, as separate architects were engaged in designing the main buildings (the leaseholder's architect), police posts (the corresponding county council architect) and maintenance depots and footbridges (Sir Owen Williams and Partners) located at each site.[48] All three organizations pushed for the appointment of a single architect to coordinate the design of each site, but despite receiving names of suitable candidates who could advise on matters of architecture and landscape architecture, the Ministry decided that none of the buildings would 'be of great importance' or 'worthy of the attention of architects of the eminence of those suggested by the President of the R. I. B. A'.[49] Britain's new, vernacular, roadside architecture would be functional and modern, but not nationally significant, as these were, 'after all, only restaurants, petrol stations, garages and hopper installations'. What's more, the service areas already had a coordinating architect, Sir Owen Williams.

Criticisms were directed at specific elements of the layout and design of both Newport Pagnell and Watford Gap service areas. The Landscape Advisory Committee were 'not very satisfied' with the layout proposed for Watford Gap service area.[50] The Committee suggested that alterations be made to the design of a concrete footbridge spanning the motorway, and

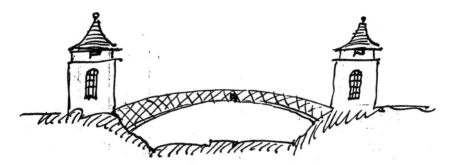

Figure 3.12   Design by Clough Williams-Ellis of a footbridge for Watford Gap service area, 1959. Reproduced from file MT 121/359 in The National Archives, Kew, by kind permission of the Trustees of the Second Portmeirion Foundation and The National Archives.

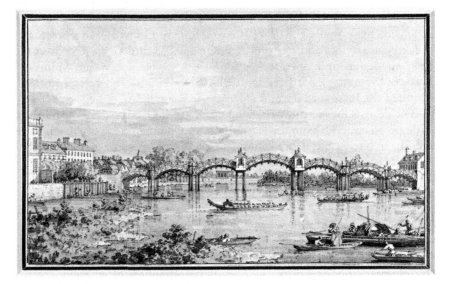

Figure 3.13   *Hampton Court Bridge*, by Canaletto, c.1754. British Museum negative PP.5–146 © Copyright of The Trustees of The British Museum. Reproduced by permission.

Clough Williams-Ellis provided a sketch of a suitable alternative, a 'footbridge supported by two elegant pavilions, after the style of Canaletto's picture of Hampton Court Bridge' (Figures 3.12 and 3.13).[51] The Ministry evidently favoured Sir Owen's functional concrete design over Clough's ornamental design, which was more suited to his Italianate village of Portmeirion than the functional, modern landscapes of M1 (on

Portmeirion, see Gruffudd 1995b). The design and layout of the service area buildings provoked more vocal concerns. The CPRE's representative on the Landscape Advisory Committee, George Langley-Taylor, expressed disquiet over the 'lack of cohesion between the different aspects' of the sites, and the use of flat roofs on buildings.[52] As he stated in a letter to the Ministry about the design of the Watford Gap police post:

> I find it difficult to comment because I fear that my objection to the long flat roof may be interpreted as an objection to modern architecture. Frankly I do not like it because I feel however right it might be as a modern building this long straight line is bound to be a jar on the landscape and I feel most strongly that in dealing with our motorways we should try to achieve a sympathy with the landscape and avoid introducing any 'shock' in our designs.[53]

Sir David Bowes-Lyon agreed, stating that pitched roofs may 'help break up the straight lines of the buildings', while Clough Williams-Ellis also disliked the designs, adding that they 'were a fair sample of the modern trend in architecture'.[54] Modern architecture *may* be appropriate for the motorway, but these committee members advocated a more conservative modernist design aesthetic; a modernism reflecting English vernacular architecture, which was more aligned with the picturesque modernism of *The Architectural Review* than the Brutalist modernism of Reyner Banham and Alison and Peter Smithson. The flat roofs remained in the plans, and architectural critics agreed with Williams-Ellis that the service areas contained average modern buildings which would not enhance the English countryside.

The main buildings at Watford Gap were designed by Harry W. Weedon and Partners. Weedon was renowned for his designs for over 150 cinemas across Britain, as well as industrial and housing projects in and around Birmingham (*The Times* 1970a), but *The Architects' Journal* disliked his 'commonplace design' for the service area buildings, which 'does not auger well for future motorways' (Astragal 1960: 417). Newport Pagnell service area – which was designed by Sidney Clough and Sons, architects of a number of Britain's largest ice-rinks of the 1920s – received similar criticisms. Writing on 'Road-style on the motorway' in *The Architectural Review*, Raymond Spurrier criticized Newport Pagnell services for its 'nondescript buildings and irresolute planning', which has the 'usual subtopian results' (Spurrier 1960a: 406). The service area had brought subtopia to rural Buckinghamshire, and Spurrier insisted that the solution was to establish a functional road-style which would become an acceptable part of the landscape, as had the age-old 'nautical tradition', the contemporary 'aeronautical tradition' and the 'hard functional engineering' of the railways (Spurrier 1960a: 407). History could provide lessons for motorway designers:

... what is important now is whether the service areas on the motorways will be as felicitous in design as they were in the stage coach era – a time when all kinds of objects, furniture, buildings, vehicles – as well as being embraced by a coherent style – were informed by a common visual excellence. Will the service area at Watford Gap, M1 be as fine as Colnbrook village was on the old Bath Road [?] (Spurrier 1960a: 408)

Spurrier insisted that the service areas needed to be coordinated in their design, with 'the guidance of a master eye trained in the art of arranging and organising shape and colour and pattern' (Spurrier 1960a: 408). Spurrier's own eyes focused on one positive dimension at Newport Pagnell: the sand-hoppers designed by Sir Owen Williams and Partners, which were located in the maintenance depots adjoining all four service areas (Figure 3.14). Spurrier upheld the towers as exemplars of an appropriate, functional, modern road-style which could influence the aesthetics of the entire motorway network: 'The promise of an heroic, functional and appropriate road-style implied by this sand-hopper tower is one of the most exciting aspects of the present programme of motorway development' (Spurrier 1960a: 406). Sir Owen's hoppers promised a great deal, but, Spurrier lamented, the layout of Newport Pagnell service area suggested that the towers 'may well be disgraced by the service areas in which they stood' (Spurrier 1960a: 406).

Sir Owen had planned each service area on a circular site (or two half semi-circles), and critics saw their forms as somewhat alien intrusions in the English landscape. Reyner Banham remarked that their shape would render 'it almost impossible to adapt the service areas and their buildings to the landscape' (Banham 1960: 786). Spurrier observed the 'disintegrative, centrifugal effect, completely out of harmony with the flow characteristics of motorway traffic, that stems from the use of radial planning' (Spurrier 1960a: 409). The motorway was a space of linearity and movement, and it was felt that the circular forms of Newport Pagnell service area were wholly inappropriate, piercing the landscape and breaking its flow. Spurrier had identified the problems of Newport Pagnell service area, but to 'illustrate and clarify' his points, and show what could be done at future sites, *The Architectural Review* commissioned architects Leonard Manasseh and Partners to 'design a model motor-way service area' based on 'an actual site on Britain's first large-scale motor-way, M1' (Leonard Manasseh and Partners 1960: 417). Manasseh's model service area was built around the two distinctive movements, or 'flow system[s]', of car drivers and commercial vehicle drivers. The two sets of consumers would require their own parking areas, which would be separated by a distinctive central piazza and service area buildings made of 'local materials'. Manasseh appears to have engaged with the distinctive, modern, picturesque princi-

The promise of an heroic, functional and appropriate road-style implied by this sand-hopper tower is one of the most exciting aspects of the present programme of motor-way development. But is it a promise that will be fulfilled? The air-photograph of the Newport Pagnell service area on M1 (page 409) with its nondescript buildings and irresolute planning, suggests that the sand towers may well be disgraced by the service areas in which they stand, unless positive steps are taken to remedy a situation in which official drift and the chances of commerce threaten to produce their usual subtopian results. In the study of motor-way service areas which begins opposite, Raymond Spurrier discusses the problem at the levels of style, planning and legislation.

Figure 3.14   A maintenance depot loading hopper designed by Sir Owen Williams and Partners. P. 406 of *The Architectural Review*, volume 128, December 1960, prefacing an article entitled 'Road-style on the motorway' by Raymond Spurrier. Reproduced by permission of The Architectural Review/Emap Construct.

ples of *The Architectural Review* 'set', and to have engaged with Spurrier's call for 'the art of townscape' to be harnessed in order to engage travellers with their surroundings and to provide them with 'enjoyment from their stop' (Spurrier 1960b: 413).

Architectural commentators and members of the Landscape Advisory Committee expressed concern about how one should categorize and subsequently treat the landscapes of a motorway service area. Do they belong to the urban, rural or suburban? Or are they a new kind of space, distinct from these three categories? Spurrier may have criticized Newport Pagnell service area for being a subtopian mess, but he was quite clear that as 'the motorway network spreads', service areas would eventually 'take the place of the towns and villages that have hitherto provided contrast, relief, shelter, refreshment, and change of scale' (Spurrier 1960b: 412). Service areas would provide a psychological break from the scale and 'cinerama views' of the motorway proper, but they must not contrast too sharply with the design of the road. As Spurrier concluded in his article 'The service-area problem', a service area 'must look right from the road, from the landscape, and from within itself. The whole ensemble must have style – motorway style' (Spurrier 1960b: 413). Sir Owen Williams and Partners appear to have developed a quite different approach to the treatment of service areas in the landscape.

At a meeting of the Landscape Advisory Committee in February 1958, Dr George Taylor expressed concern that Sir Owen's consultant, Archibald Long, had 'injected "urbanization" into his proposals for service stations',[55] and this was confirmed in July 1959 when a member of Sir Owen's staff wrote to the Ministry to outline their principle of treating the 'interior of Service Areas as partly urbanised'.[56] In the earliest plans for planting Watford Gap service area Mr Long proposed that the parking areas for private and commercial vehicles be separated by green spaces that were 'informal and more of the nature of a park' (Sir OWP 1957: 5). The exterior of the area would be planted with Limes, while it was suggested that flowering shrubs should adorn the interior along with 'more unusual trees' such as tulip tree, maidenhair tree and Wellingtonia (Sir OWP 1957: 5). As Newport Pagnell service area was originally intended to serve lorry drivers, Mr Long prepared a planting scheme which would reflect the tastes and temperaments of the largely working-class, male commercial drivers. The centre of the area would be planted with laburnum or thorn, which would be 'in keeping with the necessity for attracting and pleasing the average lorry driver who would perhaps be more stimulated by a mixture of this nature than with the commoner Ash/Elm mixtures' (Sir OWP 1957: 6). Stimulation and excitement were not the emotions the Landscape Advisory Committee wished to be associated with the landscapes of service areas. At a meeting in July 1960, Committee members expressed concern that ornamental trees and shrubs such as magnolia, liquidambar, rhodo-dendron, viburnum and Fuchsia ricartonii might excite rather than relax drivers, and that the colours of detailed 'flowering shrubs . . . may clash with that of the petrol pumps'.[57] Sir Owen's landscape consultants had treated the service areas as semi-urbanized 'island sites', separated from the motorway *and* surrounding countryside,[58] but the Landscape Advisory Committee emphasized that they must be treated as if they are part of the motorway, with large-scale indigenous trees such as elm, oak, sycamore, ash, maple, horse-chestnut and lime being planted to break up the expanse of the tarmac.[59] Motorway service areas may be traversed at very low speeds compared with the motorway proper, but inappropriate shapes, detailing and clutter would still detract from their function as spaces of relaxation, revitalization and flow. As Raymond Spurrier suggested in *The Architectural Review* in December 1960:

> The architecture of the place, right down to the design and placing, colour and texture of signs, oil cans, pumps, bollards, fences, lighting, paving, and – very important – the assimilation of the parked vehicle, will influence to a considerable extent the efficiency of the service and the pleasure in using the area. (Spurrier 1960b: 412)

During 1959 and 1960 the Landscape Advisory Committee were asked to approve the design of service area litter bins, lamp posts, roadside emer-gency telephone boxes, and a whole host of other design features. The Committee were asked to comment on the design of new traffic signs for all future motorways, but the design of these important features was over-seen by another government committee, the Advisory Committee on Traffic Signs for Motorways.

'Cutting Holes in the Landscape': Britain's Motorway Signs

With their pioneering experimental design work on the Preston Bypass and London to Yorkshire motorways, the Advisory Committee on Traffic Signs for Motorways (formed in 1957, first meeting in January 1958) paved the way for the more comprehensive redesigning of Britain's entire system of (non-motorway) road signs by the Worboys Committee between 1961 and 1963 (see Traffic Signs Committee 1963). Motorway direction and warning signs may appear to be rather mundane, backgrounded technologies we pass and quickly forget on a journey, but civil servants, engineers, typog-raphers and landscape architects held strong and often contrasting views on the most appropriate designs and locations that would enable the signs to fit into the English landscape, and the ideal designs for informing and governing the conduct of vehicle drivers. Motorway signs, then, along with all manner of other technologies and practices, which I discuss in chapter 5, were incorporated into debates about the expected conduct and experiences of Britain's new motorway drivers (cf. Merriman 2005b, 2005d).

The Advisory Committee were chaired by the shipping magnate and influential art and design advocate Sir Colin Anderson, and the membership included designer and typographer Noel Carrington, architect Sir Hugh Casson, Sir William Glanville (Director, Road Research Laboratory), motoring journalist Lord Waleran, and engineers J. S. McNeill and E. J. Powell (ACTSM 1962).[60] The Committee worked closely with scientists in the government's Road Research Laboratory to establish colours, fonts and sizes of letters and signs which would fit into the landscape and most effectively inform and influence the movements of drivers. International practice shaped a range of decisions. As the UK had signed an international agreement formulated by the Inland Transport Committee of the United Nations' Economic Commission for Europe in 1949, the nation's principal motor roads were envisioned as part of a European network as well as a national programme – the M1 was to form part of the seemingly less primary and singular route E31.[61] In March 1958 the Committee agreed that green would be the 'most distinctive colour' for the new signs, 'which would at the same time give good contrast',[62] but after Anderson, Glanville and Waleran toured the motorways of Belgium, the Netherlands and West Germany in June 1958, it was decided that a blue background, similar to that used in these three countries, should be adopted in the UK, not least because it would stand out more clearly 'against the natural greens of the countryside'.[63] In early 1958 the Committee consulted nearly thirty organizations, including the Royal Society for the Prevention of Accidents, the National Parks Commission, the Royal Fine Art Commission, the Royal Air Force and the Association of Road Traffic Sign Makers. The Road Research Laboratory informed the Committee of the results of the latest scientific studies in Britain, Europe and America, and in mid-1958 the Laboratory conducted experiments with mock-up signs at RAF Hendon in North London: 'Practical demonstrations of various experimental signs were carried out by day and night at Hendon Airfield where members of the Committee, with Lord Waleran at the wheel, drove at speed along the runways in order to test the visibility of the signs.'[64]

The Advisory Committee were keen to show that their signs were scientifically and artistically superior to anything which had been designed before. Sir Hugh Casson had initially suggested that the Committee should collaborate with members of the Royal College of Art (where Casson was a professor), but the signs were eventually designed by professional designer Jock Kinneir (1917–94) and his assistant Margaret Calvert (1936–), who had prepared signs for London's Gatwick Airport. Kinneir and Calvert went on to design the standardized sign systems for some of Britain's most frequently occupied public spaces – including signs for the National Health Service, British Railways and Sir Walter Worboys' Traffic Signs Committee – and their designs have helped to govern, speed up and bring increased efficiency to the movements of tens of millions of people nationwide on a daily basis (Hughes 2002).

The Committee's experimental signs were erected on the Preston Bypass in late 1958, and pictures of them were printed in *The Times* on 2 December 1958 to familiarize the nation's drivers with these strikingly new, modern signs, which would guide them onto and along this modern motorway (*The Times* 1958b). As the accompanying article suggested, the signs were 'a complete breakaway from present British usage':

> . . . they are much bigger (some of them are 20 ft. high), normal face letters (12 in. high) are used instead of capitals, and the letters are in white against a blue background. . . . An important departure is that route numbers on motorway signs will not be given any greater importance than place names. They will be treated as separate items of information and not attached to particular place names. (*The Times* 1958b: 10)

These design principles provoked a vigorous debate before and after the erection of the Preston Bypass and M1 signs. Concerns arose about which place names to display on signs, and a number of committee members, including Noel Carrington, argued that place names were more important than route numbers for the navigating motorist (Carrington 1976). Committee members considered the extent of the geographical knowledge of Britain's motorists and tourists: how they conceived places beyond the motorway; how they navigated between places and along routes; and what kinds of visual cues were needed to aid their movements (ACTSM 1962). More extensive and public criticisms were aimed at the size of the signs and the lettering used. The lower-case, sans serif lettering of the motorway signs was treated as a somewhat novel presence in the English landscape.[65] The majority of Britain's other road signs featured capital letters, and the one county surveyor (of Oxfordshire) who had experimented with lower-case signs in the mid-1950s was said to have been 'publicly reprimanded by the Ministry and told to remove his offending signs' (Carrington 1976: 176). No sooner had pictures of the Preston Bypass signs been printed in *The Times* than a few determined typographers and designers started writing letters of complaint. On 4 December 1958, Brooke Crutchley, the University Printer at Cambridge University Press, wrote to the Ministry to point out that capital letters and serifs were more legible and took up less space than lower-case sans serifs letters at a similar distance. Crutchley blamed the influence of modern European, particularly German, typography, which he attributed to Noel Carrington's presence on the Committee.[66] The well-respected stone carver and alphabet designer David Kindersley, a former apprentice of Eric Gill, voiced similar concerns in correspondence with civil servants and members of the Landscape Advisory Committee. To reinforce his points Kindersley devised a new type, 'MOT-Serif', which he asked the Ministry to consider as an alternative to Kinneir's Preston Bypass lettering.[67]

Sir Colin Anderson was not pleased by Crutchley and Kindersley's persistent approaches. Anderson had admired the experimental signs with lower-case lettering erected by Oxfordshire County Council, which he felt 'were in the spirit of the 1950/60 era',[68] and in a letter to David Bowes-Lyon in February 1959, he concluded that Crutchley and Kindersley were too backward-looking and anti-modern in their views: 'Personally I can't help feeling that there is an antiquarian leaning behind their attitudes and if ever there was a place where antiquarian prejudices – if indeed these do enter into it – were out of place, it would be on a modern motorway.'[69] Noel Carrington took a different stance, suggesting that printers such as Crutchley did not account for the different practices of reading associated with driving at high speed and reading a book or newspaper (Carrington 1959; cf. Crutchley 1959). In April 1959, the Advisory Committee discussed the pros and cons of both Kindersley and the Ministry's signs, while in May 1959 Kindersley, Crutchley, Carrington, Kinneir and Sir Hugh Casson joined Sylvia Crowe, Cambridge psychologist E. C. Poulton, sign manufacturer G. S. Campbell and two civil servants in a discussion organized by the Council of Industrial Design's magazine *Design* (see *Design* 1959).[70] Kindersley and Crutchley continued to criticize the Ministry's signs, and rather surprisingly Sylvia Crowe echoed their concerns, objecting to 'the "worrying squiggles" of lower-case lettering',[71] and the size of 'the vast signs cutting holes in the landscape' (quoted in *Design* 1959: 28). The Landscape Advisory Committee had forwarded their own concerns about the size of the signs to Sir Colin Anderson in February 1959,[72] while both Sylvia Crowe and Brenda Colvin disliked their blue colour. Crowe (1960) preferred black, a colour the Advisory Committee deemed 'too negative' (ACTSM 1962: 6). Colvin described 'the vast size and crude blue background of the Preston by-pass signs' as giving 'an impression of having been designed for lunatic drivers' (Colvin 1959: 246).

Design magazine reported that most of the participants at its symposium agreed that more research was needed (*Design* 1959). Three months later, and partly as a response to the vocal and very public criticisms of Crutchley and Kindersley, scientists at the Road Research Laboratory prepared a series of experiments to 'compare the relative effectiveness of upper and lower case scripts for use on traffic signs'.[73] The Laboratory's scientists would cast their judgment, and Kindersley, for one, couldn't wait for the 'truth', 'the facts', to be established (Kindersley 1960: 465). As he stated in a memorandum forwarded to the Ministry of Transport, Sir David Bowes-Lyon, Sylvia Crowe, George Langley-Taylor, Brooke Crutchley, the Road Research Laboratory and Ian Nairn in June 1960, which was later published in the journal *Traffic, Engineering and Control*: 'I believe the central mistake behind much modern industrial design is to be found in the pursuit of aesthetics rather than truth. This philosophy is without doubt responsible for the present motorway signs' (Kindersley 1960: 465). Here was a typographer who hoped that a science of signs, a semiotics, would prevent the adoption of undesirable styles of lettering. But the results of the test did not provide the authoritative and singular conclusion Kindersley desired. When the Road Research Laboratory reported on their experiments in August 1961 – tests which had involved four styles of letter, 96 signs and volunteer sign 'readers' from the RAF – they concluded that as 'there is little difference in legibility between the different types of lettering, it seems reasonable to make the choice on aesthetic grounds' (Christie and Rutley 1961: 60). The Advisory Committee and a number of graphic designers breathed sighs of relief. The pioneering modern typographer Herbert Spencer stated that, 'as a designer', he would have been 'unable to accept' Kindersley's lettering, 'which I find clumsy and which, I think, ignores both taste and tradition alike' (Spencer 1961: 61).

Modern, lower-case lettering – unfussy and with no ornamental serifs – was to distinguish the nation's motorways from its rapidly ageing, ordinary roads. These signs would reflect the modern design of the entire motorway, contributing to a motorway style, and introducing the increasingly open-minded, future-gazing motoring public to modern designs which would eventually populate spaces they occupied in their everyday lives. As the Advisory Committee on Traffic Signs for Motorways suggested in their final report in 1962, the signs would not only cultivate but also fulfil the public's aesthetic taste: '. . . we consider it likely that an increasingly cosmopolitan motoring public will encounter more and more lower case lettering on signs of all kinds and will come to expect it, among other places, on the motorway' (ACTSM 1962: 4–5). The lower-case letters were not the only new and distinctive shapes which appeared on the signs. A number of symbols were adopted for inclusion on the signs, addressing long-standing criticisms of the over-wordy and subtopian mess of the nation's uncoordinated road signs. Writing in *The Architectural Review* in 1956, Gordon Cullen had called for 'a new vocabulary' of road sign design, 'a visual code in place of a literary code' (Cullen 1956: 241), while that same year at a special 'Symposium on subtopia' at the Institute of Landscape Architects, Ian Nairn complained that 'we have had enough of the

Englishness of English road signs' (Nairn, in Crowe et al. 1956: 3). Kinneir and Calvert's less wordy, simplified motorway signs were an attempt to introduce a new modern design vocabulary into the English landscape. But would the public understand this new vocabulary, or would the government have to engage in a programme of visual education? The Advisory Committee on Traffic Signs for Motorways sounded a note of caution in their final report to Minister of Transport Ernest Marples, as it was clear that 'education' would be 'needed – even for experienced drivers – before the tempo of the motorways can be properly absorbed', and drivers would develop the competence to safely navigate the spaces of the motorway (ACTSM 1962: 2).

The new motorway signs joined a whole host of other new, modern, functional design features, or technologies, which would help to guide vehicle drivers in a safe, swift and efficient manner to their destinations. Design professionals recognized that motorway drivers experienced the landscapes and architectures of the motorway in embodied, dynamic, performative and fairly fleeting ways (cf. Lees 2001). What's more, they expressed concern that Britain's ordinary drivers lacked the skills to read and competently traverse and inhabit the landscapes of the motorway. I revisit these debates about the competence, experience, governance and education of motorway drivers and passengers in chapter 5. In the next chapter I examine the landscapes of motorway construction.

Chapter Four

Constructing the M1

In this chapter I provide an account of the mobilities, materials and practices entailed in constructing the Luton to Rugby section of the M1 in the late 1950s. I examine how the landscapes of the motorway were constructed through the movements and performances of a diverse array of official and unofficial agents of 'construction' and 'presentation' – including earth-moving machines, architectural models, maps, paintings, lorries, labourers, newspapers, films, radio programmes and a helicopter – as well as the circulations of seemingly more 'natural' entities such as rainwater and soil. The landscapes of the motorway were, and indeed still are, very much in process; shaped through the degradation and repair of materials, the growth of vegetation, as well as through the talk and movements of motorists, local inhabitants and historical geographers. The motorway has always been in a state of becoming. Its landscapes are actively 'worked' in different ways, and here I hope to expose some of the quite 'specific struggles that . . . [went] into its making' (D. Mitchell 2001: 44).

The contracting engineers John Laing and Son Limited won all four contracts to build the first 53 miles (later extended to 55 miles) of the London to Yorkshire Motorway, following their submission of an unrivalled tender of £14.7 million in January 1958 (W. K. Laing et al. 1960). In the first part of the chapter I provide a critical examination of the narratives of construction that Laing presented to the public, future clients, company employees and local residents. I focus on the different technologies which were used to construct the motorway on the ground and in the imaginations of the public. I examine how 'construction work' occurred in spaces far beyond the boundary fences of the motorway, and discuss how aerial perspectives and a military language permeated accounts of the construction process. I then show how the corporate spaces of John Laing and Son Limited were performed through the mobilities of an array of things and how different workers were incorporated into the spaces of the construction 'team' in different ways. In the second part of the chapter I focus on a one-hour radio programme about the construction of the M1. *Song of a Road* was composed by the Communist folk-song composer Ewan MacColl, BBC radio producer Charles Parker and the American folk musician Peggy Seeger in 1959. The programme incorporated sounds recorded on the construction site, extracts of interviews with motorway workers, and original songs about the construction process and the lives of the workers. While Laing's PR booklets, films, press releases and company newsletter focused on the work of the senior engineers and the construction team, *Song of a Road* focused on the lives, work and oral traditions of the largely working-class migrant labourers and tradesmen whose biographies and geographies were largely overlooked in official accounts of the motorway.

'Operation Motorway': Constructing the M1 Motorway

Technologies of construction

I begin by focusing on two technologies that were deployed during the construction of the M1, examining how their movements, materialities and durability led to the ordering of the motorway in particular spaces and times. One of these devices would not usually be thought of as 'technological': a public relations booklet that was produced to commemorate the completion of the motorway, circulated far and wide, and catalogued in libraries and archives. The other technological devices I discuss were the construction machines used to build the motorway, which clearly have very different biographies, materialities and uses to a booklet. Despite these differences I stress that the materialities, movements and aesthetics of both technologies were central to the working or ordering of the motorway as a dynamic and heterogeneous landscape, and both were implicated in the construction of the motorway as a site of modernity.

On 24 March 1958 the Minister of Transport and Civil Aviation, Harold Watkinson, inaugurated the construction of the London to Yorkshire and St Albans By-pass motorways at their meeting points at Slip End near Luton. Watkinson gave a speech celebrating the scale of Britain's first major motorway and its significance to the national economy, before being invited by Sir Owen Williams to fix the Ministry seal into a concrete plaque that was later incorporated into the parapet of a bridge.[1] This was followed by a spectacle designed to reflect the novel and technological nature of the project, as the Ministry's 'Information Branch' were keen to ensure that the tradition of 'cutting the first sod' would not be performed with a cere-monial spade.[2] Mr Watkinson pressed a button which sounded a klaxon and changed a traffic light to green, signalling two lines of excavators to start removing the top-soil. Tarmac's machines moved south along the St Albans By-pass and Laing's moved north along the London to Yorkshire Motorway, forming a 'procession' designed 'to give visible evidence of [the] immense energy of plant about to be released'.[3] The intention was to high-light the contemporaneity and modernity of construction, and to ensure that this spectacular inauguration ceremony would be appreciated by invited journalists and reported widely in the media.

While civil servants ensured that the inaugural ceremony would be appropriate for a Minister to open, John Laing and Son Limited were aware that public relations devices would provide an opportunity to promote their company as well as the motorway itself. Public relations tools were vital to the 'organizational display' of the company (P. Crang 2000: 210). In late 1959 Laing's public relations department published a booklet entitled *The London–Birmingham Motorway* that was written by the well-known engi-neering historian, railway preservationist and organicist L. T. C. Rolt (Rolt 1959; on Rolt, see Rolt 1971, 1977, 1992; M. Baldwin 1994). Rolt enjoyed writing about engineering projects in which he felt there was an 'element of the heroic in the story of their successful completion' (Rolt 1992: 175), but he found that company directors and public relations departments could be difficult employers: '. . . the executives of a large group of com-panies would commission me to write a history whereas what, in fact, they wanted was an unreal exercise in public relations in which nothing to the company's discredit had ever occurred' (Rolt 1992: 187). Rolt does not specify which of his commissions led to disagreements, but it is significant that his company history of Laing, written in 1962, was never published and exists only as a manuscript in the Laing company archive (Rolt 1962).

The London–Birmingham Motorway was aimed at potential clients, jour-nalists and company employees. It contained graphs, maps and over 60 photographs of the people, machines and sites associated with the scheme, as well as an extensive description of Laing's involvement in the construc-tion of the motorway. Rolt had just finished researching a biography of George and Robert Stephenson when he was commissioned to write the motorway booklet (Rolt 1960), and his detailed knowledge of the construc-tion of Robert Stephenson's London to Birmingham Railway in the 1830s along a similar alignment to the motorway led him to compare the chal-lenges, successes and greatness of these different, yet also surprisingly similar, engineering projects (Rolt 1959). Rolt argued that the great railway schemes of the nineteenth century – with their low gradients, embankments, cuttings and extensive earth-moving – were the only engineering projects that could be compared with the motorway, while his entwining of the histories, traditions and greatness of the two projects acted to authorize Laing's claim that this thoroughly modern motorway would assume a special place in engineering history. The 120 years that lay between the two projects had seen a 'technical revolution' that was readily apparent at the motorway's inauguration ceremony (Rolt 1959: 6). The construction of the 112-mile London to Birmingham Railway had required a monumental effort from 25,000 men equipped with 'primitive equipment', a phrase Rolt used to caption a lithograph of the railway's construction by J. C. Bourne (Rolt 1959: 7). Bourne was one of few nineteenth-century artists to depict scenes of construction (Daniels 1985), and Rolt saw his views of orderly work as showing outdated and inefficient machines and methods by modern standards. The motorway, on the other hand, was being built by a maximum work-force of only 4,700 men equipped with the most modern of machines: 'the greatest concentration of mechanical power that has ever been mustered on one contract in this country, a concentration totalling at its peak no less than 80,000 brake horse power' (Rolt 1959: 6).

The bodies shaping the landscapes of the motorway took a different form to the bodies working on the railway. The 'muscles' of Stephenson's men were replaced by the 'steel sinews' of Laing's 'dozers, scrapers and excavators', a 'giant's strength . . . put into men's hands' averaging at 20-brake horsepower per worker (Rolt 1959: 6). Rolt listed the equipment in great detail, from the number of concrete mixers, lorries and steel-bending machines, to the 'famous and familiar' names represented by the 182 different excavating machines (Rolt 1959: 25). And as the movements of their machines were largely limited to the rather isolated sites of excavation and construction work, these 'famous' firms were themselves keen to promote their involvement in the construction of this important motorway, and many of them placed advertisements in the national and local press. While the movements of their machines had shaped the landscapes of the motorway, these adverts also stressed the role of the machines and company in reconstructing the nation: physically, economically, socially. On the opening day of the motorway, 2 November 1959, Blackwood Hodge and Caterpillar placed adverts in *The Times* that were intended to appeal to the imagination of the technologically minded, progressive, responsible citizen-subject concerned with the reconstruction of the national economy and society (Figures 4.1 and 4.2). The Euclid earth-movers that Blackwood Hodge distributed had helped construct a futuristic 'highway to tomorrow', resulting in a *positive* 'brave new world' where 'super-highways will slash across the country' to provide a landscape of efficient movement and economic prosperity (Blackwood Hodge 1959: 8). For the North American company Caterpillar the motorway marked the 'beginning of a new era . . . of progress and opportunity' which would be aided by their reliable, efficient and

HIGHWAY
TO
TOMORROW

In the rush and roar of the twentieth century a brave new world is being built. And there is no time for delay.

Tomorrow, the broad white ribbons of the super-highways will slash across the country, linking factories, ports and cities, by-passing the bottlenecks and conveying food and raw materials to their destinations swiftly—safely—economically.

Tomorrow, the length and breadth of the land will be ours to travel and enjoy where and when we will. Tomorrow, the dreams of the future will be the highways of today. And the giant EUCLID Earthmovers will have played a vital part in laying the foundations of tomorrow's brave new world.

BLACKWOOD HODGE

Subsidiary Companies Branches Works and Agencies throughout the World

Figure 4.1   Blackwood Hodge advertisement, 1959. Reproduced by kind permission of Blackwood Hodge Limited and Terex Construction.

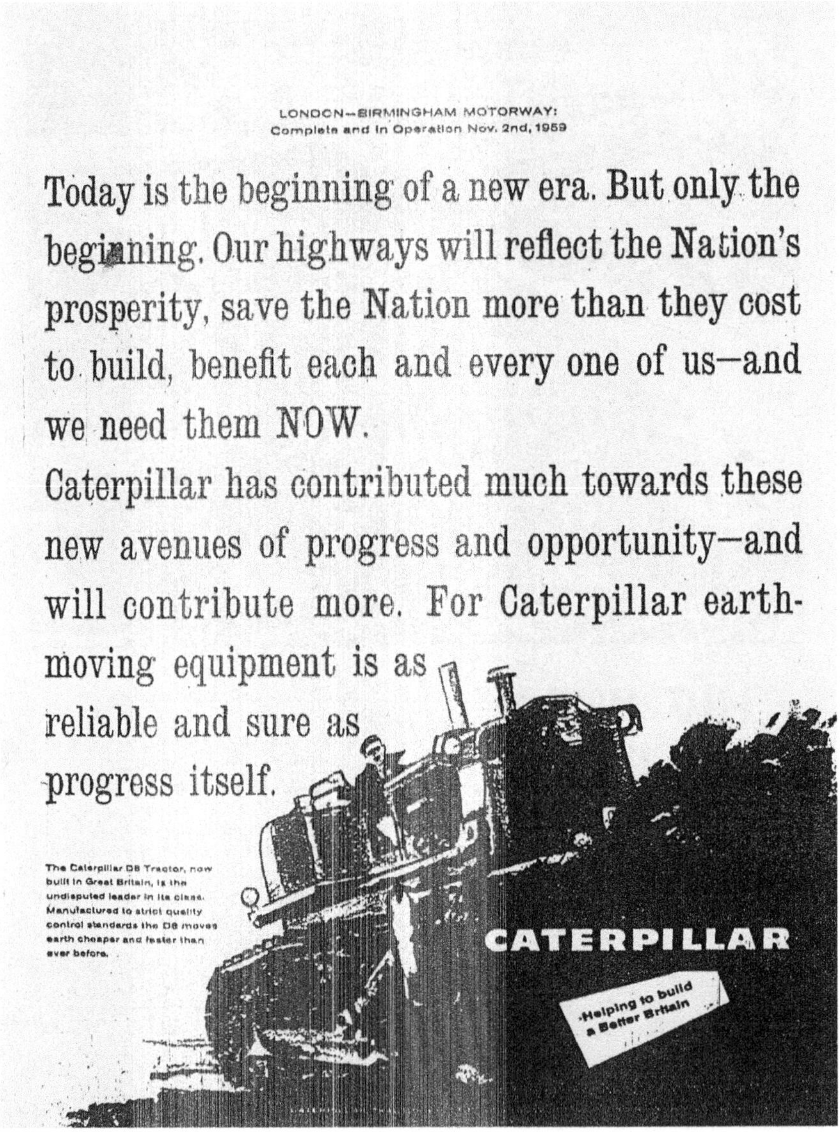

LONDON—BIRMINGHAM MOTORWAY:
Complete and in Operation Nov. 2nd, 1959

Today is the beginning of a new era. But only the beginning. Our highways will reflect the Nation's prosperity, save the Nation more than they cost to build, benefit each and every one of us—and we need them NOW.

Caterpillar has contributed much towards these new avenues of progress and opportunity—and will contribute more. For Caterpillar earth-moving equipment is as reliable and sure as progress itself.

The Caterpillar D8 Tractor, now built in Great Britain, is the undisputed leader in its class. Manufactured to strict quality control standards the D8 moves earth cheaper and faster than ever before.

CATERPILLAR

Helping to build a Better Britain

Figure 4.2   Caterpillar advertisement, 1959. Reproduced by permission of Caterpillar Incorporated.

economic D8 tractor (Caterpillar 1959: 5). During World War II Caterpillar tractors could be seen to be part of an army of imported machines turning Britain's waste-lands into spaces for the practice of an efficient, modern and mechanized farming industry, but the D8 was 'now built in

Great Britain', at their new Glasgow factory (*Road International* 1958). This was a British tractor 'helping to build a better Britain' and helping Caterpillar to fulfil their duty to strengthen the economy, society and 'the nation' (Caterpillar 1959: 5).

The mobilities of these earth-moving machines serve as important elements in constructing the landscapes of the motorway and nation. One might suggest that these machines were being controlled and choreographed by operators, engineers and civil servants, but two contemporary commentators appeared to prefigure recent debates in the sociology of science and technology by seeing them as more active and autonomous agents or beings in shaping the landscapes of construction (see, e.g., Latour 1992, 1996). In a night-time scene described by Wharton and ffolkes in *The Daily Telegraph*, the machines appeared as hybridized animal-machines which performed like robotic, futuristic 'prehistoric reptiles' with 'flaming yellow eyes': 'great earth-moving machines, belonging at the same time to the prehistoric past and the technological future, rooting and snorting in the mud . . . and to get the full atmosphere of the project you must stand and watch, preferably under a waning moon, the demoniac dance of the machines' (Wharton and ffolkes 1958). These reptile-machines are at once controlled and controlling, for while 'there are tough, experienced men driving these machines . . . it is not difficult to imagine the machines themselves taking over the project' (Wharton and ffolkes 1958). Prehistoric animality and futuristic technologies become associated with agency, inhuman sounds and movements, resistance, power, unpredictability, unfamiliar shapes, bodies and 'dances', the mythological and the sublime. This is a landscape shaped by the movements of a diverse range of technological devices and bodies, from fleshy, muscular and, at times, racialized human bodies, to the body-work of metallic machines. The motorway landscape emerges as a result of the circulations and interactions of men, construction machines, soil and public relations devices as well as other materials, knowledges and atmospheres.

Spaces of construction

While excavators and engineers played a prominent role in constructing and performing the motorway, the work performed by a text such as *The London–Birmingham Motorway* illustrates how the landscapes of construction enfolded spaces which extended far beyond the limits of the motorway, as vehicles, sounds, workers, written texts, aggregate, newspaper articles and VIPs flowed across its boundaries.

The soundscapes of construction extended out from the site, provoking a range of responses, from interest to concern. On the one hand, Buckingham's

Conservative Member of Parliament, Sir Frank Markham, complained about the disturbance to local residents caused by 24-hour working (*Rugby Advertiser* 1958a; *The Wolverton Express* 1958b), while, on the other, the Communist folk-song composer Ewan MacColl, BBC producer Charles Parker and American folk musician Peggy Seeger recorded sound effects and interviews on and around the site, which were edited, reworked and broadcast as part of their popular Radio Ballad *Song of a Road* (see below). Laing's public relations officials attempted to control access to the spaces of construction, but the leakage and extension of the construction site and work into a series of accessible public spaces – the law courts, local authority planning offices, public highways, etc. – enabled journalists to supplement articles on engineering achievements with reports of fatal accidents, local complaints, robberies, court cases involving workers, theft from the site and public inquiries – where quarry owners challenged refusals to allow the extension or creation of pits to extract gravel and sand for the motorway (*Chronicle and Echo* 1958a; *Luton News* 1958b; *Mercury and Herald* 1958b; *The Wolverton Express* 1958f). These (predominantly local) newspaper articles worked and continue to work in different ways. They spread news of the latest motorway-related sensational story into the homes and workplaces of readers, reworking people's perceptions of the project, while it was only their collation in Sir Owen Williams and Partners's press cuttings books and preservation in their archives which enabled me to encounter these otherwise unindexed journalistic accounts about the motorway.

The press cutting books provide a valuable record of local newspaper reports relating to public inquiries into the extension of gravel pits for the project. For example, in July 1958 there was an inquiry into the proposed extraction of hoggin, gravel and sand from four sites at Cosgrove near Wolverton. A representative of the Ministry of Agriculture, Fisheries and Food objected to one of the sites as it was a farm of 'national importance' with 'permanent pasture . . . [which] ranked with the arable land of the Fens', while Dowsett Engineering and a representative of Laing pointed to the need to extend local supplies of materials for Britain's largest civil engineering project, as existing supplies were likely to run out in six weeks (*The Wolverton Express* 1958c). Different registers of value were opposed by the different parties, and the appeal by Dowsetts was upheld for only one of the four sites, with the land defended by the MAFF representative remaining un-worked (*Chronicle and Echo* 1958b). When permission was granted to extract aggregate from a site, concerns were expressed about the congestion, danger, noise and damage caused by streams of lorries transporting material to and from the motorway. Newport Pagnell developed a traffic problem, a large number of country lanes had to be widened or resurfaced, and chalk dust and mud were distributed onto local roads (*The Wolverton Express* 1958d). It was argued that this was hazardous to cyclists and pedestrians, as well as being unsightly. 'Leagrave mothers' complained about conditions close to the site of a flyover near Luton, where there could be 'so much mud that it sometimes comes over the tops of children's Wellington boots' (*Luton News* 1958d). Bedfordshire and Buckinghamshire County Councils expressed concern about the conduct of lorry drivers, who would speed along narrow lanes at 50 mph, ignore halt signs, drive on verges and obliterate ditches, while H. P. Whiskin, an agent for properties along one particular lane where three lorries had crashed into a ditch, described the scene as 'just like the film "Hell Drivers"' (*Daily Telegraph* 1958; see also *Chronicle and Echo* 1958c).[4]

John Laing and Son Limited were keen to address any criticisms or concerns about the construction project. While many of the lorry drivers worked for sub-contractors beyond their control, they did employ men with road sweepers to clean local roads; a good-will gesture that was highly praised by Newport Pagnell Rural District Council and was mentioned by Rolt in *The London–Birmingham Motorway* (Rolt 1959; see also *The Bucks Standard* 1958a). Councillors were also present at a two-hour showing of Laing public relations films at Newport Pagnell in December 1958, which included the first film of the motorway, *Major Road Ahead* (*The Bucks Standard* 1958b). Laing organized public relations events on a regular basis, and the landscapes of the motorway were performed through such events and the materialities and mobilities of films, booklets, radio programmes and press releases as well as engineers, tarmac, lorries and labourers. L. T. C. Rolt confirmed that an 'essential atmosphere of goodwill was established at the outset', as Laing were the company who held a 'Motorway Ball' to raise funds for Sherington Church, and let a football team play their final league matches before the bulldozers moved in (Rolt 1959: 36; see also *Luton News* 1958c). Senior engineers and public relations officials gave lectures to local and national engineering societies, councils and Women's Institutes, while engineering students, Members of Parliament and other dignitaries were conducted on tours of the site (*Team Spirit* 1959b, 1959c). Special guests often combined a trip to the construction site and the offices and news-room at Newport Pagnell with a ride in Laing's helicopter, while local and national journalists also took to the sky to get the ultimate view and photograph of the construction site.

Aerial perspectives

While the spaces of lectures, tours, quarries and muddy roads became bound up with a series of quite distinct, embodied encounters with the landscapes of construction, helicopters enabled the production of a series of quite different depictions of the motorway and aided the movement of a broad range of things that were central to the making and performance of these landscapes. Aerial photographs of the construction site snaking into the distance were published in local and national newspapers throughout 1958 and 1959, and this gave rise to a series of quite distinctive commentaries. The motorway was seen to cut *across* the surrounding landscape, appearing as an 'ugly scar' or abject wound disfiguring the nation's face (*The Times* 1958a), but it was also argued that it would nourish and sustain the life of this landscape, functioning as a major artery in the circulatory system of the national body:

> The brown scar, which extends like a ghastly flesh wound on the face of a green and pleasant land stretches away into the haze of the summer afternoon. . . . The ordinary highways and byways appear like tiny veins, coping with difficulty with their overladen burden. The new motorway is the huge artery, oozing rich brown soil in its newly severed path (Field 1958).

The wound is coloured by local rocks and soils – white chalks near Luton and brown soils near Northampton – but while some expected it to heal, others felt that permanent scarring had already occurred (cf. Bishop 2002). A number of commentators expressed concern about the loss of valuable farmland to such developments – the end result of which might be 'starvation' (Pryke 1958: 13) – and the removal of 'the top soil, that precious national heritage' (King-Salter 1958: 11). Members of Northampton Camera Club sought to 'preserve' the local pre-motorway landscape in a portfolio of photographs which they presented to the Northamptonshire Rural Community Council (*Mercury and Herald* 1958a), while many local residents photographed and filmed the landscapes of construction in an attempt to preserve the spectacle of excavation (*Luton News* 1958a). Local journalists referred to an 'invasion' of sightseers, provoking concerns long expressed by preservationists such as C. E. M. Joad (1946), as ill-disciplined 'townsfolk' were observed crossing farmland to reach the construction site and the Newport Pagnell branch of the National Farmers Union had to instruct pedestrians to keep to footpaths, close gates and observe 'Trespassers will be prosecuted' notices (*The Wolverton Express* 1958a, 1958e).

Photographs were used by the contractors to document and preserve the construction process, and images captured from the air assumed a central place in Rolt's booklet *The London–Birmingham Motorway*. Aerial photographs were presented as mimetic and self-explanatory documents, but Rolt also provided an intensely personal description of the view from a helicopter. The flight provided him with a completely novel experience, a privileged, authoritative and seemingly detached viewpoint, and an extended 'field of vision' that had long been celebrated and promoted by planners and geographers (Daniels and Rycroft 1993: 465):

> It gave the visitor an impression . . . which no statistics, no charts, no maps however graphic could possibly convey. In long perspectives, curving easily this way and that, the line of the Motorway began to grow out of the landscape as though by magic. . . . But the effect from the air was still ill-defined and chaotic until the white-topped marginal haunches brought order, ruling precise, purposeful guide lines for the twin carriageways which the asphalt surfacing would soon ink in. Out of a huge crater of torn earth in which man and machines appeared to flounder aimlessly the disciplined geometry of a flyover junction with its overhead roundabout and connecting slip roads would suddenly emerge. . . .Whereas at ground level the sounds of mechanical plant at work were inescapable and conveyed a continuous sense of effort and urgency, from the air where no such sounds could be heard the effect was the precise opposite . . . [the impression of] road building as magical or miraculous. (Rolt 1959: 29–30)

The phenomenology of the landscape appears very different from the air, as Rolt's detachment from the machines causes the motorway to appear as if it is growing or evolving, organically or magically. In his autobiography *Landscape with Figures* Rolt (1992: 177) referred to the 'fearful mechanical commotion' of this, his first ever, helicopter ride, but the readers of Laing's public relations booklet were presented with a more detached and distanced scene – which develops visually, like a painting on a canvas. Carefully planned geometric structures bring order and discipline to the English landscape, and this description of a positive, modern landscape of orderly construction echoed the vision projected in an oil painting of the motorway which appeared on the front cover to *The London–Birmingham Motorway*.

The painting was produced by Terence Cuneo, who had worked as an illustrator and official war artist in the 1930s and 1940s, when he developed a consciously realist and 'traditional' style of painting based on what he saw as the 'three ingredients' necessary for a great work: 'colour, composition, (that is design) and draughtsmanship' (Cuneo 1977: 34). Cuneo is primarily known for his railway paintings and portraits of royalty and politicians (see Cuneo 1977, 1984), but in 1946 he started gaining commissions to paint industrial and construction subjects. By the time he visited the motorway in 1958 he had painted over twenty pictures of Laing's projects in Britain, Canada, the Union of South Africa, and Rhodesia. The paintings were hung on the walls of Laing's head offices in London and reproduced in the company calendar, bringing the spectacle of Laing and Britain's empires to the 'homes' and eyes of the company, client and senior employee; whether this be the spectacle of a half-built power station, completed office building, excavating machine or the bodies of black South

African men working in a gold mine (Cuneo 1977). Cuneo was described by Laing's staff newsletter *Team Spirit* as a 'mechanically-minded' man with the gifts of 'technical accuracy' and artistic skill, and as a somewhat heroic and adventurous painter of Laing's heroic projects: '. . . he has worked for hours by the light of lamps to capture the spirit of a busy night scene; he has clambered to inaccessible vantage points over masses of steel reinforcement, or been hauled in a bucket from one part of a job to another' (*Team Spirit* 1951: 7). A similar adventure presented itself in choosing a view for the motorway painting, as his sketches were made from Laing's BEA helicopter, which he ordered to hover over a site 'which illustrated the enterprise to best effect' (Cuneo 1977: 61). The resulting painting, *London to Birmingham Motorway During Construction at Milton Near Northampton*, depicts different levels of progress on a short section of the motorway (Figure 4.3). Traffic crosses a completed over-bridge, a machine lays the base-course between the two parallel white concrete marginal haunches, while in the foreground excavators remove earth where the Roade–Northampton railway crosses the motorway. Excavation and bridge

Figure 4.3   *London to Birmingham Motorway During Construction at Milton Near Northampton.* Oil painting by Terence Cuneo, 1958. Reproduced with the kind permission of the Cuneo Estate.

construction could not begin here until the end of the summer 1958 railway timetable, and so Cuneo was able to depict a scene of busy excavation next to near-complete sections of the motorway. One of Cuneo's favourite subjects, a steam train, crosses a temporary bridge, while the incorporation of the helicopter's windows and cockpit instruments into the picture emphasizes the viewing position and provides a modern framing device at a time when helicopter travel was fairly novel. This is a scene of mechanized construction rather than of intensely physical labour; a painting intended to please a board of directors or public relations department rather than to emphasize what Cuneo later described as a 'raw scar that stretched across the countryside like the swathe of a giant plough' (Cuneo 1977: 61). The painting of this scene by a well-known artist gave rise to an image whose symbolism and value continue to be recognized. After being displayed in the head offices of John Laing and Son Limited it was gifted to the motorway's consulting engineer, Sir Owen Williams, and it now assumes centre stage in the office of Richard Williams, Sir Owen's grandson and Chief Executive of the Owen Williams Group.

When the editors of *Team Spirit* remarked on the hovering helicopter that enabled Cuneo to produce his preliminary sketches, they pointed to the practical use of the helicopter as a vehicle for surveillance and management: 'Those working on the structure must have been mystified – and made a little anxious by the long and close scrutiny made of their work' (*Team Spirit* 1958a: 7). During a dinner speech in November 1958 S. E. Sturley, a public relations officer for Laing, went further, dismissing a rumour that 'one man who had the sack found in his pay packet a photograph – taken by helicopter – of himself leaning on a shovel!' (*Chronicle and Echo* 1958d). Whether this was true or not, the helicopter did serve as a *technology* through which senior engineers attempted to govern the movements, activities and conduct of workers and machines (Foucault 1986a, 1988; Barry 2001). As many of the construction sites were inaccessible in conventional vehicles, a 90-minute journey by helicopter to inspect the entire route seemed preferable to what was calculated to be a four-day drive along narrow lanes and tracks, with long detours to cross rivers and railway lines. As Rolt pointed out in *The London–Birmingham Motorway*, the chartering of a helicopter had been specified in the contract documents prepared by Sir Owen Williams and Partners, although its planned use as a 'flying observation platform' was carried out alongside its function as an air ambulance for workers, internal mail carrier and transporter of spare parts (Rolt 1959: 29). Far from being a sinister instrument of surveillance, Rolt saw the helicopter as a modern instrument of communication 'bind[ing] the men on the job into one enthusiastic and harmonious team' (Rolt 1959: 29). The movements of the helicopter, along with the movements and interactions of workers, machines, soil, concrete, lorries and a whole series of other things, actively constituted and performed the landscapes of the motorway.[5]

Militarized landscapes

While the movements of a range of materials and ideas were central to the continual ordering of the spaces of construction, a number of commentators felt that these movements and materials reflected a military mode of organization and aesthetic. Laing's ability to 'whistle up air support' to survey the motorway was seen by *The Daily Telegraph* journalists Wharton and ffolkes to be just one aspect of the military nature of the motorway: 'It is like a military occupation, discreet but unmistakable, of the pleasant, average, unremarkable, intensely English countryside which lies between London and Birmingham' (Wharton and ffolkes 1958). Project offices appeared like 'paramilitary huts', while the executive conducting journalists around the 'Information Room full of maps, plans and models' reminds you of someone you have met before, 'perhaps in the mess at Catterick or even Bangalore' (Wharton and ffolkes 1958). There are night-time 'sentries', short-wave radios and field telephones, 'an unofficial uniform of sweat-stained dungarees', Bailey bridges to provide temporary crossings of rivers and canals, and 'the motorway has caused the death of one decrepit house, a fish and chip shop, two sports pavilions, and half-a-dozen bungalows' (*The Bucks Standard* 1958c). While these are somewhat light-hearted references to the use of military technologies, tactics and personnel, they should not distract us from the power of military analogies and metaphors. Practices and techniques of engineering, like those of geography, have histories which are intimately bound up with practices of warfare (see Livingstone 1992; Ritchie 1997; Kirsch and Mitchell 1998; Woodward 2004). Engineers and journalists alike had personal experiences of war and a working knowledge of the armed forces, so it is not surprising that military expressions and metaphors made regular appearances in descriptions of motorway construction.[6] For example, Laing's Project General Manager, John Michie, employed a military analogy when speaking to the press in July 1958: ' "We have organised the whole business pretty well on military lines," says Mr. Michie, "as if we are carrying out reconnaissance over a battle area. With the helicopter, we can marry up the paper work with what is actually happening" ' (*The Bucks Standard* 1958c). Surveys, aerial observation and detailed plans and maps appear as vital technologies and practices for the military general, engineer and geographer alike: 'technologies of government' and surveillance through which the movements of machines and workers may be choreographed and checked (P. Miller and Rose 1990: 8).

L. T. C. Rolt used military references throughout *The London–Birmingham Motorway*. Efficient planning and engineering expertise helped to combat setbacks, enemy attacks and resulted in victory; a narrative of success that was similar to that portrayed in Ministry of Information wartime propaganda booklets (e.g. Ministry of Information 1942). The scale of the 'planning and organisation' could 'only be compared with that required to launch a major military campaign', and both the engineer and 'the military strategist' must create a flexible plan to withstand the impact of 'the enemy's counter attacks' (Rolt 1959: 9). The enemy turned out to be 'the unpredictable English climate'; the might of Laing's men and machinery heroically battling against the Englishness of the weather (Rolt 1959: 9). Construction started in March 1958 but the real battle began in June: the first of three months of heavy rain constituting 'the worst [summer] within living memory' (Rolt 1959: 37). The mobility, pooling and mixing of water with soil became a key element in the landscapes of construction. Heavy machinery had difficulty negotiating the mud, lorries sank up to their axles, and when mobile they transferred mud onto local roads. Extensive flooding occurred throughout the route, particularly at the sites for viaducts over the Nene, Lovat and Great Ouse rivers. Excavated material often proved too unstable to be used for the embankments, and a large amount of material had to be imported from quarries. The section worst effected was Contract D in Northamptonshire, where quicksand caused serious problems for engineers at Kilsby Hill just as it had for railway engineers in the 1830s (Rolt 1959: 42). Work was reduced to a minimum during the winter months of 1958–9, but in March the 'company's great spring offensive' began, which Rolt described in 'The final assault', the last chapter of *The London–Birmingham Motorway* (Rolt 1959: 63).

Rolt saw the successful and timely completion of the motorway as an inevitable outcome reflecting the careful planning of engineers who, like the 'High Command planning a major military campaign', 'determined its grand strategy', decided 'what forces to mobilise, how to deploy them in the field, and by what hierarchy of command various activities should be directed' (Rolt 1959: 17). This 'High Command' set up its headquarters at Newport Pagnell, project offices in the four contract-sections, and smaller offices for sub-sections of each project, creating an organizational structure that Rolt compared with a 'nervous system' – Newport Pagnell being 'the brain' and the project and sub-project offices 'the nerve centres' – which 'took visible form' in 'the maps and charts which lined the walls on the Planning Room at Newport Pagnell' (Rolt 1959: 27). *The London–Birmingham Motorway* contained a number of photographs of company directors, project managers and senior engineers studying maps, plans and devising strategies in this Planning Room, while Rolt himself was seemingly overwhelmed by the range of technical drawings, charts, graphs and coloured maps on the walls, especially 'the huge map' with different coloured bands which 'made progress visible at a glance' (Rolt 1959: 27). Rolt treated the map as a mimetic document, an artefact of truth, technicality and scientific calculation on which 'colour bands . . . grew and coalesced' to provide evidence of the advancement of the project; but as with other materials encountered and used by engineers, labourers or visitors such as Rolt, their value and power is constructed through practical relations with other materials and bodies (Rolt 1959: 47; cf. Harley 1988). When four Iranian journalists toured the motorway and visited the Newport Pagnell news-room in September 1958, Sir Robert Marriott's explanations of the project on behalf of Sir Owen Williams and Partners could be seen to work through such a relational performance, wherein he translated and performed his authority, expertise and knowledge through his descriptions and animation of the physical and statistical overviews provided by models, helicopters and charts (Figure 4.4). The assembling or ordering of a variety of things (human and non-human) to form the headquarters at Newport

Figure 4.4   Sir Robert Marriott and four Iranian journalists studying a model of the motorway in John Laing and Son Limited's news-room at construction headquarters, Newport Pagnell, September 1958. British Official Photograph, issued by the Central Office of Information. Crown copyright. Reproduced by kind permission of the COI.

Pagnell – including the planning and map rooms, laboratories, engineers, visiting journalists, models, offices, administrative staff, telephone lines, etc. – constituted what Law and Latour have termed a 'centre of translation' or calculation; an effect of *ordering* where decisions were generated, scientific analyses conducted, and maps and charts produced, altered, displayed and archived (J. Law 1994: 26; Latour 1996). Engineering decisions, and materials such as charts, gravel and people were then relayed through the company's communication networks, acting as essential though not necessarily that durable *presences* which directed and actively constituted the construction and materialization of the motorway.

Corporate spaces and subjects

While the motorway's landscapes of construction enfolded a broad range of subjects, technologies, materials and movements into a complex topology that was performed in different times and spaces, the contracting engineering companies and sub-contractors may *also* be characterized 'less as prior, stable, fixed entities, and more as made, dynamic, fluid achievements' (Philo and Parr 2000: 513). A company like John Laing and Son Limited may appear to be stabilized and fixed through a variety of materializations – offices and engineered structures across the world; paperwork, plans and accounts circulated and stored at different sites; narratives of construction, work and corporate identity circulated through spaces occupied by clients, employees, the press and others – but companies are constantly worked and ordered. They are dynamic achievements with different temporalities and spatialities. They are '*materially heterogeneous*: talk, bodies, texts, machines, architectures, all of these and many more are implicated in and perform the "social"' (J. Law 1994: 2). It is useful, therefore, to examine how different subjects or bodies were incorporated into, or absent from, the stories of motorway construction circulated by John Laing and Son Limited; stories in which certain kinds of worker were privileged and highlighted for their difference.

Laing's second public relations film, *Motorway*, opened with a caption which alluded to the military-style organization of construction, 'Operation motorway. D Day. March 24th 1958', before the scene switched to shots at the inauguration ceremony: of the concrete plaque, a Union Jack and Harold Watkinson's finger pressing an electric button (*Motorway* 1959). The narrator, well-known BBC television newsreader Richard Baker, compared the motorway with the London to Birmingham Railway, and described the bridges, junctions and collective horse-power of the machines. 'Fifty-five miles in nineteen months. How was it done?' he asks, as the camera cuts to a picture of Laing's team of senior engineers dressed in suits and pointing at maps: 'Expert planning and highly efficient organization on a vast scale . . .' (*Motorway* 1959). Men are shown surveying the route with theodolites, ranging rods and maps, while senior surveyors talk to a farmer; their successful negotiations being sealed by a final handshake. The remainder of the film consists of a detailed discussion of the processes entailed in actually constructing the motorway, and towards the end the narrator thanks the team of people involved in a particularly telling order. The Ministry of Transport [shot of Harold Watkinson] is followed by Sir Owen Williams and the consulting engineers [shot], the planners and administrators [shot], the project engineers [four shots], the surveyors [shot], engineers [shot], plant managers [shot], lab technicians [shot], 'the women on the lonely office sites who helped to keep the job ticking over' [shots of a typist and switchboard operator], 'the men on the machines; women too, doing a man's job' [shot of woman driving lorry], 'men from Canada, South Africa, India, Jamaica' [four shots], 'and last but not least that watchdog on progress, the helicopter' [shot] (*Motorway* 1959). This is a hierarchy of employment that emphasizes authority, expertise and otherness, with the helicopter overseeing everybody and the presence of Commonwealth nationals and women being displayed for the viewer's attention. The motorway is a collective effort, but a collective effort in which the differences of certain individuals and groups stand out.

Rolt's booklet *The London–Birmingham Motorway* contained a similar dedication, for while the motorway was seen to be a monument to a collective effort, senior engineers such as Sir Owen Williams and John Michie were deemed worthy of mention by name, while the labourers appear as anonymous figures in photographs and the 'Labour strength graph' (Rolt 1959: 44). This is not unexpected, as the booklet was distributed to long-standing Laing employees, engineering societies and institutions, councillors, local dignitaries and potential clients, and was intended to reflect the work of their permanent skilled staff rather than the temporary labour-force. The film *Motorway* was also circulated among carefully selected audiences, and after premiering to over 100 members of the technical and trade press in October 1959 it was shown to members of engineering institutions and societies, to Laing employees as far afield as Pretoria and Johannesburg, and to potential clients such as Leicester City Council, whose executives attended an evening of Laing films (*Team Spirit* 1959a, 1960a, 1960b). The films were used to display and market the work of the company and to develop a sense of collective identity, achievement and loyalty among their workers, while another key device in forging this sense of identification and corporate community was the company's monthly news-sheet *Team Spirit*.

Team Spirit combined accounts of company projects with topical stories, lists of employee movements and appointments, messages from the

Chairman, details of company sports, letters, cartoons and monthly features entitled 'Apprentice of the month' and 'Workers in the team'. The
motorway formed a major topic of discussion during 1958 and 1959. Photographs were printed of VIPs visiting the site, news was conveyed of
Terence Cuneo's painting and the films, and comparisons were made with
the construction of the London–Birmingham Railway. The newsletter was
circulated to workers on construction sites around the world, and in
November 1958 an employee in Nyasaland (Malawi) wrote to the editor
to state that he had, 'with great interest, followed all your articles so far on
the now famous London–Yorkshire Motorway' (Erasmus 1958: 7). In
March 1958 John Michie (Project General Manager) appeared in 'Workers
in the team', as number 122. As in *The London–Birmingham Motorway*,
Michie is presented as a 'very prominent member of the team' with nearly
thirty years' experience, but while Rolt had focused on his strong personality and expertise, the editors of *Team Spirit* also described his progression
through the company, family life and interests in stamp collecting, opera
and ballet (*Team Spirit* 1958d: 2). In a similar article in August 1958 it was
revealed that the Chief Planning Engineer, John Pymont, was a keen gar-
dener who 'misses his family' since he moved to a company house near
Newport Pagnell (*Team Spirit* 1958c: 2), while in March 1959 the worker
featured was a much less senior figure, a Plant and Transport Manager
named Walter Hunt (*Team Spirit* 1959e: 2). These three men were presented as experts and personalities, but also as individuals in the company
team; ordinary 'family men' who were collectively responsible for the successes of the company. The newsletter functioned as a tool through which
the company attempted to govern the movements, desires, loyalties and
social activities of company workers. It was an important medium, helping
to constitute and maintain the social networks of staff and their families,
facilitating a virtual proximity by binding together past and present friends
and colleagues (cf. Urry 2002, 2003a, 2004a).

The domestic spaces of the worker were produced through and folded
into the corporate networks/spaces of Laing. While senior engineers like
John Pymont lived in company houses, many other employees and families
parked up caravans at sites Laing established at Farley Green, Husborne
Crawley, Long Buckby, Sherington and Hanslope. The caravan site at
Sherington was the subject of an article in *Team Spirit* that was based on
interviews with the wives of workers. Despite a mobile lifestyle, these women
were seen to be able to maintain a good home and carry out domestic duties
as if they lived in 'a quiet road in suburbia', for like 'most British housewives'
they believe that 'the practical constituents of a good home' are 'attractive,
well-kept rooms, comfort, cleanliness and good food' (*Team Spirit* 1958e:
5). It was suggested that a mobile home, continually ordered and purified
by a housewife, could conform to the modern, clean, gleaming, suburban

ideals being promoted by contemporary advertisers and lifestyle magazines, and these were held in sharp contrast with what were seen to be the dated, dirty and dull type of caravan owned by gypsies (cf. Ross 1995; Sibley 1995): 'There is as much difference between a gipsy's caravan and one of these gleaming metal houses-on-wheels as between a sheep-track and a motorway' (*Team Spirit* 1958e: 5). The 'housewives' who make and manage these modern 'homes on wheels' are presented as women who fix the caravans in place (creating a stable home, comparable to a conventional house), are fixed into a pattern of mobility linked to certain spaces (they frequent the caravan and visit the village store and school), and whose mobilities are dependent on their husband's work and the decisions of Laing (*Team Spirit* 1958e: 5). There is very little freedom attached to these regularized, relational and largely unrecorded movements. Many of the labourers enjoyed the lifestyle of travelling from job to job, as did senior engineers such as Douglas Elbourne and Michael May who stayed in houses bought by Laing.[7] This mobile lifestyle helped contribute to the sense of community and of a corporate family which was promoted by Laing; where this somewhat diasporic community of workers was constantly practised and worked through talk and the circulations of mail, staff, materials, machines, company films and *Team Spirit*, which informed workers of the movements of friends they had met on previous projects.

Labouring subjects

While the caravan site is presented as a suitable domestic environment for the family of the loyal and skilled company foreman, craftsman or junior office worker, many of the casual labourers who worked on the motorway were housed in hostels or lodgings. The family life, background and mobilities of these temporary workers was, perhaps inevitably, ignored by the editors of *Team Spirit*, which was a magazine for permanent staff. Rolt did provide a brief discussion of labour in *The London–Birmingham Motorway*, but more attention was paid to the successful negotiations with unions, local labour exchanges, trade societies and hostels in Leighton Buzzard, Aylesbury, Letchworth, Baldock and Rugby than the aftermath of reduced employment levels and work targets in the winter of 1959. The workers were seen to have more self-discipline than the riotous railway navvies of the 1830s, while Rolt made sure to stress that harmonious relations existed amongst the multi-racial and multi-national work-force: 'The men of many races who made up the Motorway team became almost indistinguishable as the sun burnt English skins to the colour of the Jamaicans or the turbaned Sikhs who worked beside them' (Rolt 1959: 49). That workers of different races and nationalities were working alongside one another was seen to be positive at a time when 'race riots' had recently erupted in Nottingham and Notting Hill and the 'colour bar' was widely applied, but such statements of *presence* create a spectacle of non-whiteness as foreignness and suggest a 'colouring' of white labourers which has resonance with the nineteenth-century categorization and 'colouring' of the working classes, Irish and other groups as 'white Negroes' (McClintock 1995: 52; K. Paul 1997). What's more, Commonwealth nationals were still classed as British subjects, but politicians, the press and large sections of the public were quick to associate Britishness with whiteness: 'By simple logic, if the British population was white and the colonials were black, then the colonials could not be British. . . . The associations attached to the colonials' skin color were sufficiently strong to override the presumptive rights of their legal nationality' (K. Paul 1997: 125). The presence of non-white and non-British workers in the English countryside attracted the attention of *The Daily Telegraph* journalists Wharton and ffolkes, who compared the 'hordes of strangers', mainly Irish, who 'invaded the countryside' during railway construction in the nineteenth century with the less visible, more shadowy presence of immigrant workers in an era of increasing mechanization: '. . . here and there you see a face, often Irish, sometimes Negro, peering from a cutting or among the steel rods of an uncompleted concrete bridge' (Wharton and ffolkes 1958). References to the shifting and shadowy presences of foreign 'strangers' were not uncommon during the 1950s. In the burgeoning academic literature on 'race relations', sociologists made frequent references to the presence of 'strangers' and 'dark strangers' in Britain (Waters 1997).

Laing presented their clients and staff with a narrative reflecting upon the harmony of their multi-national and multi-racial work-force, but differences were all too frequently effaced when it came to celebrating the motorway as an example of British engineering expertise, and the 'British civil engineering operative' as 'the best of his class in the world' (J. M. Laing, quoted in *Team Spirit* 1959d: 8). The presence of workers from Canada, South Africa, India, Jamaica, Poland, Hungary and Ireland, as well as England, Scotland and Wales – coupled with the use of American construction machines and materials such as Trinidad Lake Asphalt – serves to complicate constructions of the motorway as a solely British engineering achievement. Histories of empire, migration and employment policies, and the globalization of the construction and manufacturing industries, suggest that the politics and geographies of this 'British' achievement, this motorway, are not as straightforward as some accounts suggested.

The individual biographies and mobilities of the labourers were largely ignored by the makers of the film *Motorway* and the authors of *Team Spirit* and *The London–Birmingham Motorway*, but a number of individuals did set out to document the lives of the more mobile and temporary work-force. In August 1958 a history undergraduate at Cambridge University penned an article which was strikingly different from the progress reports constructed from official press releases which peppered the local and national press (see Andrews 1958).[8] Tom Andrews had spent the summer of 1958 working as a labourer with the 'long-distance men' who con-structed the motorway, and it appears that the novelty of this experience – coupled with a rather naïve and romantic curiosity – led him to write an article about the life, work and mobilities of these 'skilled craftsmen' (Andrews 1958). Andrews remarked upon their commitment to Labour Party politics, family life and their lack of interest in world affairs, but while he admired the ability of workers of different nationalities and eth-nicities to work and socialize with one another, like other commentators, he constructed and was captivated by a division between British and non-white workers:

> A number of coloured men worked in our section, and they were completely accepted – not simply as work-mates, but as individuals. I had seen British workers getting on with Negroes amicably enough on the job before. This was the first time I saw them going off to the pub together afterwards. (Andrews 1958)

I can only guess that Andrews was a white, middle-class, young man who was attracted to the motorway by the high levels of pay which were widely reported to be available to labourers (see Schonfield 1958; G. Turner 1964). He found himself in an unfamiliar working environment with men from very different backgrounds to his own, and his enlightenment was presumably an accidental by-product of this holiday job. In contrast, three other cultural commentators – Ewan MacColl, Charles Parker and Peggy Seeger – were drawn to the landscapes of construction by the promise of cultural difference and the possibility of recording the oral folk traditions of construction workers.

Song of a Road: Folk Song, Working-Class Culture and the Labour of a Motorway

Folk culture, industrial heritage and the oral tradition

During 1958 and 1959 Ewan MacColl, Charles Parker and Peggy Seeger prepared a one-hour radio programme about the construction of the M1 motorway.[9] *Song of a Road* was broadcast on the BBC Home Service on 5 November 1959 – three days after the motorway opened. The programme was the second of eight Radio Ballads that the trio produced for the BBC between 1958 and 1964. The first, *The Ballad of John Axon*, had portrayed the life and work of a Stockport engine driver who was posthumously awarded the George Cross after he sacrificed his life to warn bystanders that his train was out of control. The idea for the Ballad came from Charles Parker, Senior Features Producer for BBC Midland Region, who commissioned Ewan MacColl and Peggy Seeger – two of the most prominent figures in the post-war British folk-song revival – to write, compose and arrange the programme. The programme received widespread praise, and in late 1958 the team started work on *Song of a Road*. The Ballad presented a very different account of motorway construction to the public relations booklets and films of Laing, and to understand the approach of MacColl, Parker and Seeger it is important to provide a brief sketch of their role in the British folk-song revival.

Ewan MacColl was born Jimmie Miller in Salford in 1915 (see MacColl 1990; Parker 1965; Harker 1980, 1985; Watson 1983; Samuel 1989, 1994; Boyes 1993). His Scottish parents, Will and Betsy Miller, were keen folk singers, and his father was actively involved in union politics. Jimmie joined the Salford Clarion Players and the Young Communist League aged 14, and became actively engaged in writing political poems, songs and street theatre in the 1930s. He formed the Red Megaphones in Salford in 1931, Theatre of Action in Manchester in 1933, and would go on to found Theatre Workshop with his first wife, Joan Littlewood, in 1945 (MacColl 1990). Theatre Workshop performed in community halls, miners' welfare clubs and theatres in Britain and Europe, and Ewan MacColl (as he renamed himself in the 1940s) wrote 11 plays until his departure in 1952 (MacColl 1985, 1990; MacColl and Goorney 1986). It was then that he began to get more involved in folk song: collecting ballads, writing songs and preparing radio programmes for the BBC, including the series *Ballad and Blues* in 1952–3 (MacColl 1990). MacColl collaborated with individuals such as the Communist folk singer and writer A. L. Lloyd and his good friend, the American folklorist Alan Lomax. All three became key figures in an emerging folk-song revival that was fuelled by the increasing popularity of American political song, jazz and blues, and the British craze of Skiffle (Boyes 1993). MacColl, Lomax and Lloyd saw this second revival as one which must be concerned with industrial song and the folk culture of the urban working classes, and they constructed an 'urban pastoral' which was held in sharp contrast to the 'English [rural] pastoral' of the turn-of-the-century folk-song revival cultivated by Cecil Sharp, Vaughan Williams and others (A. L. Lloyd 1961; Watson 1983; Samuel 1989: 24; Boyes 1993; Revill 2000). In the preface to his 1954 collection of industrial folk ballads, *The Shuttle and Cage*, MacColl stressed that:

There are no nightingales in these songs, no flowers – and the sun is rarely mentioned; their themes are work, poverty, hunger and exploitation. They should be sung to the accompaniment of pneumatic drills and swinging hammers, they should be bawled above the hum of turbines and the clatter of looms for they are songs of toil, anthems of the industrial age. . . . The folklore of the industrial worker is still a largely unexplored field and this collection represents no more than a mere scratching at the surface. A comprehensive survey of our industrial folk-song requires the full collaboration of the Trades Union movement. (MacColl 1954)

In the early 1950s MacColl and Lomax spent nearly six months visiting trades unions to make them 'aware of their cultural responsibilities' to help preserve and archive workers' song, while in 1951–2 Lloyd persuaded the National Coal Board to fund a competition which resulted in the collection *Come All Ye Bold Miners* (A. L. Lloyd 1952; Dallas 1989; MacColl 1990: 272; Boyes 1993). Lloyd, Lomax and MacColl were revivalists seeking to discover, preserve and articulate a working-class industrial folk heritage. As Raphael Samuel (1989, 1994) has pointed out, this predated the rise of other industrial preservation movements (including industrial archaeology and railway preservation) by a number of years (see also Newton 1963). But while Lloyd was concerned with reviving a tradition of *English* song – albeit an Englishness which was very different to that evoked by Vaughan Williams's *Greensleeves* – MacColl was concerned with the revival of (regional) *British* songs – notably Scots ballads, which he had learnt as a boy from his parents (Howkins 1989; Samuel 1989; MacColl 1990). This was a Britishness which was inclusive and progressive, which could combine American traditions of jazz and blues; bringing about an alignment of different oral and musical traditions that would provide one of the defining features of the BBC Radio Ballads (Samuel 1989; MacColl 1990; Boyes 1993). Nevertheless, MacColl and others were keen to ensure that what they felt to be foreign music and song (particularly American music and song) should not destroy British folk culture and produce 'a kind of cosmopolitan, half-baked music which doesn't satisfy the emotions of anybody' (MacColl 1961: 20). One result was that the folk clubs associated with MacColl, notably the Ballad and Blues Club (founded in 1954) and later the Singers' Club, adopted a policy that 'residents, guest singers and those who sang from the floor should limit themselves to songs which were in a language the singer spoke or understood' (MacColl 1990: 288).

Concerns about what was seen to be the erosion of a British working-class culture by an Americanized mass culture of glossy magazines, television, milk bars and pop song were most notably expressed by Richard

Hoggart in his 1957 book *The Uses of Literacy*. Hoggart diagnosed the problems, but key figures in the folk-song revival felt that they could provide solutions and document and revive vernacular folk cultures (Fisher 1986). With the advent of the portable EMI Midget tape recorder in 1952, radio producers with access to these new technologies, as well as folk-song and oral history collectors, could approach their subjects in new ways. As Charles Parker reflected in 1975: '. . . you could go . . . into "life" and not do a formal interview of only three minutes – because you were recording on disc . . . – but you could go out with a little box, sit down, and talk. Now, this to me, was a revelation' (Parker 1975: 98). It was with such a tape recorder that Parker arrived in Stockport to meet the friends and relatives of John Axon.

Charles Parker had joined the BBC in 1948 (and the Midland Service in 1954) after serving as a submarine lieutenant-commander during World War II and studying history at Queens' College, Cambridge (Fisher 1986). Parker was a devout Christian, conservative, 'liberal bourgeois' BBC producer (Parker 1975: 98), but during the early 1960s his politics began to change. His collaboration and friendship with committed Communists such as Ewan MacColl and George Thomson (Professor of Greek, University of Birmingham), coupled with the impact of the people he met during the production of the Radio Ballads (notably the coal miners he interviewed in 1961 for the *The Big Hewer*), led him to become an avowed Marxist, proponent of folk song and campaigner for social justice (Fisher 1986; on Thomson, see Burns 1999; Enright 1988). Parker became actively involved in MacColl's Critics Group, the Birmingham and Midland Folk Centre, the Grey Cock Folk Club and the West Midlands Gypsy Liaison Group. During the 1960s he taught courses on folk music at the Birmingham Workers' Educational Association and the Department of Extra-Mural Studies at the University of Birmingham (Fisher 1986). His increasingly radical opinions did not fit well within the BBC. After refusing to accept downgrading, following BBC restructuring, Parker was forcibly retired in December 1972 – a move which brought the threat of union action, was widely reported in national broadsheets, and was condemned in a letter to *The Times* by prominent academics with whom Parker had collaborated, including Fred Inglis, Denys Thompson, George Ewart Evans, Peter Abbs and Stuart Hall (Abbs et al. 1972; *The Times* 1972; Parker 1973; Fisher 1986). In the 1970s Parker continued to promote the 'actuality ballad' as a vital tool for preserving working-class, oral folk tradition. He conducted seminars and guest lectures on folk song, popular culture and radio in schools, universities, and at education conferences, and in 1974 he built upon his earlier work in documentary folk theatre by co-founding the Banner Theatre of Actuality (Fisher 1986).

Interviewing and recording 'the motorway men'

After the success of *The Ballad of John Axon*, Denis Morris – Head of Midland Region Programmes (BBC) – suggested to Parker that the construction of the M1 might provide a suitable topic for a Radio Ballad.[10] At first Parker was sceptical about whether a suitable story-line could be found,[11] as he was concerned that Morris had a particular image of construction in mind:

> [His image of construction was] dominated by documentary film techniques ... a bulldozer coming over the skyline with pulsing symphonic music behind it, while a voice throbbing with history talks of the titanic nature of the undertaking, of ... the tradition of the great engineers ... , and you are made to feel that something is really happening in the hearts of the men you see driving their bulldozers and preparing their projects and plans and that all are fired with a tremendous sense of epic achievement! (Parker 1964: 2)

This was the image presented in Laing's films and *The London–Birmingham Motorway* (Rolt 1959), but during their visits to the construction site Parker and MacColl encountered engineers who they felt had little appreciation of history or of the epic nature of the project, farmers who were only concerned about their properties, and workmen who were largely in it for the money. A further problem emerged when Parker's superiors requested that the team produce 'a less subjective approach than that used in *John Axon*', as the motorway was 'a national affair' (MacColl 1990: 316). MacColl felt that Parker agreed with this view, while he favoured a ballad based upon the views of the working-class men engaged on the project (MacColl 1990). Nevertheless, a draft plan reveals that Parker was well aware of the critical possibilities of this new ballad:

> ... one is tempted to conceive a programme exposing: (i) the ruthless exploitation of labour which still remains the feature of civil engineering, albeit the assumption is now that better wages more than recompense for long absence from home and brutalising conditions of work and living [*sic*]. ... (iii) the arrogance which technological advance has planted in man and a consequent failure to conceive spiritual values which are the fruit of human personality as sacred. ... the over-riding atmosphere is one of a war-time operation. There is a battle in progress but a battle in which the ultimate objectives are profits and not engineering achievement.[12]

Parker knew that such a treatment would not receive the cooperation of John Laing's Public Relations Department or the Ministry of Transport and Civil Aviation. The team decided to take a 'less subjective' approach

(MacColl 1990: 316), but in December 1958, after a month of preliminary recordings, Parker and MacColl were refused further access to the site by John Laing and Son Limited.[13] Parker suggested that this was due to reports on the television programmes *Tonight* and *Panorama* which had drawn attention to the long hours worked by men on the project.[14] Senior Ministry engineers, including J. F. A. Baker, also refused to participate, but with the full support of Sir Owen Williams, Laing finally allowed Parker and MacColl to speak to the workers. The pair were accompanied at all times by a public relations assistant 'who sought to dictate the persons to be interviewed, and the terms in which they were interviewed',[15] and MacColl felt that their presence meant that, 'in the minds of the workers', he and Parker became identified with 'the management'.[16] It was only after negotiations between the BBC and Laing that they were given more freedom on site. In any case, Laing's public relations team could not have prevented the team from going to the independent workers hostels, the pubs the men frequented, the offices and employees of Sir Owen Williams and Partners, or the sections of the St Albans Bypass Motorway built by Tarmac – who cooperated fully with Parker and MacColl.

In total, MacColl, Parker and Seeger recorded between 80 and 90 hours of 'actuality'.[17] They recorded the sounds of pouring concrete, aggregate and asphalt; the noise of earth-movers, tractors, concrete trains and Tarmac's aeroplane; and mouth organs, office typewriters and short-wave radios.[18] Interviews were conducted with senior engineers and technicians – including Sir Owen Williams, Owen Tudor Williams, Mr Price (Chief Resident Engineer), J. C. Cryer (Project Manager, Contract A), G. R. Aspinall (Tarmac Civil Engineering), Tarmac's Chief Chemist and Laing's surveyors and soil analysts – who were asked technical questions about the design and construction.[19] The interviews with less senior workers and labourers were very different, and they formed the bulk of the recordings. Parker appears to have conducted much of the interviewing, and would usually start by asking the worker his name and what he was doing. Fairly quickly the questioning was directed to more personal matters: where the worker came from, whether he was married and had children, why he chose that line of work, and whether he enjoyed the mobile lifestyle of a construction worker. The recordings provided Parker and MacColl with the raw material (the 'actuality') for the Radio Ballad. Parker's passion led him into quite extensive discussions with interviewees, and MacColl remarked that it was common for him to 'talk more than the person he was recording' (MacColl 1990: 329). In a number of interviews Parker also comes across as someone who was speaking down to the workers. During a recording in a public house frequented by Irish labourers, which MacColl and Parker visited in an attempt to record traditional Irish songs, Parker's tone is evident:

Charles Parker: What, do you mean having machines means that you don't write shanties and things; you don't need them?

Irish man: No

Charles Parker: That's very true. But you still need to sing. We all need to sing. Your need to sing is greater than theirs. Because if you're not going to be . . . [interruption]

Irish man: You can't sing swinging a shovel, can you?

Charles Parker: Well, your grandfathers did!

Irish man: er . . .

Charles Parker: And the Negroes. A great deal of the Negroes' tradition is work songs; swinging an axe, you know, in the forests and swamps.

(Transcribed from cassette, BCA MS4000/LC84)

That many of the Irish labourers didn't sing, know the traditional Irish songs that Parker and MacColl knew, or match their stereotypes of Irish labourers was held to be another example of how the increasing use of machines was destroying working-class folk cultures (Parker 1959a). Nevertheless, after further enquiries they managed to record quite a few work songs.

The large number of Irish workers on the motorway were of great interest to these two folk-song enthusiasts, but their fascination with working traditions and folk culture also extended to workers of other nationalities. English workers from a number of different regions were interviewed, although the accents of local employees – who are said to have comprised up to 50% of the total work-force (Rolt 1959) – are few and far between on either the Radio Ballad or the interview tapes. On the other hand, the chance to talk to an Indian, Hungarian, Pole, Italian or West Indian provided an opportunity not to be missed:

Charles Parker: Where are they from, mostly? Are they mostly from Ireland?

Irish foreman: Mostly from Ireland, yes. There's two Englishman, and there are two West Indians.

Charles Parker: How do you find them?

Irish foreman: Alright. Very satisfactory.

Charles Parker: It might be an idea to have a word with a West Indian man.

Ewan MacColl: Yes, that would be a good idea.

(Transcribed from cassette, BCA MS4000/LC84)

Who it was that was working satisfactorily is unclear. When they did speak to a West Indian they asked him what he thought of the British winters, while an Indian man was asked what he had done for a living in India and why he had come to England. One of the most telling recordings featured a number of short songs – a Jamaican man singing 'Saviour and friend', an

Irishman singing 'Lonely I wander', Ewan MacColl singing a nineteenth-century navvy song, and a man singing 'McAlpine's Fusiliers' – after which can be heard a discussion between an Irishman and a Jamaican labourer which did not make the final programme.[20] The recordings suggest that while official observers presented optimistic accounts of harmonious relations between managers and workers of different races, the 'colour bar' (and racism in general) may well have been experienced by motorway workers, whether in fairly subtle or overt ways:

Jamaican man [talking about bulldozers]: I can work it . . . It was my job in my country.
Irish man: Why the hell don't you drive them here then?
Jamaican man: I write in many times for a job, but just because I've been there, and they see that I'm a different nationality, they don't give it to me.
Irish man: That shouldn't count.
Jamaican man: It did.

(Transcribed from cassette, BCA MS4000/LC86)

This account is typical of the experiences of 'de-skilling and discrimination endured by black Britons' during the 1950s (K. Paul 1997: 120). The British government attempted to address what it characterized as the 'problem' of black, imperial (and, by law, British) subjects emigrating to Britain, while they deemed the immigration of white men and women from Ireland and other European nations to be more acceptable and less noticeable, and their difference more assimilable (K. Paul 1997).

Visits to the pub to record songs and less formal conversations were not the only excursions MacColl and Parker made away from the construction sites. Recordings were made in 'Bob's Café' on the A45 in July 1959, which Parker had read about in an article by W. J. Morgan in *The Observer* in November 1958.[21] Morgan (1958) had interviewed regular visitors to Bob's Café on the changing practices and traditions of lorry driving, an exercise which may well have had an appeal for MacColl and Parker, two enthusiasts of vernacular life (on lorry driving and transport cafés, see chapter 5). Recordings were made in canteens and Tarmac's Wolverhampton depot, while on a number of occasions Parker and MacColl interviewed men in their hostels. Parker later described the independent hostels as 'positively Orwellian', their condition being made especially evident after he and MacColl spent a 'harrowing night in a hostel' that accommodated two or three hundred workmen:

It had, I imagine, been an army camp at one time, and it was for me a nightmare of dank concrete floors, disused static tanks and heaps of coarse food shovelled into men, some of whom had spent twelve hours at work with only a snack to sustain them. (Parker 1964: 3)

The camps provided Parker with a lesson about the conditions in which some of these working-class 'folk' lived and worked, but MacColl has suggested that Parker still 'maintained a fairly formal relationship with his informants' (MacColl 1990: 329). It was during the recording of *The Big Hewer* in 1960–1 that Parker began to see his interviewees as equals, developing a great respect and, occasionally, friendships with individuals whom he had previously seen as somewhat quaint and inferior (Fisher 1986; MacColl 1990).

'Singing of a motorway'

When the recordings were complete, Ewan MacColl and Peggy Seeger (the American folk musician and singer, and Ewan's third wife) edited the tapes to produce about 30 to 40 minutes of 'actuality' – which was separated from the questions of Parker and MacColl, and integrated with songs written by MacColl and Seeger.[22] The songs drew upon information conveyed in the recorded interviews, as well as facts and figures obtained from booklets, maps and articles collected by Charles Parker.[23] Parker developed a particular interest in the geology of the route, collecting information from the contractors and consulting engineers, articles from the press, and purchasing Ordnance Survey geological maps for the entire length of the motorway. Geological investigations had accompanied the excavation phase – as occurred during railway construction in the nineteenth century (see Freeman 2001) – and Dr R. Casey of the Palaeontological Department of the Geological Survey of Great Britain discovered a large number of ammonites (up to 18″ across) along the line of the motorway (see *Rugby Advertiser* 1958b; Smith 1958; Geological Survey of Great Britain 1959). Parker requested an interview with Dr Casey, but was refused by the government's Department of Scientific and Industrial Research.[24] The team recorded very little actuality about the geology of the route, and Parker felt that this was a major fault with the Ballad (Parker 1964).

Song of a Road opened with a song – variously titled the 'Muck-Shifter's Song' or 'Cats and Back Acters' – about the ancient rocks and soils, and the modern earth-moving machines, which resembled 'pre-historic monsters' (MacColl and Parker 1959: 5; Seeger 2001). This song, as with many of those in the Ballad, was a rousing folk-style number celebrating the epic nature of the project, into which sections of recorded actuality were interspersed where Sir Owen Williams, Owen Tudor Williams and other engineers talk about geology and earth-moving:

22. *Singer*: In the beginning there was the land.

23. *Tape A Sir Owen*: Now when you dig the ground you realise what a real
bundle of dirt we live on.

24. *Singer*: Earth moved, mountains, plains and seas formed.
Swirling gases rolled away,
Sun in the sky burned down on the saw-toothed mountains

25. *Chorus*: When you're up on the seat,
In the cold and the heat,
You never think what you're lifting
You're bashing away every hour of the day,
You're working at the old muck-shifting,
Cats and back acters,
Rubber-tired tractors,
Derricks and cranes for lifting
Whatever your rig, bulldozer or jib
You're doing the old muck shifting
(MacColl and Parker 1959: 4)

The 'Muck-Shifter's Song', and the Ballad as a whole, may be seen as
examples of MacColl's tendency to celebrate a specifically masculine heroic
industrial folk culture (Boyes 1993). As Georgina Boyes has observed, 'the
industries' that folk revivalists thought 'worth researching and writing about
were never catering or nursing, hairdressing or office work' (Boyes 1993:
240). Women did appear in later sections of the Ballad. MacColl, Parker
and Seeger used the large number of recordings of male labourers talking
about their life away from their family and friends to write songs about
their mobile lifestyle. They contrasted the mobilities of the men with the
immobility of the wives, girlfriends and families back 'at home', but while
celebrating the heroism and masculinity of these mobile working men, the
songs reinforced 'conventional' gendered, heterosexist attitudes to family
life (cf. Cresswell 1993; McDowell 1996). The Ballad stopped short of
explicitly celebrating the freedom of these men and their mobilities, and
their absence from home was presented as essential to maintain their family
life, economically and socially. This was particularly noticeable in the
team's portrayal of the mobilities, masculinities and family life of the Irish
migrant workers:

185. *Singer*: Why do you live your life alone.
And leave your wife and kids at home?
Why don't you stay, why must you roam
And work on the open road

MUSIC SEQUENCE 9A ENDS

186. *Tape A Belfast man*: It's a man's place to be at home with the wife but unfortunately there's no work in our part of the country now.

· · ·

188. ['The Exile Song'] *Singer*: Just a note, for time is short, dear,
Hard the work and long the day,
But my heart is with you Mary,
Though I'm many a mile away.

Kiss the children for me, Mary,
Do not let them pine or grieve,
Tell them how I'm working for them,
Why our home I had to leave.

Building dams, airfields and factories,
Moving concrete by the load,
I'll be with you in November,
When I'm finished on the road.
(MacColl and Parker 1959: 25–6)

These final three verses formed 'The Exile Song', which was based upon the words of James Graham, a labourer from Kilkeel in Northern Ireland, and sung to a traditional Scots tune accompanied by the Uileann pipes (MacColl and Seeger 1963; Seeger 2001). After this there followed the only song sung by a woman, 'Come, Me Little Son', which was also performed to a traditional Scottish tune (MacColl and Seeger 1963; Seeger 2001). A woman sings to her son, telling him about the father he has never really known, who is working hard to feed and clothe them. The song-writers locate the woman in a distant home, while 'her man' is forced to assume a mobile lifestyle due to lack of work. This portrayal of a lonely grieving wife was criticized by contemporary journalists as inaccurate and romantic. As David Paul wrote in *The Listener* in November 1959, 'the rigours of social-realist balladry demand that' the wife of a worker 'sits at home, rocking the cradle with her toe and crooning a lullaby', rather than 'click[ing] the needles in front of [the] telly' or 'slip[ping] out with Sadie for a quiet drink with two nice chaps at the local' (D. Paul 1959: 847).

'The Exile Song' and 'Come, Me Little Son' were just two songs in the lengthy section of the Ballad on the life and work of the 'long-distance men'. The sequence began with recordings of workers stating where they come from. The answers of men from different parts of England, Poland,

Hungary, Ireland, India and the West Indies were edited together into a short montage, at the end of which Jack Hamilton from Cork assured the listener that everybody got on just fine: 'There are black men and white men, and they're all colours up there . . . , oh! they're a good old crowd; oh yes, they are yes; oh they're a good old crowd; oh we get on good.'[25]

While this was seen to be a harmonious, multi-cultural and multi-national work-force, the fact that many of the workers were Irish reinforced MacColl's decision to draw upon a number of traditional Irish (as well as Scots) ballads. Francis McPeake from Belfast, 'one of the greatest living traditional singers in Britain', sang 'The Exile Song' – accompanying himself on the Uileann pipes – and 'The Driver's Song' – accompanied by Peggy Seeger on the banjo (Parker 1961: 4). The other singers included MacColl himself, A. L. Lloyd, Jimmie MacGregor and Isla Cameron from Scotland, Isabel Sutherland and Seamus Ennis from Ireland, Lou Killen from Gateshead, Cyril Tawney from Plymouth, and two West Indians, John Clarence and William V. Thomas. But while a large number of the songs were derived from traditional British and Irish songs, MacColl and Seeger drew their inspiration from other traditional folk musics – notably jazz. Peggy Seeger, on banjo and autoharp, and Alf Edwards, on concertina and ocarina, were joined by Bobby Mickleburgh on trombone, the Calypsonian guitarist Fitzroy Coleman and Bruce Turner's jazz band (consisting of Turner on clarinet, Jim Bray on double bass, John Chiltern on trumpet and John Armitage on drums).[26]

Broadcasting the Ballad

The musicians and singers recorded *Song of a Road* in October 1959, and the Ballad was broadcast on the BBC Home Service on 5 November 1959 at 8 pm. The *Radio Times* included an advert for the programme and an article by Parker which emphasized the rich traditions of civil engineering and song that the Ballad drew upon and preserved (Figure 4.5) (Parker 1959a; *Radio Times* 1959a). Parker wasn't happy with the *Radio Times* advertisement, as he had requested an illustration by Eric Fraser, who had drawn two abstract sketches to promote *The Ballad of John Axon*.[27] Instead of reflecting the dynamism and 'awesome size and powers' of the machines, Parker complained that the illustration 'achieved the effect of a static Dinky Toy'.[28] The situation was remedied before the programme was repeated on 29 December 1959, and Parker felt that the new advertisement featured an 'absolutely first class' illustration of earth-moving machines and men on a musical score/road (see *Radio Times* 1959b).[29] *Song of a Road* received mixed reviews in the press. Under the title of 'popular art for a cultural minority', *The Times*'s (1959b: 4) correspondent stressed that the ballad

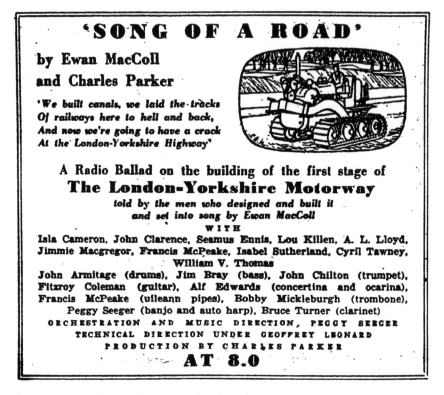

Figure 4.5   *Song of a Road*. Advertisement from the *Radio Times*, volume 145, 30 October 1959, p. 42. Reproduced by kind permission of the BBC Written Archives Centre.

was 'experimentally playing in a highbrow way with its theme'. In *The Sunday Times*, Robert Robinson (1959) described the songs as having very little to do with 'a contemporary workman on a contemporary road' and as being more 'folksy-song' than 'folk song'. Writing in *The Listener*, David Paul criticized the Ballad for being 'portentously unreal' (1959: 847), but Parker (1959b: 938) responded by suggesting that they had never attempted to convey 'reality'.

Song of a Road was broadcast in at least fifteen countries (Winter 1961), and in October 1960 it was played to Laing's staff in Rhodesia and Nyasaland (*Team Spirit* 1960c). At the suggestion of K. G. Gerrard, Laing's Public Relations Officer, The Institution of Civil Engineers requested a playback of the Ballad on 3 May 1960 at the 'tea interval' of their symposium on the London to Birmingham Motorway. Laing's request was somewhat surprising, as their public relations team had unsuccessfully requested that two sections of the Ballad be cut, and Parker explained how they had apparently tried to 'prevent the programme being heard at all in the most embarrassing exchange' at the press preview.[30] The section of the Ballad which offended both Laing's public relations team and Phyllis Faulkner at the Ministry of Transport and Civil Aviation was a piece of actuality recorded in a hostel in Leighton Buzzard, in which Jack Hamilton from Cork compared the accommodation with a concentration camp:

> *Hamilton*: The bed I'm lying in! There's humps and hollows in it b'jabers it's like a camels back. It is – I thought I was up in the desert [the] first time I slept inside on it. – looking at camels' humps and hollows. Me arse was all blisters and carbuncles! I could hardly walk – I thought I wouldn't be able to go to work. Its an awful joint – concentration camp! All they want there now is some gas chambers and they'll smother us, that's all they want there.
>
> (MacColl and Parker 1959: 25)

Parker felt that a private company such as Laing had no right to obstruct a public institution such as the BBC in conveying the story of this 'national undertaking' to the taxpayer and citizen.[31] The BBC were, after all, an Establishment institution (Fairlie 1959), but MacColl and Parker's social and political agenda, and the subjects of the Ballads, meant that their work diverged from the largely conservative stance and principle of impartiality associated with the BBC (see Fairlie 1959). Although Hugh Carleton-Greene (Director General of the BBC between 1960 and 1969) sent a 'congratulatory note' to Parker following the broadcast of *Song of a Road*, both Parker and MacColl maintained that the BBC disapproved of the Radio Ballads (Parker 1973: 134). Parker suggested that it was *these* programmes which gave rise to accusations that he made 'disproportionate demands upon resources' (Parker 1973: 134), while MacColl later suggested that the BBC saw the Radio Ballads as expensive working-class epics which were 'full of value judgments' and 'took sides' (MacColl 1990: 330). More importantly, MacColl remained a member of the Communist Party until the early 1960s, and this did not go unnoticed by the press and BBC executives (Samuel 1989; MacColl 1990; Denselow 1996).[32]

What is clear is that the scripting and editing of the Radio Ballads enabled Parker, MacColl and Seeger to juxtapose sections of speech and song and raise questions about the social and political situations associated with different projects and subjects:

> 131. *Tape A Mr Price*: I feel now in my own mind that we are building something which is well worthy of the name of a road, something really good for the country.

MUSIC 8A ENDS

132. *Tape A Welsh farmer.* Well I don't know, I'm not worried about the country I'm worried about the farm, they've cut the farm in two you see – left some bits here and there – got a bridge – must make the best of bad job! Its not too bad, but there you are what can you do! (laughs).

(MacColl and Parker 1959: 19)

The juxtaposition of such extracts of speech may have caused the listener to question the impact of the motorway on the lives and work of labourers, farmers and local residents, but MacColl saw their *broad* approach – particularly the celebration of the work of senior engineers – as unnecessary (MacColl 1990: 316). The inclusion of middle-class managers as well as working-class labourers and tradesmen produced a 'thoroughly confusing – and at times boring – programme', which taught Seeger and MacColl a 'valuable lesson about the precise nature of the relationship of speech to class and to traditional songs' (MacColl 1990: 316).

> In the case of playing back the road builders' actuality, we had observed that there were basic differences in the way in which words were used by the manual workers on the one hand and by the planners and white-collar staff on the other. The latter, though educated and 'articulate', were tedious to listen to. . . . To our 'uneducated' speakers, however, we could listen for long periods without any decline in concentration. (MacColl 1990: 316–17)

MacColl appears to exoticize and be captivated by the idiosyncractic speech of the motorway's workers, but he set out to analyse the differences in a more formal and objective manner. MacColl and Seeger made a 'rough analysis of the speech in a number of tapes chosen at random', and while the managers used 'an extremely small area of the vocal-effort spectrum', spoke at a consistent tempo, and rarely used metaphors and similes, 'the labourers' used a large area of the vocal-effort spectrum, changed tense and tempo regularly, and used 'similes and metaphors with obvious enjoyment' (MacColl 1990: 317). The labourers' voices had a dramatic effect which captivated listeners, while the speech of the managers was 'impassive' and 'dull' (MacColl 1990: 317). Their voices and actions were seen to be out of place in a Radio Ballad about 'heroes of labour' (Samuel 1989: 24).

Song of a Road was later described by Ewan MacColl as 'such a butchered work' that neither he nor Peggy Seeger was keen to have his or her name 'associated with it' (MacColl 1990: 327). It taught them that the Radio Ballads should 'only be about people, not about processes', and the six subsequent Radio Ballads focused on the lives and work of a range of groups, including individuals whose lives were far-removed from the male heroes of industry that critics associated with the folk-song revival (Parker 1964: 4). Ballads on fishermen, coal miners and boxers (*Singing the Fishing*, *The Big Hewer* and *The Fight Game*) contrasted with programmes on teenagers (*On the Edge*), polio sufferers (*The Body Blow*) and contemporary gypsies and travellers (*The Travelling People*) (see Fisher 1986; MacColl 1990). Nevertheless, in all eight Radio Ballads, MacColl, Parker and Seeger challenged prevailing stereotypes of these individuals and groups. *Song of a Road* populated the landscapes of motorway construction with the voices and stories of migrant labourers and their families, who received little or no mention in the official press releases, films, newsletters or booklets produced by John Laing and Son Limited and the Ministry of Transport and Civil Aviation (*Major Road Ahead* 1958; *Motorway* 1959; MT 1959; Rolt 1959). Laing's texts celebrated the achievements and expertise of senior engineers and staff members, and individual labourers tended to be subsumed into the construction 'team'. The nationalities and ethnicities of the different workers, although highlighted in an attempt to emphasize the progressiveness and harmony of the employment practices, were easily effaced in celebrations of the motorway as a British engineering achievement. Public relations officers shadowed MacColl and Parker during their visits to the construction sites to ensure that their movements, questions and answers conformed with official accounts, but they could do little to control their access to the leisure spaces of workers. Parker, MacColl and Seeger's Ballad celebrates the construction of Britain's first major motorway, but their interest in the lives of the workers, their socialist politics and their reworking of traditional jazz and folk music (particularly Scottish and Irish folk music and song) presents motorway construction as a complex achievement. This was an achievement not simply of a British construction company, but of 'a thousand men from the earth's four corners' (MacColl and Parker 1959: 2).

The BBC Radio Ballads inspired broadcasters, educationalists and arts groups to experiment with recorded actuality throughout the 1960s and 1970s, and in 1965 MacColl, Parker and Seeger worked on a 50-minute black and white television film which revisited themes raised in *Song of a Road*. *The Irishmen: An Impression of Exile* was produced and directed by Parker's friend and BBC Birmingham colleague Philip Donnellan (1924–99), with advice from Irish singer Seamus Ennis, who had performed on *Song of a Road* (*The Irishmen* 1965; Pettitt 2000a).[33] The film 'gives powerful and proud voice to the Irish emigrant experience in Britain' (Pettitt 2000b: 85), focusing largely on the experiences of male migrants, interspersing and intertwining footage of: Irish communities in north-west London and western Ireland; the journey of a fisherman from Connemara to work in London; men constructing the new Victoria Line of the London Underground; and men working on Cementation's southern extension of the M1 at Brockley Hill interchange (Junction 4) on the northern outskirts of London (*The Irishmen* 1965; Pettitt 2000a). Footage of reinforced concrete bridges being lowered into place, and of men digging the earth, collecting their wages, driving Euclid earth-movers, sitting in huts and laying steel reinforcement cables was overlain with songs such as 'The Rambler from Clare', 'Dublin Jack of all Trades', ' The Tunnel Tigers' and 'The Rocks of Bawn', and interspersed with footage of the bare bodies of men drilling through the London clay to construct the Victoria Line.[34] *The Irishmen* portrays the embodied actions, hard work, machines and displacements required to construct a motorway or tube line, a landscape which in Don Mitchell's words might be considered to be 'both a work and an erasure of work' (D. Mitchell 1996: 6; 2001). Donnellan's background, political commitments and radical vision was strikingly similar to that of Charles Parker, and later in life he described how his approach to film was heavily influenced by the Radio Ballads and the interview and recording techniques of Parker (Pettitt 2000a). Indeed, one of the most striking features of *The Irishmen* is the soundtrack, which contained 'songs to Irish traditional melodies' arranged and in many cases written by Ewan MacColl and Peggy Seeger (these are reprinted in Seeger 2001). Paul Lennihan and well-known Connemara Gaeltacht singer Joe Heaney sang many of the songs, while the entire soundtrack was edited by Charles Parker (*The Irishmen* 1965; Seeger 2001). *The Irishmen* explored important subjects that were largely unspoken or ignored in Establishment BBC circles, including the discrimination and racism experienced by many immigrants (including the Irish). The film appears to have been too frank, realist and political for the BBC, and it was dropped from the broadcasting schedule on the grounds that it was 'shapeless, pretentious . . . and boring' (Pettitt 2000a: 358; 2000b).

Chapter Five

Driving, Consuming and Governing the M1

The first section of the London to Yorkshire Motorway, and the greater part of the St Albans Bypass Motorway, were completed on schedule on Saturday, 31 October 1959, in time for the official opening on Monday, 2 November. In this chapter I examine how the motorway was experienced and consumed by drivers and passengers in late 1959 and the early 1960s. As the M1 was Britain's first major stretch of motorway – which opened eleven months after the 8¼-mile-long Preston Bypass Motorway in Lancashire, North-West England – it was seen by commentators not only to be strikingly new, but also to be a distinctively modern force cutting through the rural landscapes of South-East England. While Prime Minister Harold Macmillan and the press celebrated the opening of the Preston Bypass as Britain's first motorway, the M1 attracted more widespread publicity. Firstly, the M1 was a far longer section of motorway than the Preston Bypass, but secondly, and more significantly, the M1 was located far closer to London and the offices and homes of the majority of national journalists, politicians and cultural commentators. The M1 was easy to visit, observe and write about. Here was the first southern English motorway, a motorway with no speed limit, an experimental space, which would require close scrutiny and regulation to ensure that vehicle drivers passed safely, efficiently and pleasurably through the landscape. In this chapter I focus on attempts to predict, control, criticize and celebrate the movements of vehicles, drivers and passengers who passed through the spaces of the M1 motorway in the late 1950s and 1960s.

In the first two sections I examine the attempts of politicians, journalists, the police and motoring organizations to predict and govern the conduct and movements of motorway vehicle drivers before and after the opening of the M1. Three key theoretical ideas underlie my argument, although I do not explore these in great detail here (see Merriman 2005b, 2006b for more detailed theoretical discussions). Firstly, drawing upon the writings of Michel Foucault, Nikolas Rose and others on 'government' and 'governmentality', I attempt to show how practices of subjectification and self-government are vital to all manner of programmes of government, including seemingly large-scale (state) programmes concerned with governing others at a distance (Foucault 1985, 1986b, 1988, 1991, 1997; P. Miller and Rose 1990; Gordon 1991; N. Rose and Miller 1992; N. Rose 1996, 1999; Dean 1999; Joyce 2003; Merriman 2005b, 2006b). Secondly, I loosely draw upon ideas developed by actor-network theorists, anthropologists of material culture and sociologists such as Andrew Barry, Nikolas Rose, Peter Miller and others to examine how a broad array of more or less mundane material devices or technologies are vital to the articulation and translation of programmes of government, particularly those concerned with governing at a distance (Foucault 1986a, 1988; Dean 1999; Latour 1992, 1993; N. Rose 1999; Barry 2001; Joyce 2003; Merriman 2005b, 2006b). Thirdly, and relatedly, I draw upon writings by sociologists to examine how the motorway vehicle driver was constructed by experts and authorities as a complex, heterogeneous and hybrid figure (Michael 1998, 2000, 2001; Lupton 1999; Sheller and Urry 2000; Merriman 2006b; cf. Latour 1992, 1996). In the first two sections I draw upon these theoretical ideas to examine how a range of experts and authorities attempted to educate vehicle drivers about good motorway driving through the production and distribution of a series of *technologies of government* that would enable drivers to translate these recommendations into embodied and habituated techniques for conducting oneself and one's vehicle safely along the motorway. I examine how motorway driving was seen to produce distinctively new experiences, sensations, subjectivities and ways of being, and I show how experts concerned themselves with enhancing the abilities, competence and performance of both driver *and* vehicle. I study the reactions of Minister of Transport Ernest Marples to the conduct of the first motorway drivers, and I reveal how commentators argued that motorway driving ushered in new senses of space, distance and speed. Finally, I examine the work of the motorway police and Automobile Association in monitoring and regulating the movements of vehicle drivers on the motorway.

In sections three and four I examine how the M1 was constructed and experienced as a space of modern consumption. In section three I show how the M1 caught the public's imagination, becoming a popular tourist destination and a cultural reference point and marketing tool for pop groups, toy manufacturers, children's authors and a playwright. I examine how motorway driving became entwined with debates about masculinity, before looking at the maps and guides which were produced of the M1, and the ways in which the M1 acted as an organizing device in the landscapes of rural Bedfordshire, Buckinghamshire and Northamptonshire. In section four I examine how the first two motorway service areas at Newport Pagnell and Watford Gap emerged as spaces of regulated consumption. I show how Newport Pagnell service area became a popular 'hang-out' for teenagers, and discuss the reluctance of lorry drivers to leave their transport cafés for the modern service area cafés. In the final section, section five, I show how the government approached the M1 as an experimental space, a space of scientific inquiry, statistical calculation, accidents and death. I discuss the work of economists and statisticians in the government's Road Research Laboratory and at the University of Birmingham who undertook one of the first large-scale British cost-benefit analyses on the economic impact of the construction of the M1, before going on to examine studies by Road Research Laboratory scientists of the first motorway accidents, which revealed the difficulty of attributing blame to the different actors involved in the complex movements and moments of driving.

Motorway Driving, Embodiment, Competence

Governing the conduct of motorway drivers

During 1957, 1958 and 1959, politicians, journalists, motoring organizations and the police started to express concern about the ability of drivers and vehicles to cope with the speeds and conditions which were expected to prevail on Britain's new motorways. Would the nation's motor vehicles withstand the high average speeds ushered in with motorway driving? Would British vehicles break down? How fast would motorists drive? Would they be able to comprehend and successfully negotiate the modern flyover junctions? How would motorists react to the choice of two or three lanes? Would they stay in one lane, or straddle two? In July 1958, Minister of Transport and Civil Aviation Harold Watkinson introduced a series of experimental motorway regulations to the House of Commons (*Parliamentary Debates* 1958). The new regulations detailed the vehicles and drivers who would be prohibited from travelling along the motorway, and the focus was quite clearly on the combined abilities and performance of, or the 'distribution of competences' throughout, the simplified and hybridized vehicle driver (Latour 1992: 233; 1996). Motorway vehicle drivers had to be motorized, fully trained and licensed, and of an appropriate size and weight. Hence, cyclists, mopeds, animal-drawn vehicles, excessively large and heavy loads, agricultural vehicles, pedestrians, learner drivers and invalid carriages were, and are, all excluded from these spaces. The motorway regulations were displayed on prominent signs at the entrance to the motorway (visible on Figure 5.5), but while these regulations were legally binding, a more enabling quasi-legal code of conduct, the Motorway Code, was drafted in 1958 (*Parliamentary Debates* 1958; Merriman 2005b). The Motorway Code was issued to motorists who used the Preston Bypass Motorway in December 1958 (MTCA and COI 1958), and in July 1959 it was incorporated into a new look Highway Code (MTCA and COI 1959). The Motorway Code was comprised of 'statements of the criminal law . . . mixed up with rules of conduct' (*Parliamentary Debates* 1959b: 802), and it functioned as a moral contract, a Foucauldian 'technology of government' that would persuade drivers to translate its coded recommendations into embodied and habituated techniques for conducting oneself and one's vehicle safely along the motorway (Merriman 2005b, 2006b). Drivers were informed of how to join and leave motorways, which 'have no sharp bends, cross-roads, roundabouts or traffic lights' (MTCA and COI 1959: 14). Advice was provided on the new practices which had to be learned and performed, whether practices relating to lane discipline and overtaking, or to mirror usage, emergency procedures and exiting strategies: 'To leave a motorway at one of the intermediate exit points, get into the left-hand lane in good time, stay in it, and give a left-turn signal well before you reach the slip road' (MTCA and COI 1959: 18). The Code was designed to address concerns about how drivers might behave on the new motorways, and these uncertainties surfaced in both serious *and* satirical commentaries on the 'unknown' spaces, situations and movements of motorway driving.

In a satirical article, 'M1 for Murder', published in *Punch* magazine five days before the opening of the motorway, H. F. Ellis predicted chaotic scenes, in which the nation's drivers would be driven mad by the freedom, speeds and strikingly different experience of motorway driving:

> The knowledgeable say that motorways drive men mad. . . . Quite sensible, respectable drivers . . . become corrupted on a motorway. They intend no evil . . . in a mile or two the sense of freedom from interruptions, the heightened tempo of the road, the *whizz-whizz* of passing vehicles, infects the blood. The needle creeps into the sixties, into the seventies, to the ultimate middle eighties. Without conscious volition they are driving at speeds clean outside their experience and skill, happily unaware of the paralysing effects that, let us say, unbalanced wheels can produce at higher speeds. . . . (Ellis 1959: 362–3)

Ellis accepted that 'the young and ardent' would and could drive their sports cars and motorcycles at speeds of over 90 mph, but such velocities were not suitable for the average or below-average driver, who lacked the necessary vehicles, experience and/or skill (Ellis 1959: 362). Throughout the twentieth century a range of groups were criticized and stereotyped for their driving practices, from women, the middle classes and the elderly, to the *nouveau riche* and ethnic minorities (O'Connell 1998; Katz 1999). Ellis's satirical target was the lorry drivers 'released from the constraints of A5', the 'normally rational people in unbalanced saloons', and the 'old fool in a worn-out soap box' (Ellis 1959: 363), whom he expected to move beyond their physical and mental abilities. These vehicle drivers would literally drive themselves into a state of emotional and mental, as well as physical, breakdown.

During 1958 and 1959 motoring journalists as well as politicians provided advice which was intended subtly to reconfigure the 'embodied dispositions' of prospective motorway drivers (Edensor 2003; Sheller 2004: 228; Thrift 2004). Following the publication of the Motorway Code in November 1958 Northampton's *Chronicle and Echo* newspaper expressed its faith in the powers of the Code to regularize the embodied practices and conduct of motorway drivers, such that their very being would be transformed when traversing these distinctively different spaces: '. . . the motorist on the new highway will only become a being apart while he is actually on the motorways. When he leaves them he will automatically be transformed into an "ordinary" motorist . . .' (*Chronicle and Echo* 1958e). The modern spaces of the motorway would shape the very being, the ontologies, of vehicle drivers, requiring new kinds of skill and spatial awareness. New techniques of driving, looking and concentrating would be required, and although the *Chronicle and Echo* suggested that these habits might come automatically or naturally, other journalists suggested that drivers would need to consciously develop such skills. As the Road Haulage Association magazine *Road Way* advised its lorry driving readers in December 1959: 'Many drivers . . . will have to condition themselves to motorway driving – to adopt new techniques quite foreign to those they have habitually used on normal highways' (McLintock 1959: 22). Motorway driving becomes identified as a practice apart, and to complement the Motorway and Highway Codes, and provide further guidance for motorists, national and local newspapers, the motoring press and motoring organizations published leaflets, supplements and special articles before the M1 opened.

The Royal Automobile Club issued a 'Know your motorway' booklet to its members (Brendon 1997), and the Automobile Association provided a colourful 'Guide to the motorway' for motorists entering the M1, which contained a map of the route, pictures of the new signs, extracts from the Motorway Code and details of breakdown services (Figure 5.1).[1] In an article entitled 'Motorway One: new way from the Midlands to London' published on 30 October 1959, *The Autocar* provided information on how to get to the motorway, the design of the new signs, maps of the route, photographs of the junctions, explanations of the Highway Code, and

diagrams illustrating 'flyover drill' (*The Autocar* 1959a). On 4 November *The Motor* inserted a special pull-out guide into its magazine – its cover photograph emphasizing the monumental scale and modernity of the motorway, with a low-level close-up shot through the arches of one of Sir Owen Williams's skewed over-bridges (Figure 5.2). The pull-out guide featured an annotated map of the route and junctions, a list of 'motorway dos and don'ts', details of how to access the motorway 'from the South', and an 'A to Z on the motorway' containing 'pertinent extracts from the Highway Code' (*The Motor* 1959a). Motorists were urged to detach the supplement and take it on their first motorway outing, as it would 'prove indispensable to users of this new highway' (*The Motor* 1959b: 339). However, the supplement was not intended as a tourist guidebook, for the motorway was 'not [intended] for sightseers' (*The Motor* 1959a). Rather, the guide was presented as a functional, educational tool, and the motorway was presented as an economic necessity, an 'essential link in the industrial life of the country' (*The Motor* 1959a). National newspapers such as the *Daily Telegraph*, *The Guardian* and *The Times* published special motorway supplements on 2 November 1959, as did provincial and local newspapers such as the *Birmingham Post* and *The Herts Advertiser and St Albans Times*. The attention of these journalists focused on educating potential motorway drivers about the design and construction of the motorway, and its appropriate use, but a number of journalists, motoring organizations and commercial companies also focused their attention on the condition and performance of the nation's vehicles.

Technological enhancements for the vehicle driver

John Urry has suggested that motor vehicles insulate drivers 'from the environment', reducing 'the sights, sounds, tastes, temperatures and smells of the city and countryside' to 'the two-dimensional view through the car windscreen' (Urry 2000: 63). This observation appears to match the ideals of many car manufacturers and advertisers, but in both the past and the present vehicles have enabled quite different, embodied, multi-sensual engagements with the landscape (cf. Edensor 2003, 2004; Sheller 2004). Motorway driving would have been a noisy, cold, draughty and rattling experience for many car travellers who traversed the M1 in November 1959, and politicians, motoring organizations and journalists were well aware that a high proportion of the nation's 7.9 million vehicles (including 4.5 million cars) were not built to travel at consistently high speeds (Eason Gibson 1961; Plowden 1971). The Automobile Association's 'Guide to the motorway' contained a special section entitled 'Your car on the motorway', which informed drivers that as their 'vehicles will be subjected to

Figure 5.2 Cover of *Motorway*, a pull-out guide to the M1 issued with *The Motor*, volume 116, 4 November 1959. Reproduced by kind permission of Haymarket Publishing.

unaccustomed stresses' it would be essential to keep them in 'first-class condition'.[2] The editors of *The Engineer* speculated on 'the influence that motorway conditions . . . will have upon the future design of vehicles' (*The Engineer* 1959: 583), and 'J. L.', writing in the 'topical technics' section of *The Motor*, wondered whether an 'alert British car manufacturer' would be 'enterprising enough to introduce a "Motorway Cruiser" model in time' for the opening of the M1 (*The Motor* 1959c: 135). The Cruiser would be more aerodynamic, insulated against noise, and have front-wheel drive, overdrive gear facilities, powerful headlights, good windscreen washers and a 'decent car radio' (*The Motor* 1959c: 135). J. L. even remarked that the increasingly popular American-style 'tail fins may justify themselves as aids to stability' for motorway driving 'on windy days' (*The Motor* 1959c: 135).

While no Motorway Cruiser was launched, manufacturing companies used the publicity surrounding the opening of the M1 to associate their vehicles and automotive products with the spectacle surrounding this new space of high-speed driving. Advertisements peppered the national and local press. On 3 November 1959, Ford placed illustrations of its 'Thames Articulated Traders' over an aerial photograph of the new motorway, urging freight operators to 'be first on *the road* with Ford': ' "Rarin' to go", Ford salutes the opening of Motorway M1 with transport you can trust to keep up the pace. For Thames Articulated Traders are designed as fast movers in a fast-moving age. They fit the shape of the roads to come. . . .' (Ford 1959: 8). In an advert in *The Times* on 2 November India Tyres invited motorists to purchase their Super Multigrip and Super Tyres, as 'the "motorway outlook" demands a new outlook on tyres' (Figure 5.3) (India Tyres 1959: 5). India Tyres were manufactured 'For the man who *really* drives!' and the company suggested that although Britain's (male) motorway driver may be able to raise his performance on the motorway, India Tyres would be required to 'make the most of your car's power' and to enhance the capabilities of your machine. Ability becomes distributed across, and located in, the masculinized bodies of vehicle *and* driver (cf. Latour 1992, 1996; Lupton 1999), and all manner of accessories, as well as vehicles, are promoted as tools to reinforce the masculinity and enhance the performance and speed of 'the man who *really* drives' (India Tyres 1959: 5). In another advert, printed in *The Times* on 2 November, Automotive Products Associated Limited attempted to persuade drivers to buy their high-performance brakes, clutch, steering system and filters, as there was a risk of motorway drivers being let down by their inferior vehicles: 'You need more than skill behind the wheel' (Figure 5.4) (Automotive Products Associated Limited 1959: 9). The company attempted to flatter potential motorway drivers, confirming that while they might have the skills and expertise, their vehicles might not have the ability to perform at sustained

Figure 5.3   India Tyres advertisement, 1959. Reproduced by permission of Goodyear Dunlop UK.

Figure 5.4 Automotive Products Associated Limited advertisement, 1959. Reproduced by permission of the AP Group Limited.

high speeds. Motorway vehicle drivers are presented as 'tightly coupled' assemblages (Perrow 1984), and motorway driving itself is underlain by networks of reliance, faith, competence and trust linking motorway engineers, civil servants and vehicle manufacturers with British motorists and their vehicles (Hawkins 1986; Giddens 1990; Latour 1992, 1996; Lupton 1999; Lynch 1993; Beckmann 2004; Featherstone 2004). As Automotive Products Associated phrased it:

> On this new motorway, you need complete *faith* in your vehicle to back every ounce of your driving skill. You can put your *trust* in LOCKHEED disc and drum brakes . . . *rely* on a BORG & BECK Clutch to withstand increased transmission strains . . . *be sure of* steering accuracy with THOMPSON Tie Rods and Ball Joints – and *depend* on PUROLATOR Filters for utmost efficiency. These vital components are designed and developed by the [Automotive Products Associated] group to match the performance of the modern car. (Automotive Products Associated Limited 1959: 9, itallicization mine)

During the 1950s a significant number of motorists were driving and maintaining pre-war cars, and the Automotive Products Associated advertisement reflected the booming trade in new and used parts, and the popularity of do-it-yourself car maintenance, at this time (*The Times* 1959c; Samuel 1994). The message to aspiring motorway drivers, a new breed of consumer, was loud and clear: the average car would need to be improved and customized by its discerning owner, in an attempt to improve the performance of both driver and vehicle (on car customization see Moorhouse 1991).

'Motorway Madness': Driving, Governing, Expertise

'A power for good or evil': Ernest Marples warns the public

As I discuss below, a range of experts and authorities, from senior police officers and AA patrolmen, to racing drivers, issued advice to the 'ordinary' motorist before and after the motorway opened, but some of the strongest words of warning were provided by the new Minister of Transport, Ernest Marples, during a speech at the opening ceremony of M1 on 2 November 1959.[3] Ernest Marples (1907–78) replaced fellow Conservative Harold Watkinson as Minister of Transport following the General Election of 8 October 1959. Marples had been an active and successful Postmaster-General since July 1957, and few were surprised when his close ally Prime Minister Harold Macmillan promoted him to the Cabinet as Minister of

Transport. Marples stood out from the Conservative crowd. He was a working-class Mancunian who had won a scholarship to his local grammar school and went on to become a successful accountant, property developer and 'construction tycoon' (Sampson 1967: 175; Dutton 2004). Macmillan described him as one of only two 'self-made men' in his new cabinet (Macmillan 1972: 19), but others criticized him for being a somewhat 'cocky', 'talkative', 'exuberant' (Sampson 1967: 94, 175), 'flashy' and 'slick' minister (Evans 1981: 127) with an 'unerring flair for publicity' (Ramsden 1996: 69).[4]

Marples assumed centre-stage at the opening ceremony of the London to Birmingham and St Albans Bypass motorways. In his speech at Slip End near Luton, Marples celebrated the design, construction and completion of this innovative, monumental, modern, scientific motorway, but he urged caution amongst the nation's drivers:

> This motorway starts a new era in road travel. It is in keeping with the bold, exciting and scientific age in which we live. It is a powerful weapon to add to our transport system. But like all powerful instruments it can be a power for good or evil.[5]

What kind of place and force would the motorway become? Would it be 'a power for good or evil'? Well, Marples continued, this would be determined by Britain's drivers:

> It will bring immense benefits if drivers use discipline, common sense and obey the rules. But disaster and tragedy may descend on those who drive recklessly or selfishly. For on this magnificent road the speed which can easily be reached is so great that senses may be numbed and judgment warped.[6]

Marples emphasized the need to govern the conduct and movements of vehicle drivers on the motorway, but while official regulations, codes and policing would form important elements of a regime of good governance, he was quite clear that habituated, everyday techniques and practices of self-governance and self-discipline by vehicle drivers would underlay the production of a safe, efficient and social driving environment (Merriman 2005b). Marples stressed that 'new motoring techniques must be learnt', and as a final piece of advice he offered the public 'two mottoes': 'Take it easy motorist' and 'If in doubt – don't'.[7]

After his speech, Marples used a radio in a Ford Zephyr police car to order the motorway to be opened at every junction along its 72 miles. The ceremonial party left the carriageway, travelling a few hundred yards south to a bridge at the Pepperstock interchange (Junction 10), from where they observed the first vehicles passing along the motorway. Marples was not happy with what he saw:

I was frightened when I saw the first drivers using the road. I have never seen anybody going so fast and ignoring the rules and regulations. Out of the first four cars I saw three were not keeping to their traffic lanes – they were straddling them. Another car came along and broke down, apparently with an overheated engine and no oil. . . . I was really appalled at the speed at which some of the cars were travelling. Drivers must not forget that it is not a matter of skill alone on the motorway. There must be skill and judgement and discipline. . . . If necessary we shall have to consider such things as a maximum and minimum speed limit.[8]

After he had observed the first vehicles passing the Pepperstock Junction, Marples travelled to London to attend a commemorative luncheon hosted by John Laing and Son Limited and Tarmac Civil Engineering Limited in the Lancaster Room at the Savoy Hotel (*Team Spirit* 1959f).[9] In a speech at the luncheon (see above), Marples expressed his shock and horror at the conduct of the first drivers to use the motorway, and his comments hit the headlines of both tabloid and broad-sheet newspapers on 3 November 1959. 'Motorway 1 opens – and Mr Marples says: "I was so appalled",' announced the *Daily Mirror* (Mennem 1959: 5). 'Minister "appalled" by new motorway driving' was the strap-line in *The Times* (1959d: 8). 'I'm so scared!' shouted the headline in the *Daily Sketch* (Fothergill 1959). Motoring organizations and journalists hit back, criticizing the Minister for being '"a bit hasty" with his remarks' (*The Times* 1959d: 8; see also *The Motor* 1959d). *The Motor* explained how it was an unrepresentative minority of overly keen sports-car owners who were the first to pass Marples at the Pepperstock interchange; the first three cars being a Rover, a Bentley and a Jaguar 'travelling very fast' (*The Motor* 1959e: 518).

Vehicle drivers of all shapes and sizes had queued up at the terminal junctions to be the first to experience the new motorway. The car at the head of a three-quarter-mile queue at the northern terminus at Dunchurch was a modest Ford Consul carrying the Walsh family from Birmingham (*Birmingham Mail* 1959a). Journalists also queued up to test out and review the new motorway, and it was they who occupied a number of the sports cars which passed Marples at high speed. A reporter from *The Herts Advertiser* was a passenger in one of the first cars on the motorway, a 3.4 litre Jaguar which travelled between Park Street and Pepperstock at over 120 mph (*The Herts Advertiser* 1959a). A journalist from *The Autocar* (1959b: 582) stated that 'it was a dramatic event to cruise' in the 'staff car . . . at up to 100 m.p.h. along the deserted carriageway', while a reporter from the *Daily Telegraph* stated that he was perfectly safe at 75 mph, as were the 'fast, well driven cars' which passed him at well over 100 mph (*Daily Telegraph* 1959a: 1). These journalists were quite clear that the problem drivers they encountered were not those travelling at high speeds. It was the drivers who had not learned the Motorway Code, ignored the advice of journalists, civil servants and motoring organizations, and moved through the spaces of the motorway in an inappropriate manner who represented a danger to the skilled motorway driver.

The spatialities of motorway driving

The lack of lane discipline concerned many journalists. Basil Cardew, writing in the *Daily Express*, suggested that 'the line drivers may become the scourge of the motorway' (Cardew 1959a), and a party of foreign journalists expressed similar concerns: 'Lane drill today was very bad,' stated Herr Veidermann from the German Tourist Information Bureau to a *Birmingham Mail* reporter (*Birmingham Mail* 1959a). A journalist writing about advanced motorway driving techniques for *The Autocar* went further, stating that lane discipline was, 'unfortunately, practically unknown in this country' (*The Autocar* 1959c: 560). The *Daily Telegraph*'s reporter focused his criticisms on those who 'showed a blithe disregard of common-sense overtaking rules' by pulling out without indicating or checking mirrors (*Daily Telegraph* 1959a: 1). Drivers must use their vehicle's technologies to express their intentions (Katz 1999; Featherstone 2004). The motoring correspondent from *The Times* stressed that electric indicator lights and mirrors should be made compulsory fittings for vehicles using the motorway, and that 'overtaking is really the essence of motorway driving technique' (*The Times* 1959e: 4). The Automobile Association emphasized the vital importance of 'mirror discipline' and safe overtaking (McKenzie 1959a), and throughout these different commentaries there is the suggestion that motorway driving necessitates a different set of driving practices, technologies, accessories and spatialities: new kinds of spatial and kinaesthetic awareness, new spatial practices, new techniques and sensibilities when driving. High-speed multi-lane motorway driving is seen to be strikingly different to driving on Britain's ordinary single-lane, single-carriageway roads. As motoring writer John Eason Gibson stated in an article for *Country Life* on 'The pros and cons of M1':

> . . . the motorway calls for a completely different type of skill. Because one's vision both forwards and to the rear through the mirror is greatly extended on the motorway, one can easily be faced with the task of judging the relative speeds of four cars in front and the same number visible in the mirror. This is far from being as easy as it might at first appear. (Eason Gibson 1959: 1089)

The attention of the motorway driver must be refocused, and spread between the front, rear and side of the vehicle (on attention and modern culture, see Crary 1999). As BBC producer David Martin suggested in an article in the *Radio Times* about his television programme about the M1, 'driving techniques must be altered. The motorist will have to realise that what is coming behind him is of more importance than what is in front of him' (D. Martin 1959).[10] *The Autocar* went further still, stressing the importance of gauging the speeds and positions of all surrounding vehicles:

> This means a full knowledge at all times of exactly what is going on astern – what cars are there, whether they are of the type likely to be travelling much faster than you are, their rate of approach, and where they are placed on the road (from which you can deduce any intended manœuvres). (*The Autocar* 1959c: 560)

Motorway drivers must learn to read the ever-changing geographies of the motorway, and these new habits were felt to place new demands on the bodies of the motorist (cf. Edensor 2004; Sheller 2004; Urry 2004b). High speeds and increased stopping distances meant that the driver's 'critical vision must be extended proportionally', as they would have to be 'capable of spotting a potential accident at anything up to a mile' (*The Autocar* 1959c: 561). The *Daily Telegraph*'s motoring correspondent, W. A. McKenzie, wondered whether 'the law relating to physical fitness [to drive] may have to be amended' (McKenzie 1959b: 13). These concerns were summarized nicely in *The Autocar*:

> Fast driving – even in a straight line – is completely strange to some people, particularly those who have not been fortunate enough to take their cars abroad. There are some who are neither physically nor mentally equipped to cope with the increased tempo, or possess the greatly reduced reaction times that high speeds demand – let alone having eyesight in keeping with the demands of safe, fast driving. (*The Autocar* 1959c: 561)

Wing mirrors – absent from many vehicles in the 1950s – could open up and reflect the spaces alongside and behind one's vehicle, helping drivers to divide their attention, but they could not compensate for the poor vision or poor reaction times of motorists. Motorway drivers, then, must assess their bodily capacities and their fitness to drive on M1, and they must learn to look in new ways (on athleticism and driving, see Sachs 1992).

Distance starts to be perceived and experienced in new ways. Writing on the 'Problems of motorway cruising' for *Country Life* in June 1961, John Eason Gibson made a series of observations familiar to geographers writing about time-space compression and the impact of transport technologies on conceptions of space and distance (see, e.g., Schivelbusch 1986; D. Harvey 1990; Thrift 1995, 1996):

> The ordinary motorist is quickly finding that the new express motorways are producing a startling change in his conception of distance. Many motorists to whom I speak have stopped regarding Birmingham and Coventry as so many miles from London; now they regard those cities as so many minutes away. (Eason Gibson 1961: 1490)

Despite the suggestion that the motorway heralded new conceptions of distance, Eason Gibson was observing changes which had accompanied the increasing use of timetabled stage coaches and railways in previous centuries (Schivelbusch 1986; Thrift 1990, 1995). Motoring journalists also suggested that motorways enabled the speed of vehicles to be sensed and interpreted in new ways. A reviewer from *The Autocar* stressed that the 'striking scale' of the motorway gave a 'completely altered impression of speed' to the motorist (*The Autocar* 1959b: 582), while in the same issue another journalist suggested that 'two senses of speed . . . actual speed. . . . and relative speed' are essential for this new kind of driving (*The Autocar* 1959c: 561). These comments can usefully be related to recent academic understandings of speed. Firstly, the ways in which the speed of oneself is experienced relative to the movements of other vehicles and the motorway landscape itself echoes Derrida's assertion that 'there can be no sensation of speed without a difference in speed, without something moving at a different speed' (Mackenzie 2002: 123). Thus, while commentators observed that 'the wide and treeless . . . expanse of the motorway' may confuse one's sense of speed (Cardew 1959a), differences become apparent and are accentuated when one is passed by a vehicle moving at a markedly different speed. Speed, then, is seen to be experienced in different ways in the spaces of the motorway. My second point relates to the style of academic writings on speed. While social scientists, including geographers, insist that we must focus on the social dimensions of space and time – the diverse spatialities and temporalities which are performed in the world – speed is still largely approached as a calculus of distance travelled over time, and as a simple matter of speeding up (although see Kern 1983; Schivelbusch 1986; Thrift 1996; Millar and Schwartz 1998; Schnapp 1999; Laurier 2004). In contrast to the influential yet somewhat sweeping writings of Paul Virilio on speed, it is useful to approach speed as a 'machinic complex' (Thrift 1996: 263), which – like time and space – is technologized, experienced and sensed in different ways in different cultural contexts (cf. Virilio 1986; Armitage 1999a, 1999b). With the opening of the speed-limitless M1, the social acceptability of driving at high speed changes, and commentators are quite clear that drivers must develop a new sense of speed, and gauge the speed of others in new ways.

Journalists identified overtaking, lane discipline and exiting as spatial practices which needed to be perfected on the new motorway, but many of the M1's first drivers appeared to be ignoring the advice laid out in the Motorway Code. Writing in the *Daily Telegraph* nine days after the opening, W. A. McKenzie showed no concern, revealing his faith in the power, authority and normalizing effect of the Motorway Code, which would function as a Foucauldian 'technology of government':

> All these offenders will be conditioned by the new Highway Code, designed to include the tenets of good driving on motorways. In a short time, as in other countries where motors-only roads came as a new aspect of road travel, the public will learn the few fundamental principles of safe conduct on M1. (McKenzie 1959c: 13)

Civil servants, the police, motoring organizations and journalists felt that additional programmes would be needed for educating and policing motorway drivers. At Laing's ceremonial luncheon at the Savoy, former Minister of Transport and Civil Aviation Harold Watkinson expressed his hope that motorists would improve their driving technique and take the Institute of Advanced Motorists test, 'a passport to safe driving on motor roads'.[11] In mid-November 1959, a series of posters containing extracts from the Motorway Code were displayed on the back panels of a fleet of Bedford lorries which used the motorway on a regular basis (*The Autocar* 1959d). In late 1959 and 1960 public information films providing advice on lane discipline, drowsiness, parking and 'the correct use of hard shoulders' were broadcast on television, including 'five filmlets on Motorway Behaviour produced by Shell-Mex and BP Ltd' (*The Times* 1959f: 14). All of these different propaganda exercises were designed to prevent the kind of illegal manoeuvres and stops witnessed by journalists and police patrols during the first few weeks and months of operation: the drivers who had 'pulled up to consult maps or . . . sight-see' (*Daily Telegraph* 1959a: 1), were parked up 'happily nibbling at their sandwiches' (Hay 1959), or who made U-turns across the central reservation.

Policing the motorway

The first sections of the M1 came under the jurisdiction of five different county police forces (Bedfordshire, Buckinghamshire, Hertfordshire, Northamptonshire and Warwickshire). In May 1959 senior police officers observed motorway patrols in the Netherlands and Germany, and in September 1959 Chief Superintendent John Gott (Hertfordshire

Constabulary) organized a two-week training course for the officers who would patrol the 72 miles of motorway (St Johnston 1959).[12] This new driving environment required new kinds of policing techniques and specialist training, and senior officers passed on their skills and advice to the ordinary motorist through the media. Lancashire Chief Constable Eric St Johnston penned an article on motorway policing for *The Guardian*'s special motorway supplement of 2 November 1959 (St Johnston 1959), while Lancashire Chief Superintendent J. J. Wren and Hertfordshire Superintendent F. Pritchard discussed the problems facing both police and motorists in a special discussion on the BBC radio programme *Motoring and the Motorist* on 30 October (*The Herts Advertiser* 1959b).

Motorway patrols were conducted in specially equipped Ford Zephyr estate cars, which were painted white (rather than the usual black) for increased visibility. In June 1960 white B. S. A. Flash motorcycles were trialled by Bedfordshire police (*The Motor* 1960; Dixon 1961), and the aim of deploying both sets of patrols in white vehicles was to provide a highly visible police presence which would deter motorway drivers from committing offences and force them to reflect upon and govern their own conduct and speed: 'It has already been seen that [when police] supervision is relaxed drivers stop on the hard shoulder or even on the carriageway itself, lane discipline deteriorates and infringements of the Motorway Code increase.'[13] During the first week of operation the police took a fairly lenient approach to minor infringements, but on 8 November they issued their first summons (resulting in a £5 fine) to a driver who had stopped without due cause (*The Times* 1959h). During the following months members of the public were reported for a range of offences – from walking along the motorway and driving backwards, to dangerous driving and driving without a full licence – but the patrols were not solely concerned with maintaining order and stopping mischievous motorists. The Home Office and Ministry of Transport instructed motorway police to compile daily logs of motorway driving standards, traffic flows, breakdown numbers and the circumstances surrounding accidents.[14] Motorway police were enrolled into a broader programme for understanding, calculating and governing the movements and conduct of drivers, collecting data that would enable civil servants, government scientists and transport economists to assess the successes and failures of this new, experimental motorway (see section five of this chapter, below).

In the early 1960s concern was expressed that the M1 might encourage criminal activity. At the Council of Europe's Third European Conference of Directors of Criminological Research Institutes in November 1965, Terence Morris – the sociologist and historian of crime – pointed out that the M1 had brought about a 'fantastic increase in the mobility of criminals', as 'Birmingham gangs in high-speed cars were now able to make hit-and-run raids on

London, and vice versa' (*The Times* 1965a: 11). Writing in 1971, Northamptonshire Chief Constable John Gott argued that neighbouring police forces needed to work closely to prevent motorways being used as getaway routes (Gott 1971). During the early 1960s lorry hi-jacking had become increasingly prevalent (see, e.g., *The Times* 1961c), and one film-maker linked the geography of such crimes to the M1 and the London to Coventry/Birmingham corridor. In *The Hi-Jackers* (1963), written and directed by Jim O'Connolly, a lorry is hi-jacked *en route* for the M1, and when the driver and his travelling companion are interviewed by the police, they are informed that the frequency of this crime, combined with the proximity of M1 and high volume of traffic, made it unlikely that they would catch the criminals or recover the load. The M1 corridor is presented as a difficult space to police, and a space where criminals can slip away undetected.

The Automobile Association and the vehicle driver

In the months leading up to November 1959 the Automobile Association recognized an opportunity to promote their breakdown services and gain free publicity as the media spotlight focused on the opening of the M1. On 19 October the AA invited journalists from the press and television to view the Association's special fleet of Land Rovers and Ford Escort estate cars, and fly over the motorway in their new de Havilland Rapide spotter aircraft (Figure 5.5). The motorway patrols appear as an orderly, uniformed force equipped with modern vehicles, and the use of a spotter aircraft secured widespread coverage in the national press (see, e.g., *The Times* 1959g). The aircraft served as a technology of surveillance, a 'hovering eye' that could radio the position of stationary vehicles to the Association's 'Super Mobile Office' at Newport Pagnell (*Daily Express* 1959a). Two new radio transmitters helped to form a comprehensive 'radio "umbrella"' spanning the entire length of the M1,[15] and when a stationary vehicle was located, a patrol would be despatched to address the problem. AA patrolmen were instructed to undertake a range of tasks. Breakdown and accident assistance formed the core, but in the first year patrols were also required to report both ordinary and irregular events occurring on the motorway, including information on:

Accidents and breakdowns.
General standard of driving.
Lane discipline.
Average speeds.
Types of vehicles.
Density of traffic in general terms.[16]

Figure 5.5 Automobile Association patrol vehicles, spotter plane and patrolmen on parade at junction 14 on 19 October 1959. A sign displaying the types of vehicle and driver which are 'prohibited on the motorway' is visible in the background. Image supplied by The Automobile Association Limited. All rights reserved.

AA officials undertook a traffic census at the Broughton interchange (junction 14), and all of this 'progressive publicity' was fed to the media through the Association's 'Press Information Centre'.[17] As with the motorway's police patrols, the AA stressed that their patrolmen were the best they had, 'specially selected for their personality, mechanical knowledge, first aid qualifications, keenness and general adaptability'.[18] The patrolmen attended an intensive four-day course which included practical exercises on the motorway, as well as lectures on motorway regulations, first aid, accident procedures, towing, vehicle maintenance and AA publicity.[19] The AA patrols, like the motorway police patrols, were specially trained and equipped to regulate the irregular movements and practices of the nation's motorway drivers.

Despite persistent warnings to the public, over 100 breakdowns were reported to the AA on 2 November 1959. In referring to the scenes, one Bedfordshire police spokesman stated that some 'drivers have gone absolutely mad' (*The Guardian* 1959c), and the press were soon referring to the occurrence of 'motorway madness' (*The Herts Advertiser* 1959c: 3). Drivers of older, more modest, saloon cars seemed to get carried away by the freedom ushered in with this speed-limitless road, and many attempted to push their vehicles to breaking point. Fan-belts and clutch pedals broke, tyres burst, cars overheated and ran out of petrol, and 'the engine of one car dropped out' (*Daily Telegraph* 1959b). The high speeds which were sustainable on the new motorway put new kinds of stresses on vehicles, and one leading manufacturer, Rootes, banned its distributors from delivering new cars to customers using the M1 (Hay 1959). Calls to the AA's breakdown control centre dropped dramatically after the first month, but as traffic flows increased in Easter and the summer months of 1960, the number of breakdowns rose (Figure 5.6).[20] Press reports of 'motorway madness' soon died down, but the occurrence of several serious accidents in late 1959 fuelled further debate about the abilities of Britain's drivers *and* vehicles to perform in a safe, orderly and efficient manner on motorways (on these accidents see section five of this chapter, below).

Motorway Modern: Consuming the M1

M1 and the popular imagination

> M1 is still news. It is surprising that this should be so, since the world is lined with motorways of one sort or another. But Britain's first route of the kind has come to this insular nation as an innovation as foreign as a ski-jump course or a bull-fight ring, and every day the public, the police and the

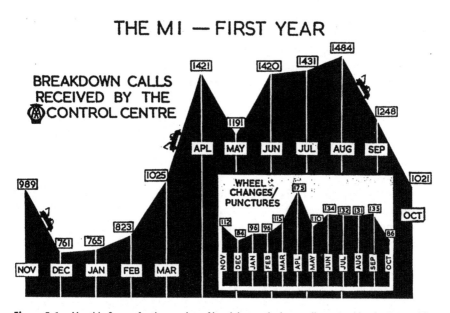

Figure 5.6   Monthly figures for the number of breakdown telephone calls received by the Automobile Association, November 1959 to October 1960. Supplied by The Automobile Association Limited. All rights reserved.

producers of the motorway are discovering something new about it. (McKenzie 1959b: 13)

Writing on 25 November, almost one month after the opening of the M1, W. A. McKenzie emphasised the novelty, innovativeness and foreignness of Britain's first major motorway. The motorway emerges as a space of discovery and experimentation, an exotic space which had not been fully incorporated into English society and provided a new experience of the rural English landscape: 'The absence of villages, sharp bends, cross roads, roundabouts, speed limits and traffic lights leads to a curious elation' (*Birmingham Mail* 1959b). Driving on the motorway is seen to be like 'driving in a new world', 'a dream world' (*Birmingham Mail* 1959b), and reports highlighting the *difference* of the motorway continued to appear in the national press, television and news reels throughout 1960.

This modern motorway caught the press and the public's imagination. A large number of motorists queued up at the junctions to try out their vehicles on Monday, 2 November, and 3000 vehicles used the motorway in the first hour. The flow rate fell to 1500 vehicles an hour for the remainder of Monday (*The Times* 1959d), but the busiest day by far was the first

Sunday after opening, when the flow rate peaked at 5000 vehicles an hour (*The Times* 1959i). Sunday drives had been a popular pastime amongst car-owning families since the 1920s and 1930s (O'Connell 1998), and the motorway quickly became established as a popular destination and route for 'heavily laden family cars' (*The Times* 1959i: 10):

> As soon as it opened the M1 acted as a magnet for drivers from miles around. For the first month or so the Sunday afternoon pastime was a run on the M1, family cars mixing with high-speed sports models and even with London buses on sightseeing tours. (M. V. Rose 1960)

London Transport ran its special Sunday afternoon motorway bus trips throughout November 1959 (*The Times* 1959f), while 'thousands of sight-seers lined the motorway's bridges and flyovers' to enjoy the free entertainment provided by passing traffic (*The Times* 1959i: 10).[21] The motorway emerges as a modern spectacle, and it becomes firmly located as *the* site of modern motoring, *the* driving environment for the 'modern motorist'. In 1960, Manchester publisher Tom Mellor launched a new magazine, *Motorway: an International Magazine for the Modern Motorist*. Successive covers featured aerial perspective sketches of the M1's carriageways, confirming that it was the new motorway which made the difference, while the contents included articles on motorway driving, new vehicles and road construction news (*Motorway* 1960, 1961).

'Motorway culture' starts to spread in 1960 and 1961, extending far beyond the movements of motorists and journalistic statements. The motorway becomes a common reference point in a broad range of more or less-popular cultural forms, from plays and films, to children's books, toys and popular music songs. In David Turner's 1962 play *Semi-detached* – a satirical comedy about the social mobility and relationships of the affluent lower middle-class Midway family, who live in the fictional Midlands suburb of Dowlihull – the M1 makes an appearance as *the* new route from the Midlands to London; a busy, fast, modern road which fills the Midway's well-to-do son-in-law Nigel Hadfield with feelings of nervousness and anxiety when he travels on it (D. Turner 1962: 36).[22] Turner, a writer of the long-running popular radio series *The Archers*, playfully feeds off contemporary concerns about the spaces of the motorway and practices of motorway driving; the M1 becoming associated with feelings of nervousness and fear, as well as elation and excitement (cf. Edensor 2003; Sheller 2004; on anger and driving see Michael 1998, 2000, 2001; Katz 1999; Lupton 1999).

In 1960 the London-based instrumental group the Ted Taylor Four released their new 7 inch single 'M1' (Figure 5.7).[23] The band had formed in 1952, playing 'strictly tempo dance music' on the 'Mecca Ballroom circuit', before undertaking session work and occasional television

Figure 5.7 Cover of the sheet music for 'M1' by the Ted Taylor Four, published by Meridian Music Publishing. Reproduced by permission of Peer Music (UK) Limited.

performances in the late 1950s, and releasing their first single, 'Son of Honky Tonk', on Oriole Records in 1958 (Burke 1992: 3). Their second single, 'M1', is a rather unusual high-tempo instrumental track, resonating with the sound of Ted Taylor's clavioline (an early electric keyboard).[24] The song was originally to have been titled 'Left Hand Drive' to appeal to the audiences they performed to at American air bases in Britain, but the widespread publicity associated with the opening of the motorway persuaded them to change the title: 'The building of Britain's first "super road" the M1 and all its publicity made the change of name a sound commercial idea' (Burke 1992: 6). 'M1' joined other amusingly titled songs in the band's repertoire, including two that were inspired by the spaces of the modern road – 'Cats Eyes' and 'Flyover'. Whether these titles were serious or not, the group hoped that their new single would attract some of the excitement which had surrounded the opening of the M1. The song received 'three hits' and 'a miss' on BBC Radio's *Juke Box Jury* in October 1960, but it hovered 'around on the brink of the charts without ever quite making it' (Burke 1992: 6).

Two toy companies, American firm Marx and British firm Tri-Ang, featured artistic impressions of the M1 and Sir Owen Williams's distinctive bridges on the covers to their independently produced, British figure-of-eight race-track sets, associating the excitement and modernity of this *new* motorway with their fairly standard toys.[25] Earlier versions of both sets had sported different automotive scenes on their box covers, but the motorway enabled both Marx and Tri-Ang to re-brand their sets, bringing the aesthetics, movement, excitement, freedom and speed of the new motorway to the young child, who could race *their* cars on *their* own motorway, whenever they desired. For those children without a motorway toy set, the BBC Home Service for Schools aimed to bring the excitement of the new motorway into their homes in a series on *Current Affairs*:

> *Announcer*. . . . Today's programme is about M1, the new motorway. Listeners will find it useful to have a map of England before them.[26]

The narrator, well-known Welsh preservationist and BBC broadcaster Wynford Vaughan-Thomas, described the key features of the motorway as he travelled at 70–80 mph through the landscape. This is a landscape narrated at speed, a new landscape, a new motorway, which had yet to feature on the majority of maps. Vaughan-Thomas encouraged the listener to map this space:

> Now take a look at your map, if you haven't already done so, and see exactly where the motorway is. . . . it runs right up through the heart of England. . . . you can roughly trace its route by drawing a line from London to Birmingham which passes to the west of all these towns.[27]

The child/listener is encouraged to act as a pioneering geographer-cartographer, charting a new space, a new road, on their own map. The motorway was to be permanently charted, firmly located on the map of England, but it remained a somewhat exotic, undiscovered or at least rarely visited place for many children (and adults) throughout the 1960s. In 1968 another BBC Radio for Schools programme, *Exploration Earth*, described a journey 'Along the M1' as the first in a series of journeys 'Around Britain'.[28] The narrator, Paddy Feeny, described his drive from London to Leeds along the now much lengthened M1, holding interviews with motorway police, the manageress of Watford Gap service area, motorway engineers and maintenance staff along the route. Each child could hear about, but also see, the sights of the motorway without travelling on it, turning to pages 20 to 23 of their Summer 1968 *Exploration Earth* booklet, which featured aerial photographs of the motorway, service areas and junctions, as well as pictures of a police van and the inside of a service area café (BBC Radio 1968).

Masculinity, driving, expertise

> Driving skill was constantly identified as a natural masculine quality and the act of driving was often fetishised as something of an art form. As such, it was something to be savoured. Driving could re-energise a man. (O'Connell 1998: 53)

In his discussion of gender and motoring in early twentieth-century Britain, O'Connell shows how driving and masculinity were continually entwined by cultural commentators (see also Scharff 1991; Sachs 1992; McShane 1994). The majority of motorists were and are men. In 1964 only 13% of British women held a driving licence, compared with 56% of men (O'Connell 1998: 43). Single women had lower average disposable incomes than single men, while the predominance of single-car households and the gender dynamics of many 1960s families meant that it was frequently the male head of a household who would control access to, and drive, the family car for work and/or leisure (O'Connell 1998; R. Law 1999). As the India Tyres advert I discussed earlier in this chapter revealed (see Figure 5.3), the motorway driver was frequently constructed as a male and distinctively masculine figure. The presence of all-male AA and police patrol teams reinforced the construction of the motorway as a space of male expertise, but despite this ongoing construction of the motorway as a space of masculinity, motoring journalists, motoring organizations and civil servants appear not to have explicitly reflected upon gender *differences* in their discussion of motorway driving and the conduct of drivers.[29] Motorway driving was frequently associated with masculinity, but it appears not to have been identified as an unfeminine activity or skill.

In 1961, two children's books were published in which the M1 appears as an important cultural reference point, the space for a plot to unfold, a space of masculine adventure, freedom and speed. The first of these, *The Mystery of the Motorway*, was written by Robert Martin, author of at least twelve other 'mystery' titles (Figure 5.8) (R. Martin 1961). The M1 becomes the location of a mysterious car crash, in which a high-powered light-silver Mercedes overturns at over 100 mph. The driver dies, but a passenger escapes and is picked up by a large black Humber saloon. The crash is witnessed by Mike Dance, a young engineer in his late teens/early twenties who is driving home from Birmingham to London. Mike, his young brother Jim and Jim's friend Birdie become amateur investigators in an attempt to solve the mystery of the motorway crash. The three discover the hideaway of an international crime syndicate who are responsible for the crash, a syndicate who are led by the shadowy foreign financier Aldoise Mendoza, whom the boys castigate for profiting from the Second World War. Arrests are made and the events surrounding the crash start to

Figure 5.8   The front cover of *The Mystery of the Motorway*, by Robert Martin. Illustrated by D. G. Valentine. Published in London by Thomas Nelson and Sons Ltd. Reproduced by permission of Nelson Thornes from Robert Martin's *The Mystery of the Motorway*, 1961.

unravel. Throughout the book the M1 serves as a space of excitement, modernity and danger, as well as of mystery. The book acts as a social manual for boys entering their teens. Robert Martin emphasizes the distinctiveness of motorway driving in his discussion of Mike's drive down the M1, explaining how Mike gives adequate signals, keeps his distance, maintains a steady cruising speed and keeps in the left-hand lane. Driving skill and expertise must be equal to the capabilities of the vehicle. As a policeman informed Mike and the book's readers after the fatal crash, 'those who drive the fastest cars are not necessarily people who know *how* to drive a fast car' (Martin 1961: 14). The author repeats warnings made by Ernest Marples (see below); nevertheless the book draws positive associations between motorway driving and speed, youth and masculinity. In one paragraph, Mike's friendly neighbourhood 'bobby' asks him whether he uses the motorway to go to work in Birmingham, chuckling that this was 'a silly question – can't see a young feller like you missing a chance to drive on a fast road' (Martin 1961: 59). In another section, differences are revealed between the attitudes of two generations. For Mike and Jim the motorway is a space of excitement, but for their uncle and aunt the motorway emerges as a space of fear and concern, and both are worried that Mike uses the motorway: ' "Oh dear, oh dear!" His aunt's voice came from behind them. "That dreadful Motorway! I'm sure I'm scared to death every time I know you're driving along it, Mike" ' (Martin 1961: 40). The fear is that the motorway is becoming a dangerous, disorderly and risky space of uncontrolled movement; a space of youthful abandon not unlike the rock 'n' roll venues and cinemas which had dominated the headlines four or five years earlier.[30] Despite Aunt Mary's concerns, the narrator reveals Mike to be a safe and competent driver, a sensible and suitable role model, who emerges as a victorious amateur British detective fighting against suspect and shadowy foreign criminals. As with several centuries of books aimed at young boys, masculinity and adventure become closely entwined in the story (Phillips 1997; on driving and adventure see Sachs 1992).

The second story was Bruce Carter's *The Motorway Chase*, a 1961 book published in Hamish Hamilton's series of Speed Books.[31] *The Motorway Chase* was a much shorter and simpler story than Robert Martin's book. This book was aimed at a younger readership, and the opening lines immediately connect the reader with the dreams and ambitions of the hero: 'Clive Green wanted to be a racing driver. He knew he could drive fast on the track and win races. One day he would get the chance to show that he could drive. One day Clive would be World Champion driver' (Carter 1961: 6–7). Clive drives a car-transporter lorry for a living, and he is on his way to London, travelling at 40 mph along the motorway before leaving to stop for fuel at 'the big garage . . . run by Sam Miller, the famous racing driver' (Carter 1961: 11–12). Sam is standing by a red Ferrari, but he has a problem. He needs someone to drive his 'supercharged' Mini to the Silverstone racing circuit, as his mechanic has broken an arm. By luck, Clive has space on his lorry, so he sets off for Silverstone along the motorway, 'dream[ing] of the day when he would be World Champion driver' (Carter 1961: 19). Looking ahead under one of Sir Owen Williams's bridges, Clive sees that the police are stopping vehicles – they are on the look out for an escaped convict. Suddenly, Sam Miller's Ferrari breaks through the check-point at break-neck speed. The convict has stolen the car! The police set off in pursuit, as does Clive in the racing Mini which he had on his transporter. Soon Clive reaches 110 mph, 'it was unbelievable that the little car could travel so fast' (Carter 1961: 29). The Ferrari turns off the motorway, and this is Clive's chance to catch up. He passes the police cars, 'steering the car round [*sic*] the corners brilliantly with the skill of a great racing driver' (Carter 1961: 33), and before long he catches up, as illustrated on the cover of the book (Figure 5.9). Clive overtakes the Ferrari, forcing it off the road; thus allowing the police to arrest the convict. Clive's skill as a driver, combined with the engineering excellence of Miller's British rally car – which would in real life win the Monte Carlo rally between 1964 and 1967 – sees him recapture the convict and the Italian sports-car. As a reward Sam Miller gives Clive an opportunity to try

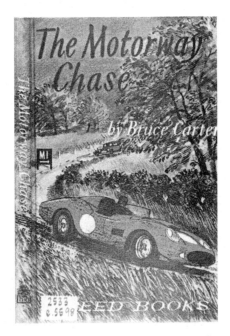

Figure 5.9   The front cover of *The Motorway Chase*, by Bruce Carter. Illustrated by Bernard Wragg. Published in London by Hamish Hamilton. © Copyright 1961 Text and illustrations Hamish Hamilton Ltd. Reproduced by permission of Penguin Books Ltd.

out for his racing team and drive the Ferrari back up the M1. As he joined the motorway Clive 'let the Ferrari full out', approaching '130 miles an hour, 140, 150 miles an hour', as he thought about racing for Sam Miller's team (Carter 1961: 53). On the final page, the reader is reminded of Clive's dream, which appears within reach, providing a link between everyday life and champion status for the ordinary boy at home, reinforcing and reconstructing his masculinity and his desires: 'After that chase, Clive knew he was a good driver. One day Clive would be World Champion driver' (Carter 1961: 55). The motorway is constructed as a democratic or meritocratic space of masculinity, 'where all men . . . can demonstrate their superior masculinity and self-worth' (Walker et al. 2000: 163). Every boy and man is presented with the possibility of moving from the speed-limitless M1 to the race track, and from the status of amateur to professional race driver. This was the golden age of that heroic figure of masculinity, the British grand-prix racing driver. Stirling Moss, Britain's most successful racing driver to date, was a household name (Brendon 1997), but racing drivers were not infallible. They too could make mistakes.

In his speech at the Savoy Hotel on 2 November 1959, Minister of Transport Ernest Marples urged caution, stating that even young, male British racing drivers needed to take it easy and be aware of their mortality: 'Skill alone is not enough at high speed' (Marples, in *The Guardian* 1959b). To reinforce the point, Marples referred to the tragic death of Mike Hawthorn, Britain's first ever motor-racing world champion. Hawthorn had crashed his Jaguar on the Guildford Bypass in January 1959, only three months after winning the 1958 Drivers' World Championship with Ferrari (*The Times* 1959k, 1959l). Despite such warnings the press and the public still revered the authority and expertise of the formula one racing driver.

On Sunday, 8 November 1959 Ferrari's 27-year-old grand-prix racing driver Tony Brooks wrote a review of a journey along the M1 for *The Observer* newspaper. Brooks was a world-class driver who had finished second in the 1959 Drivers' World Championship behind Jack Brabham (Brooks 1959; Owen 2000). *The Observer* presented him as an extraordinary driver, an expert with quick reaction times who was far safer than the average motorist – while Tony Brooks had taken 0.2 seconds to react to an emergency on the British School of Motoring reaction test machine, the average driver took 0.7 seconds (Brooks 1959). Who better, then, to review the new speed-limitless motorway? The opening paragraph of the article reflected on other matters, providing a striking celebration of the novelty, difference and detachment of the M1 motorway from the English landscape:

To drive up M1 is to feel as if the England of one's childhood, the England of the British Travel and Holidays Association advertisements, is no more. This broad six-lane through-way, divorced from the countryside, divorced from towns and villages, kills the image of a tight little island full of hamlets and lanes and pubs. More than anything – more than Espresso bars, jeans, rock 'n' roll, the smell of French cigarettes on the underground, white lipstick – it is of the twentieth century. For all that, it is very welcome. (Brooks 1959: 5)

Brooks describes how Britain's newest motorway brings an international-metropolitan modernity into the English countryside, and this vision of a highly modern motorway 'divorced from the countryside' was far removed from the Landscape Advisory Committee's vision of a motorway which *was* different but could also be easily integrated into the landscape (Brooks 1959: 5). For Brooks, the M1 was as modern, different, contemporary and 'other' as such Americanized products as jeans and rock 'n' roll, but also fashionable continental European forces, including the Espresso bars which were springing up in London during the late 1950s, forming part of the 'Italianization of London style' (Gardiner 1999: 107; Maddox 2003; see section four, below). Brooks suggested that the motorway enfolded these modern urban forces into the rural English landscape, creating new mixtures, contrasts and views. He was not disappointed. Rather than becoming nostalgic about a past England, Brooks actively welcomed these modern influences, but his opening observations soon gave way to a discussion of driving practices on the motorway.

Brooks compared the M1 with roads he had experienced abroad, stating that it is 'infinitely superior to the autobahnen and autostrada in Germany and Italy' (Brooks 1959: 5). The M1 may be good, but he was clear that there was no point in being overly positive, for 'if this is to be the prototype of future motorways it is worth being highly critical about it' (Brooks 1959: 5). We ride in Brooks's Aston Martin north from Marble Arch. Slow progress through the outskirts of North London is signified by the 100 gear changes he records in a 40-minute period. We then move on to the M10 branch of the St Albans Bypass at Park Street:

> At first everything was wrong: an L-driver teetering along uncertainly in the middle of the road, a van broadside on and reversing on a feed-off, many cars sitting complacently astride the fast lanes. But then we all seemed to settle down. . . . The three smooth-surfaced lanes, good standard of driving by every-one (lorry drivers in particular), the good conditions and the effortless cruising of the Aston Martin were hurrying us along. (Brooks 1959: 5)

The drive settles down into an effortless collective performance, as the Aston Martin, other vehicles, other drivers and the spaces of the motorway are gathered and placed into an orderly flow. Brooks expressed his elation as he covered 18 miles in 12 minutes, with 'constant use of the mirror, indicator and headlight flasher', but no gear changes (Brooks 1959: 5). Brooks emphasized that *he* was safe driving at such speeds, but other drivers must beware, as 'different cars and different drivers have different safe limits' (Brooks 1959: 5). His most critical comments were aimed at the road itself, particularly the width of the hard shoulders, and the absence of a safety barrier and anti-dazzle fencing (Brooks 1959: 5). The anti-dazzle protection was felt to be crucial, for Brooks had been much troubled by dazzle on his return journey to London in the dark and rain. The drama of 'motorway dazzle' was captured in a grainy black-and-white photograph of the view from Brooks's Aston Martin by *The Observer*'s picture editor Peter Keen, which showed how 'two cars on the opposite carriageway throw a blinding light on to the wet windscreen' (Brooks 1959: 5; on Peter Keen's photojournalism see Harrison 1998). Brooks soldiered on, safely reaching his destination.

Racing drivers, AA patrolmen, policemen and the heroes of children's books were not the only individuals whose masculinity was seen to be reinforced through heroic actions and performances in the spaces of the motorway. Later in this chapter I discuss how the male-dominated culture of lorry driving was affected by the opening of the new motorway, but other professional drivers were also heroized by the press. One example was Mr Fennel, the driver of one of Midland Red's new non-stop 'Motorway Express' coaches. The first journey of the new coach, leaving Birmingham at 9.45 am on Monday 2 November, proved to be one of the journalistic highlights of the opening day. The coach, which was packed with special guests and journalists, was advertised as 'the first motorway express in Great Britain', and although timetabled to take 3 hours 25 minutes, it broke all expectations, arriving in London's Victoria coach station after just 2 hours 51 minutes (*The Guardian* 1959d; *The Times* 1959m).[32] The journalists on the coach found it hard to contain their excitement. Writing in the *Daily Sketch*, Norman Aubury (1959) reflected on his 'astonishing journey' along the motorway, while a journalist from *The Times* explained how they 'were almost the fastest thing on wheels between Birmingham and London today': 'In the first 57 miles, in which we touched 80 mph at least twice, we were only overtaken by four cars – two Jaguars, an Aston Martin and an Austin Healey' (*The Times* 1959m: 8).

The new motorway and coach service brought much-needed competition to inter-city public transport. The coach cost half the return rail fare and took only slightly longer, at a time when British Rail had closed its main London Euston to Birmingham line for modernization and electrification (Beane 1959). What's more, Midland Red's coaches provided unrivalled comfort. These modern coaches were air-conditioned and carpeted, and they had headrests, 'imitation Jaguar skin' upholstery and a toilet (*Daily Telegraph* 1959b). One company, British Industrial Plastics (BIP), were so proud of their role in helping to manufacture the light-weight polyester and glass-fibre bodywork for this modern hybrid coach that they placed an advert explaining their work in national newspapers on 2 November (Figure 5.10) (BIP 1959: 9). BIP had manufactured 'over 100 polyester/glass fibre body components in Britain's fastest coach' (BIP 1959: 9), which had reached 85 mph during tests at the Motor Industry Research Association's track near Nuneaton (*The Times* 1959n; on the coach, see Richards 1994). This was a well-tested, high-performance coach, but this did not prevent a tyre from bursting during the return journey to Birmingham on 2 November. As a *Daily Express* journalist observed, disaster could have struck, but Midland Red's experienced hand-picked veteran driver saved the day: '. . . the front off-side tyre exploded. Veteran driver Fennel fighting the wheel, forced the swaying 10-ton vehicle on to the safety "hard shoulder" bordering the road' (*Daily Express* 1959b). Fennel is presented as a strong, experienced coach driver, a heroic, masculine figure who 'coped effectively with a minor crisis' (*The Times* 1959m: 8). Unlike Brooks and other grand-prix racing drivers, Fennel is not celebrated for his youth, but as in other cultural commentaries, motorway driving, expertise, physicality and masculinity become entwined.

Guiding motorway travellers

The British public inhabited, consumed and traversed the spaces of the M1 motorway in a broad range of ways. Newspaper readers, television viewers and radio listeners expressed excitement, ambivalence and horror at the modernity of the motorway and journalistic accounts of breakdowns and lawlessness. Children consumed the spectacle of this exciting new space of travel through books, toys, pop songs and radio programmes. Vehicle owners were encouraged to buy the Highway Code and specialist accessories to enhance their skills and the performance of their vehicles. In this section I continue this story, examining how the opening of the M1 led to the launch of maps and guides to the motorway and local 'sights', which in turn reconfigured and recentred the surrounding landscape around the geographies of the motorway.

When the M1 opened in November 1959, one of the first challenges facing aspiring motorway drivers was to *find* the nation's new motorway. The majority of motorists would not have purchased the latest revised small-scale Ordnance survey maps (see P. R. T. Newby 2004), and as the motorway was not shown on older maps, many motorists relied on the special guides which were issued by motoring organizations, petrol companies and the press. As a late 1950s Shell Foldex Great Britain map I possess reveals, some motorists drew the line of the M1 onto existing road

Figure 5.10 British Industrial Plastics and Midland Red's Motorway Express coach. Advertisement reproduced by kind permission of BIP (Oldbury) Limited.

maps – improving their accuracy, extending their useful life, saving them money. Map manufacturers saw such new features of the English landscape as key marketing tools in the early 1960s. In 1961, the Shell Touring Service issued a road map of 'South West England and South East England', which listed its important features on the cover: 'In full colour on tough paper clearly showing roads with Ministry of Transport road numbers, railways, rivers, canals, mileages between principal towns and main built-up areas. All motorways are shown with exits and entrances, those under construction in a dotted line' (Shell Touring Service 1961). The M1 and other motorways appear in bright yellow, outlined with thin red parallel lines. The motorway's line stands out on the map, highlighted as *the* route to take, the new route to the North.

In 1959 and 1960 the Royal Automobile Club and Automobile Association started to convey information on the new motorway in their members handbooks. The AA's 1959–60 handbook contained a full reprint of the Motorway Code, and showed the motorway and its junctions on its national maps (AA 1959). The Royal Automobile Club's 1960 *Guide and Handbook* went further, printing pictures of the new motorway signs, a detailed map of the route and a list of 'where to stay and where to eat in the vicinity of the London–Birmingham Motorway' (RAC 1960: 100). The Royal Automobile Club sought not only to direct their members *to* the motorway, they also helped inform motorway travellers of the most proximate accredited services and amenities. In this and subsequent guides we see the geographies of the surrounding landscape, particularly the geographies of local hotels, restaurants and tourist attractions, become gathered around the motorway and located in relation to its junctions. The motorway becomes a way of organizing or relating features of the landscape, enacting the kinds of corridoring effects discussed by Peter Bishop in his work on the Alice Springs to Darwin railway (Bishop 2002). This was particularly evident in successive editions of Raymond Postgate's *The Good Food Guide*, which was rapidly joining 'the AA or RAC handbook and maps' as 'part of the usual travelling library of numerous middle-class motorists' (J. Postgate and Postgate 1994: 269). In 1963 Postgate commented fairly favourably on the food on sale in the Grill and Griddle restaurant at Newport Pagnell service area (R. Postgate 1963; see below), but as early as 1961 he started to mention the propinquity of the M1 in reviews of restaurants in villages such as Newport Pagnell, Dunchurch and Kislingbury:

The Swan Revisited, Newport Pagnell . . . It has a genuine interest in food and it is well worth turning off the racket and rush of M.1 to dine here or sleep in a Victorian bedroom. (R. Postgate 1961: 42)

Dunchurch, Warwickshire – Dun Cow Hotel . . . Usefully placed near the M1, this is a fourteenth-century coaching inn which offers a warm welcome and good food. The glossy menu is not remarkable for its diversity. . . . (R. Postgate 1965: 187)

Villages, restaurants, inns, hotels and 'sights' all become worth visiting due to their proximity to the M1. They are gathered around the M1's junctions, linked to this national transport corridor. The Duke of Bedford's Woburn Abbey is a case in point. In 1958 it was announced that the M1 would cut across 100 acres of the Duke's land at Ridgmont, but when asked if he was concerned, he pointed out that the motorway would help bring important revenue to his fledgling tourist venture: 'It will mean that Woburn will be within an hour's run of London and that will mean more visitors. . . . Besides, the Government are paying me well enough for the land' (quoted in *Luton News* 1958e). The Duke of Bedford had opened Woburn Abbey to the public in 1955 to help pay death duties, and in guides to the Abbey then and since, maps show quite clearly the route between the Park and the M1 – locating the M1 in relation to the Park, just as today's tourist road signs enfold the Abbey into the spaces of the motorway.[33]

During the 1920s and 1930s, publishers, petrol companies and motoring organizations started to produce a vast array of publications and maps aimed at motorists (Liniado 1996; Matless 1998; O'Connell 1998; Gruffudd et al. 2000). As the number of motorists increased in the late 1950s and 1960s, publishers expanded their activities, producing cheaper and more popular guides (Roscoe 1996). In 1968 Shire Publications produced a new kind of guide book for the nation's motorists. Earlier guides had tended to focus on single counties or regions, but this new guide book, Margaret Baker's *Discovering M1*, used the linear transport corridor of the M1 as its organizing or gathering principle. *Discovering M1* was written as 'a glove-compartment guide to the motorway and the places of interest that can be seen from it' (Baker 1968: 1). It repeated the style of earlier guides and itineraries published for stage coach and railway travellers – such as the Great Western Railway's *Through the Window* guides (e.g. GWR 1924). Descriptions of landmarks and topographic features – for example, the aerials at Daventry and Rugby, and ridge and furrow fields near Crick – were printed opposite maps, photographs and sketches of the route (Baker 1968). All of these were designed to enliven and illustrate the journey and the landscape for passengers: 'It is written for passengers – perhaps bored by the apparent monotony of a road devoid of strip development and place-name signs – and is arranged for easy assimilation at around 60 mph' (Baker 1968: 3). Drawing upon the arguments of Marc Augé, one could suggest that the guide was designed to animate, enliven or translate a monotonous non-place; rendering the landscape legible for passengers (Augé 1995). I have problems with such a simplistic reading, and instead I prefer to think of the guide as a tool which helped to perform this place, enabling travellers to encounter and consume the landscapes of the motorway in different kinds of ways (Merriman 2004b). This small-format guide was designed to reside in a glove compartment, and be taken out and read at a particular speed. At 60 mph, landscape features should appear and disappear as the passenger reads, glances and returns to the text.

Motorway Service Areas and the Motorist-Consumer

In chapter 3 I discussed the debates which emerged in the late 1950s about the location, design and landscaping of the M1's first two service areas at Newport Pagnell and Watford Gap. In this section I examine how they were designed and experienced as spaces of consumption; exposing the geographies of sites which have been largely overlooked by social scientists (although see D. Lawrence 1999; R. Green 2004).[34] The main buildings at Newport Pagnell and Watford Gap service areas were not opened until August and September 1960, respectively, but 'emergency facilities' – including toilets, fuel and breakdown services – were provided for motor-way travellers at both sites from 2 November 1959.[35] Extensive discussions lay behind the planning of the service areas, and the Ministry of Transport were careful to regulate the activities of both operators and visitors. Architectural and landscaping plans were scrutinized by civil servants, Royal Fine Art Commissioners and members of the Landscape Advisory Com-mittee. Commercial advertising was banned from outside areas, signage was strictly controlled, and developers were instructed that they must advertise and enforce a 'code of conduct for service area users'. The Code included six 'rules relating to the circulation and movement of vehicles', and twelve 'rules relating to the behaviour of the general public':

> . . . 4. Drive carefully and do not exceed 30 mph within the service area or on the slip roads. . . . 11. Do not obstruct or otherwise interfere with free passage on any road, path or footbridge. 12. Do not offer anything for sale or hire. . . . 13. Place all litter in the receptacles provided. 14. Do not walk on any flower beds. . . . 15. If you have brought any animal into the service area in your vehicle, keep it under control at all times. . . . 18. Do not play games within the service area.[36]

Politicians were keen to ensure that the movements and conduct of visi-tors were carefully regulated, but they also focused their attention on the consumption habits of motorway travellers. Food and drinks menus and prices were checked to ensure that operators did not abuse their monop-oly,[37] while the Ministry refused to allow the provision of single-brand (or 'solus') petrol stations, for although they would have brought in consider-ably higher rents, civil servants felt that the nation's 'motorists would expect a choice'.[38] Blue Boar sold Regent, Fina, Shell and BP fuels at

Watford Gap service area, while Motorway Services Limited (Forte and Blue Star garages) provided Regent, Fina, Esso and Mobil to motorists at Newport Pagnell (*The Autocar* 1959b). The nationalities of the petrol companies sparked debate and criticism in the House of Commons. Despite Blue Boar's clear advertisement of Shell and BP fuels at Watford Gap from November 1959 (see Blue Boar 1959), one Conservative Member of Parliament, Thomas Dugdale, criticized the government for the lack of choice, pointing out that the nation's motorists had been prevented 'from buying British petrol from stations on the M.1.':

> Is the Minister aware that it was about a year before Shell or B.P. sold any petrol at all there, and that it was necessary to buy American petrol because there was no other available? Was that part of the campaign to help the dollar or was it merely due to incompetence on the part of Shell and B.P.? (*Parliamentary Debates* 1960a: 1293)

Concerns about the national economy, particularly the weakness of the pound, became entwined with debates about national identity and the importance of selling *and* buying 'British petrol'.[39] Dugdale felt that British consumers, British motorists, would not only demand choice, but they would also want petrol and oil which was drilled, refined or marketed by British companies. Marples agreed, pointing out that Shell and the UK Petroleum Industry Advisory Committee had refused to cooperate with the service area operators due to their objecting to the government's decision to provide multi-brand (rather than 'solus') petrol stations.[40]

Petrol consumption was not the only fluid activity to be regulated and scrutinized by official bodies. The government decided that pubs and bars would have no place on this high-speed motorway, but service area operators were permitted to apply for 'table licences' from local magistrates courts – which would enable them to serve alcohol with meals (*The Manchester Guardian* 1958; *Parliamentary Debates* 1959c).[41] Blue Boar decided not to apply for a table licence, but Motorway Services Limited applied for one for their Grill and Griddle restaurant. The application was rejected at a court hearing in February 1960 (*The Times* 1960a). The company presented evidence from 'the British Holidays and Travel Association' arguing that foreign tourists would 'be disappointed' if alcohol was not available, and criticizing their opponents for trying to 'apply prohibition to the motorway' and make 'the M1 extra-judicial territory outside the laws of this country' (*The Times* 1960a: 6). The application was opposed by the police, two local breweries, the local congregational church, the Temperance Council of the Christian Churches, the Methodist Circuit, the Baptist Union and several thousand petitioners (including many lorry drivers and locals) (*The Times* 1960a: 6). A representative of the National British Women's Total

Abstinence Union expressed her concern that the sale of alcohol might increase accident and fatality rates: '. . . it would be terrible if the wonderful M1 were to become Murder Incorporated' (cited in *The Times* 1960a: 6). Roger Harman, a solicitor acting for a number of local temperance bodies, expressed his concern that the sale of alcohol might turn the restaurant into a fashionable destination: 'A roadhouse on the M1 might become a place to which people drove out from London for dinner in the summer because it took virtually no longer than driving from London to Maidenhead' (cited in *The Times* 1960a: 6). Harman's predictions were not that inaccurate. While the Ministry of Transport and Civil Aviation were adamant that motorway service areas must not serve as *destinations* for motorists, Newport Pagnell and Watford Gap very quickly became established as exciting, popular, modern destinations for motorists and their families.

Writing in *The Guardian* on 2 November 1959, Eric Hartwell, Joint Managing Director of Forte's and Company, identified four types of motoring-consumer whom he expected to visit Newport Pagnell and other service areas:

> The business traveller.
> The pleasure motorist.
> The motor-coach party.
> The commercial vehicle driver. (Hartwell 1959: 13)

These different consumers are defined by the vehicles they travel in and assumptions about their budget and the kinds of food they would want to eat. Two distinct 'types of catering service' were planned to meet their needs: 'quick snack' cafeterias would provide 'a quick, comfortable service of popular cooked meals and snacks in modern surroundings', while the Grill and Griddle Restaurant would provide 'a licensed restaurant and snack bar service . . . for those who wish to eat in a more leisurely manner' (Hartwell 1959: 13). Hartwell appeared to be less concerned about the demographics of his customers, and it is likely that he was not expecting as much interest from one type of customer – widely seen as a new social phenomena – who flocked to Newport Pagnell after it opened.

'The Teenage Consumer' hits the road

During the late 1950s, newspapers and magazines contained a large number of articles, features and photographic surveys on the fashions, transgressions and consumption practices of Britain's increasingly affluent youth (M. Harrison 1998). In his 1957 book *The Uses of Literacy*, Richard Hoggart criticized the impact on working-class culture of such Americanized media, practices and spaces as glossy magazines, jukeboxes and milk bars (Hoggart 1957). In July 1959 market researcher Mark Abrams's report *The Teenage Consumer* revealed how newly affluent teenagers had increased earnings and spent a high proportion of their money on clothing and entertainment (Abrams 1959). In more colourful prose, Colin MacInnes' 1959 novel *Absolute Beginners* depicted the consumption practices and spaces occupied by the now somewhat mythologized 'hedonistic, working-class teenager' (Hebdige 1988: 69; Bugge 2004). As the work of all of these different commentators attests, and as the earlier comments of Tony Brooks remind us, American, British and continental European practices, products and styles were being continually interlaced – as they had been for decades. Coffee bars, rock 'n' roll music, Vespa scooters and other commodities and spaces rendered the consumption practices of the young visible to the public gaze, and it was in the midst of such changes in the late 1950s and early 1960s that Newport Pagnell's service area cafés emerged as fashionable, youthful, modern (and provincial) extensions of Soho's popular coffee bars. As one regular visitor reflected somewhat nostalgically in 1985, Newport Pagnell service area quickly became a 'place of pilgrimage for teenagers hoping for instant glamour':

> For young people, the new road was a concrete escape to a new kind of excitement. Along it, on a Saturday night, would swarm the Morris Minors, XK Jaguars and Norton motor bikes, eating up the miles at incredible speeds in search of the bright lights. Their destination? Mr. Forte's snack-bar on the M1 . . . this cosy man-made island called out to Britain's youth, the generation of teenagers who did not know there was anything special about being young but forsook the coffee bars of Soho to spend Saturday night 'doing a ton' on this long straight road. (Greaves 1985: 8)

Charles Forte had become known for the large number of small, modern, 'popular' cafés, milk bars and restaurants which he operated in central London, but during the late 1950s and 1960s he branched out into motorway service areas, motels, up-market restaurants and hotels (Forte 1986: 88; Maddox 2003). The popular 'quick snack' meals appealed to Forte's young crowd, who could order a hamburger for 2s or a Knickerbocker Glory for 3s/6d, while the rich and famous are said to have flocked to the 'Grill and Griddle' restaurant for more up-market fare: 'From the steamed-up windows of the snack-bar you could watch the Bentleys and Rolls-Royces streaming into the car park' (Greaves 1985: 8). Famous faces, 'Establishment' and 'pop', are said to have frequented the 'Grill and Griddle' restaurant, from the Beatles, Tom Jones and Mick Jagger, to future prime ministers Harold Wilson and Edward Heath (Greaves 1985). I suspect that such stars were less frequent visitors than Suzanne Greaves's nostalgic reminiscences suggest, but it seems clear that famous people did indeed visit the restaurant to experience the atmosphere of the service area and motorway, dwelling in a restaurant designed to be 'bright, modern and comfortable, to blend with the exciting conception of the motorway itself'

(Hartwell 1959: 13). I am unclear whether the famous visitors liked the food, and a review of the 'Grill and Griddle' in the 1963–4 edition of *The Good Food Guide* suggests that it provided good, standard – but not exceptional – fare:

> This Forte's snack bar on the M.1 (London to Birmingham side) provides cooked-while-you-wait meals like eggs, bacon, steak, etc. or you can have sandwiches. If you are hungry and bored with driving on the motorway, you can stop and eat well for 4/- and upwards. It keeps open quite late. (R. Postgate 1963: 237)

As new and larger service areas were opened during the 1960s, a few operators hit the headlines with their plush restaurants and unusual culinary creations.[42] In 1966 frozen fish company Ross opened their Terence Conran-designed 'Captain's Table' restaurant at Leicester Forest East service area on the second section of the M1 (*Interior Design* 1970). The restaurant's eight AA stars, and the presence of 'Jamaican Fish Jambaraya' (*sic*) and '*flambé* curry' on the menu, caught the attention of the culinary correspondent of *The Times* (1967: 23). At a time when service areas were receiving increasing criticism from journalists, consumer groups and food critics, Raymond Postgate gave a half-hearted thumbs-up to the 'Captain's Table': ' "The best on the motorways" is no high compliment; however, this nautically-decorated restaurant . . . does its best with what its owners, the Ross Group, send it. Fish, as you might expect, is the best buy . . .' (R. Postgate 1969: 186). In chapter 6 I discuss some of the criticisms which were aimed at motorway service areas – particularly those on the M1 – from the late 1960s, but for now I want to return to the early 1960s and the spectacle of the motorway and its service areas.

As Britain's motorists revelled in the modernity of the M1 and its service areas, some visitors bought postcards of the motorway and service areas from vending machines and service area shops (Hartwell 1959; see Parr 1999). In the early 1960s, a black and white postcard was produced featuring four views of 'The M.1. Motorway' (Figure 5.11). Three views featured Motorway Services Limited's Newport Pagnell service area, while the fourth photograph showed the St Albans Bypass section of the M1, looking north, at junction 9. Newport Pagnell service area had become a destination, a place to visit, eat, hang out and buy a souvenir postcard. The modernity of the motorway's structures was attracting the attention of postcard manufacturers, who recognized that profits could be gained from the novelty of this modern spectacle. One colour postcard from the early to mid-1960s featured a photograph of the flyover, embankment and rather barren roundabout at junction 11, where the M1 meets the A505 Luton to Dunstable Road (Figure 5.12). Today, this postcard might well be

Figure 5.11   'The M.1. Motorway', black and white postcard, c.1960–3. Source: Author's collection.

Figure 5.12   'M1. Flyover, Dunstable Road, Luton'. Colour postcard of a bridge, embankment and roundabout at junction 11 of the M1, published in the M and L National Series, postmarked 18 July 1966. Source: Author's collection.

labelled a 'boring postcard' (Parr 1999; see chapter 6), but when it was produced it would have been seen to reflect the modernity of the motorway and its structures. The motorway emerges as a modern spectacle or event, and this postcard enabled it to transcend its local, regional and even national geographies, as it was consumed by more or less distant friends and relatives. This modern landscape, 'M1. Flyover, Dunstable Road, Luton', was and is placed and enlivened through the production and consumption of these modern postcards.

Lorry drivers and transport cafés

The opening of the M1 was not welcomed by the owners of transport cafés and garages located on the old A5 road. Traffic levels on the A5 dropped dramatically as lorry and car drivers switched to the nation's new motorway, but after their curiosity was fulfilled many lorry drivers returned to the A5. *The Times* and *The Economist* suggested that this was due to 'their attachment to familiar transport cafés' and the absence of catering facilities in the motorway service areas (*The Times* 1959e: 4; *The Economist* 1960). Giles expressed the same view in a cartoon published in the *Daily Express* on 5 November 1959 (Figure 5.13) (Giles 1959). Two lorry drivers return to 'Ye olde "A5" transport cafe' having driven their lorries on the M1 the previous day. The café owners are not happy, and the punishment for these formerly loyal customers is to receive handle-less cups. Giles's transport

"Given us cups with no 'andles just because we used the M1 yesterday"

Figure 5.13   'Ye olde "A5" transport cafe'. Cartoon by Giles in the *Daily Express*, 5 November 1959. Reproduced by permission of Express Newspapers.

café is a place occupied by familiar, loyal, working-class lorry drivers; cloth-capped and sheep-skin-coated men who would regularly call in at their favourite café for a fry-up or bacon sandwich, a cup of tea, a cigarette and a chat (on the urban landscapes and characters of Giles's cartoons, see MacInnes 1961).[43] This was a distinctive environment, fostering distinctive consumption practices and social relations. Transport cafés provided an opportunity to chat to other lorry drivers; a way of overcoming the loneliness of long-distance lorry driving (Hollowell 1968). But would the new motorway service areas bring about a new café culture? Would service area cafés appeal to the tastes of Britain's lorry drivers? The motoring correspondent of *The Times* expected a natural progression, a shift, when the service area cafés eventually opened in late 1960: 'As for the lorry drivers, doubtless they will become attached in time to the cafés being built for them at the service areas on the motorway' (*The Times* 1959e: 4).

When the main buildings at Newport Pagnell and Watford Gap service areas finally opened in August and September 1960, commercial drivers started to visit their transport cafés for cheap meals and drinks. Sir Owen Williams and Partners had originally intended that Newport Pagnell would be a 'minor' service area aimed largely at lorry drivers and Watford Gap would be a 'major' service area catering for all road users (Sir OWP 1957: 6, 5; see chapter 3), but it was Watford Gap service area which quickly gained a reputation amongst lorry drivers for its transport cafés. Lorry drivers were already familiar with Blue Boar's cafés on the A5 near Watford Gap, the A45 at Dunchurch, and on the A5 near Lutterworth (Blue Boar 1959). Blue Boar was a local firm, familiar to lorry drivers, but such associations were not welcomed by one senior civil servant, who hoped that the redesign of the catering facilities and provision of waitress service restaurants at Watford Gap in 1964 would enable it to 'lose at least some of the stigma of having been started in 1959 as primarily a commercial drivers' facility'.[44]

During the 1950s, the number of lorries on Britain's roads increased rapidly. In July 1959 a survey published by the Ministry of Transport revealed that 'for the first time in our history more goods are now carried by road than by rail (56 per cent. by road, 44 per cent. by rail . . .)' (MacEwen 1959: 258), and in the early 1960s sociologists and cultural commentators began to take an increasing interest in the life and work of the long-distance lorry driver (Seymour 1964; Bugler 1966; Hollowell 1968). In a 1966 article in *New Society*, Jeremy Bugler described his trip with 'the lorry men' from London to Liverpool, recalling conversations with them about the arts of lorry driving, pay, prostitutes and transport cafés (Bugler 1966). Bugler travelled north with a driver named Graham in his 'Ford 1800 articulated lorry', 'winding through the small North London streets towards the M1', cruising at 40 mph along the motorway, then stopping at 'the Blue Boar, near the Watford Gap . . . the type of plastic-bright, glass-sided café that the motorway has brought' (Bugler 1966: 181). In this account, the truckers section of the Blue Boar was accepted as a modern transport café, an example of 'the lorry drivers' social centre' (Bugler 1966: 181). University of Southampton sociologist Peter Hollowell observed a different trend. In *The Lorry Driver*, a detailed book-length sociological study of the working environment and practices of British lorry drivers, Hollowell revealed how a number of the drivers he had 'met had stopped their patronage of the motorway cafés because of the "lack of life" in them, as well as the greater expense' (Hollowell 1968: 52). Hollowell's research participants preferred the variety, cheapness and character of older transport cafés. In these different accounts we see that lorry drivers incorporate the consumption spaces of the new service areas into their own personal geography of lay-bys and transport cafés, with drivers opting to stop at specific cafés depending on their route, tastes, company policy and the sites frequented by their colleagues, friends and acquaintances.

Assessing the M1's Performance: Cost-Benefit Analysis, Scientific Experiments, Accidents

In this section I examine how the first sections of the M1 emerged as a space of scientific experiment, economic calculation and death in 1950s and 1960s Britain. As early as 1937, civil servants suggested that it would be appropriate to build an experimental British motorway between London and Birmingham (see chapter 2), and once the government had announced its decision to proceed with construction in February 1955, scientists, engineers, designers and economists realized that they would need to pay close attention to the events and actions that unfolded on the nation's first major motorway. In the first part of this section I examine how economists undertook one of the first large-scale cost-benefit analyses in Britain in an attempt to predict the traffic flows and economic benefits resulting from the construction and use of the M1. The motorway emerges as a space of calculation, a space where performance could be predicted and enumerated. In the second and third parts of the section I examine how, after opening, government scientists, engineers and road safety experts scrutinized the performance of the motorway and its motorists, focusing their attention on the irregular *and* disorderly movements of construction materials and motorists, identifying and rectifying failures, and investigating the causes of accidents.

Calculating economic performance

In February 1955 Gilbert Walker (Professor of Commerce at the University of Birmingham) and Dr William Glanville (Director of the Road Research

Laboratory) asked the Ministry of Transport and Civil Aviation for £5000 – to add to £3000 they had received in US Conditional Aid – to undertake a study into the economic impact of the construction of the London to Birmingham Motorway.[45] Civil servants expressed some reservations. Earlier that month the government had committed themselves to the construction of the M1, and doubt was expressed about the need for such a government-funded scientific study and the possibility that it might suggest minimal economic gains or even losses. S. R. Walton explained:

> . . . it is perhaps a little curious that an investigation into the need for such a motorway shall continue when a decision has in the meanwhile been taken to go ahead with it. This would be more serious if there was any possibility that the investigation were going to demonstrate the lack of need for a motorway, but Mr Baker and I are agreed that, knowing Dr Glanville's views on motorways and without casting any aspersions on the objectivity of the research, there is no likelihood of this.[46]

Walton and other civil servants were clear about the political implications of undesirable statistical results, and the ways in which the 'apparent facticity of the figure obscures the complex technical work that is required to *produce* objectivity' (N. Rose 1999: 208). Nevertheless, faith was placed in Dr Glanville, the Road Research Laboratory and the expected results of this 'objective' study.

The Ministry of Transport and Civil Aviation approved the funds for the study in July 1955. In the meantime the Road Research Laboratory had started their research (on the survey, see RRL and UB 1960; RRL 1965).[47] In June and July 1955 Britain's largest traffic origin-and-destination survey was conducted at 23 points on different routes between London and Birmingham. In total, 41,000 drivers were interviewed about their points of origin, destination and purpose of travelling in an attempt to build up a statistical picture of the movements of different type of vehicle, which were classified as: private car, coach, light van or one of three classes of lorry. Journey times and speeds were calculated for different routes, which were compared with data collected during trips to France, Holland and Belgium. The movements of vehicles and their contents were costed by economists, who calculated the savings to be made from motorway construction and the rate of return on the government's investment. They did this by weighing construction, maintenance and increased vehicle mileage costs against the projected savings in working time, fuel consumption, vehicle wear and tear, and accidents. There were no limits to what could be enumerated and predicted; rendered 'representable in a docile form' (N. Rose 1999: 198). Fatal casualties were estimated to cost £2500 per 'unit' (i.e. person), with serious casualties estimated as £650 and slight casualties at £50. Using these figures it was estimated that the motorway might save £215,000 a year using 1955 traffic estimates – equating to a saving of 520 casualties a year (including 20 fatalities) (RRL and UB 1960: 57).

The University of Birmingham and Road Research Laboratory's economists approached the motorway as an experimental space; a space where traffic flows and their economic impacts could be predicted, quantified and costed. The economic performance and productivity of this space could be projected, with every single bodily and automotive movement being 'flattened' to a statistic – whether the press on the accelerator, the change of route, or the journey to a hospital or mortuary (on this 'flattening', see de Certeau 1984). This was an 'academic' exercise. The decision to construct the motorway had already been announced, but research papers could be written, academic researchers employed, statistical formulae developed. The preliminary findings of the study were released in August 1959, but a degree of critical press attention brought a swift end to the project, as one civil servant's pen revealed: '. . . this report got a bad press. I hope we shall terminate this research in due course.'[48] Senior civil servants expressed concern about the relatively low traffic flows and economic gains which had been predicted – this was partly due to the use of 1955 traffic flow rates – but they agreed that the full report could be published as it 'is long and therefore will only be read by statisticians'.[49]

The 89-page report – Road Research Technical Paper No. 46, *The London–Birmingham Motorway. Traffic and Economics* (RRL and UB 1960)[50] – received little attention in the national press, but within a few months, and over future decades, it was to be widely referenced and frequently praised by statisticians, economists and transport geographers as one of the first large-scale British cost-benefit analyses (see, e.g., RRL 1965; B. R. Williams 1965; Boorman 1966; Bamford and Robinson 1978; Starkie 1982). Dr Michael Beesley (1924–99), a Lecturer at the University of Birmingham – who co-authored the final report – went on to work with Christopher Foster on the highly influential cost-benefit study of the Victoria Line of the London Underground, and he later became founding Professor of Economics at the London Business School and a highly influential government adviser (C. D. Foster 1999). Both studies, of the M1 and the Victoria Line, became influential reference points for statisticians and politicians. As economist G. H. Peters stated in the third edition of *Cost-Benefit Analysis and Public Expenditure*, published by the free-market think-tank the Institute of Economic Affairs in July 1973, the 'largely experimental' motorway study raised a whole series of dilemmas about how to predict, calculate and most importantly *delimit* the social and economic costs and benefits of the motorway:

> There was one externality: the savings to persons who would continue to use existing roads. . . .There was no question of attempting to include any

far-reaching external effects of the motorway on, for example, the economies of the towns most affected by it. Nor did the study group discuss whether the improved transport network would give a further twist to the spiral of increasing concentration of population in London and the Midlands, which is sometimes held to be detrimental to the development of the economy in general. This question is probably unanswerable by speculation and analysis, of the cost-benefit or any other kind. Indeed at this juncture it is worth remarking on a judgement of Mr C. D. Foster who held that a full social cost-benefit study – in which *all* costs and benefits are included – is by its very nature an impossible practical ideal at which to aim. The line must be drawn somewhere; and in transport studies to date it has been drawn to include only the benefits to users of transport facilities. Indeed Mr Foster described full 'social appraisals' in which all costs and all benefits to whomsoever they may accrue are estimated as a 'piece of utopianism'. (Peters 1973: 28, 29)

Would the motorway strengthen the regional economy surrounding the London to Birmingham axis, the lower portion of Eva Taylor's 'industrial coffin'? Would this be to the detriment of more peripheral regions and the national economy as a whole? In any event, would it be possible to measure such effects? The problem here is not only one of quantifying complex qualitative effects and events, but also one of 'cutting the network', deciding how to delimit and represent spaces, effects, practices, movements (Strathern 1996: 517; Callon 1998a). Foster recognized that quantification and calculation necessitated partiality, the drawing of lines, the making of cuts, bracketing, the creation of 'externalities' (Callon 1998b). As Daniel Miller has recently shown, the literature on the 'externalities' associated with driving continues to flourish, 'forcing us to raise our sights and imagination to bring in all that the car implicates: aggregate effects, landscapes of roadways, patterns of work and patterns of leisure' (D. Miller 2001: 15). The problem is that this literature frequently assumes *a priori* cuts and divisions between an interior and an exterior, vehicle and landscape. What's more, the social and cultural dimensions of driving and driving spaces tend to be 'downplayed', and at worst quantified (D. Miller 2001: 16). Instead, I would argue that we must focus on automobility/car cultures as 'system[s] of fluid interconnections' (Urry 2003b: 69).

Engineering performance

After the M1 opened on 2 November 1959, scientists at the government's Road Research Laboratory took the opportunity to study the performance of this experimental motorway and its drivers. As with the cost-benefit study, government statisticians attempted to categorize and quantify the rich and varied movements and presences of an array of vehicle drivers, 'flattening' their 'trajectories' into regular, 'homogeneous' and static figures or types (de Certeau 1984: xviii). Automatic traffic counters registered and summated the movements of vehicles passing three points on the motorway, and Road Research Laboratory scientists tried to explain daily and monthly fluctuations in these figures – reconstructing the desires, decisions and movements of motorway travellers from registers of their presence or absence (G.R. Green 1961). Experiments were conducted on the differences in travel time and fuel consumption for journeys between Park Street and Dunchurch using (variously) the M1 and A5/A45 (Dawson 1961), but alongside these studies of the regular movements of the average driver, government committees, statisticians and engineers also focused their attention on the irregular and undesirable movements and performances of both motorists *and* the motorway: the irregular movements of vehicles involved in accidents, the presence of broken-down vehicles, and movements or failures in the construction materials.

On 12 November 1959 Ernest Marples decided to set up a committee 'to assess the lessons to be learned from the London–Birmingham Motorway and to consider their application to other motorways'.[51] The Motorways Panel, latter called the Motorways Working Party, included representatives of the Ministry, the Road Research Laboratory, the AA and the RAC, while in January 1960 the Working Party recommended the formation of a second committee, the Traffic Engineering Committee, to consider 'traffic and safety matters relating to the motorway'.[52] The Motorways Panel held their first meeting on 20 November, and in the eighteen days since the motorway had opened, a number of significant events had occurred. The first problem was with the narrow eight feet-wide hard shoulder at the edge of the carriageway – of which 50 miles was comprised of stone covered in tar and 57 miles of stone covered with sand and grass seed (Sir OWP 1961). On 3 November, 36 hours after the opening, a lorry pulled over onto the grassed hard shoulder, causing the embankment below to partially collapse (*The Times* 1959o). Over time, the hard shoulders became rutted and water-logged, and in the first three months there were six reported accidents resulting from heavy lorries sinking into mud. The Traffic Engineering Committee observed that by January 1960 'much of the hard shoulder' had become 'unusable',[53] and during late 1960 and early 1961 the 57 miles of grassed hard shoulder were widened and reconstructed at a cost of £558,000 (Sir OWP 1961).[54]

The motorway was an ongoing experiment, a linear field-site, and during 1960 and 1961 the Motorways Working Party and Traffic Engineering Committee examined different aspects of the motorway, from the merits of lighting particular sections, different styles of painted white line and coloured cat's eyes, to the design of direction signs and emergency warning signs, the siting of emergency telephones, the layout of junctions and the effects of speed on accident rates.[55] As early as 1957, and throughout construction, engineers at the Road Research Laboratory expressed their concerns about the design and construction quality of the motorway.[56] Indeed, in a public statement issued in September 1962, the Ministry of Transport admitted that there had been significant compromises in the motorway's design:

> At the time the M1 was designed, and in the absence of any practical experience of motorways in this country, certain compromises in the application of the basic specification were accepted in the interests of economy both in cost and the use of land. Experience since the motorway was opened has shown that some of these compromises went too far, but, after account has been taken of the cost of improvements and repairs now being put in hand, M.1 will remain the cheapest motorway that this country has built or is likely to build in the future.... The new motorways now being built incorporate the lessons learnt from M.1.... (quoted in *Roads and Road Construction* 1962: 288)

The M1 was a cheap compromise, an austere structure built in a period of increasing affluence and rapidly rising traffic levels. Lessons had been learned, and future problems were not expected to occur, but the *Daily Telegraph*'s motoring correspondent, W. A. McKenzie, was not happy. In November 1962 he accused the Ministry of Transport of ignoring the lessons learned by engineers in 'Germany, Italy, Holland, France and the United States' (McKenzie 1962). The Ministry appeared to have 'ploughed on its own way', producing a motorway which was 'showing signs of disintegration within three years of completion'. McKenzie was quite clear: the specifications for the design of the M1 were inadequate and 'the blame lies with the Ministry'.

The M1 and other motorways have continued to act as spaces of experimentation, whether for new styles of crash barrier, anti-dazzle fencing, warning sign, emergency telephone or automatic traffic sensors. One of the most controversial experiments was the imposition of a temporary 70 mph speed limit in 1965. When the motorway was opened, journalists celebrated the freedom and exhilaration of driving along this speed-limitless road. After watching the first cars drive along the M1 in November 1959, as noted above, Ernest Marples had suggested the possibility of imposing a speed limit. In the early 1960s, sports-car and racing-car manufacturers openly tested their high-performance cars on the motorway,[57] and in June 1964 Ernest Marples held talks with the President of the Society of Motor Manufacturers and Traders after it was revealed that 'the M1 motorway has been used to test cars at speeds approaching 200mph in preparation for the Le Mans race' (*The Times* 1964c: 7). Marples was reported to be satisfied that a speed limit was not necessary to control such activities, but following a series of fatal pile-ups in fog, including one on the M6 in November 1965, the National Road Safety Advisory Council and senior police officers advised the new Minister of Transport, Tom Fraser – who had succeeded Marples following Labour's election victory in October 1964 – to introduce an experimental 70 mph speed limit for all motorways during winter months (*Parliamentary Debates* 1965; *The Times* 1965b, 1965c). The Road Research Laboratory monitored the experiments throughout 1966 and 1967, and after their studies showed a decrease in the accident rate, Minister of Transport Barbara Castle elected to make the 70 mph speed limit permanent from September 1967 – despite extensive opposition from the roads lobby and 'drive-up and drive-in' protests at Newport Pagnell service area by the short-lived militant group 'Motorists' Action' (Nightingale 1967: 789; Castle 1990). Government statistics had 'proved' that the speed limit saved lives, and Castle felt justified to make this temporary measure, this governmental *technology*, permanent.

Accidents, death, blame

> ... to invent the sailing vessel or steam ship is to *invent the* shipwreck. To invent trains is to *invent the* derailment. To invent the private car is to produce *the motorway pile-up.* ... If we take the area of private motoring, for example, ... the multiple security systems with which our vehicles are equipped will do nothing to change this fact: in the course of the twentieth century, accidents became a heavy industry. (Virilio 2002: 24, 25)

Academic writings on car crashes/accidents have tended to focus on the design of safety mechanisms, the injuries resulting from crashes, road safety propaganda, perceptions and calculations of risk, or on distinctive cultural engagements, whether in film – e.g. Jean-Luc Godard's *Week-End* (1967) and David Cronenberg's *Crash* (1996); art – e.g. the 1960s disaster paintings by Andy Warhol (Crow 1990; H. Foster 1996); or literature – e.g. J. G. Ballard's exploration of the sexual desires, injuries and deaths resulting from the deliberate and accidental 'crashing' together of human bodies and cars, most famously in his 1973 novel *Crash*, which Cronenberg filmed in 1996 (Ballard 1973; see also Schnapp 1999; Sinclair 1999).[58] In contrast, as Roger Cooter and Bill Luckin (1997) have shown, relatively few social historians have examined car accidents, particularly the changing ways in which statisticians and actuaries have quantified events, calculated risks and have contributed to the normalization, de-personalization and secularization of accidents. In this section I aim to explore attempts to understand *both* seemingly uneventful *and* shocking or unusual crashes on the M1 in 1959 and the early 1960s. Following Paul Virilio's comments above, I want to examine how the motorway crash was approached as a potentially new type of accident associated with the distinctive spaces, movements and speeds associated with motorways and motorway driving (on the social dimensions of the car accident, see Schnapp 1999; Brottman 2001; Arthurs and Grant 2003; Beckmann 2004; Featherstone 2004).

Satirists set the tone before the M1 opened. In the titles of two *Punch* articles published in October and November 1959 – H. F. Ellis's 'M1 for murder' and Lombard Lane's 'That "M for murder" road' – both authors made a play on the title of the popular 1952 play/1954 film *Dial M for Murder*, associating the new motorway with wilful violence, fear and death. Journalists picked up the baton after the first serious crashes, and the Ministry of Transport and Road Research Laboratory began collating statistics and investigating the events surrounding accidents. Individual accidents, injuries and deaths become de-personalized and sanitized in statistical tables and graphs. In February 1961 Ernest Marples revealed the numbers of people killed and injured on the M1 in the first 12 months, along with comparative statistics for the A5 and A45 before and after the motorway opened (*Parliamentary Debates* 1961). Traffic levels had increased, complicating the comparison. Nevertheless, Marples revealed that the number of injured crash victims had reduced from 724 on the A5 and A45 in 1958–9, to 685 on the A5, A45 and M1 in 1959–60 (*Parliamentary Debates* 1961: 23–4).[59] In July 1961, W. F. Adams, Chairman of the Traffic Engineering Committee, declared that the accident rate on M1 was 'satisfactory' compared with that on European motorways and US toll roads, but despite such reassurances Adams and other government scientists still identified matters of concern (W. F. Adams 1961: 178). After the occurrence of a number of major accidents in late 1959 and early 1960, journalists and government scientists started investigating the effects of different factors on their severity. The problem became one of attempting to identify causes or attribute blame to particular actors (J. Law 2001). Indeed, some scholars might argue that motorway accidents are an *inevitable, normal* or *systemic* effect of thousands of performances and movements by different drivers in vehicles of varying condition, trying to avoid other vehicles while driving on roads of differing states of repair in particular weather conditions: 'If interactive complexity and tight-coupling – system characteristics – inevitably will produce an accident, I believe we are justified in calling it a *normal accident*, or a *system accident*' (Perrow 1984: 5).[60] I would suggest that, just as the emotions and desires associated with driving need to be understood in relation to the 'tight-coupling' or hybridization of drivers and vehicles (cf. Katz 1999; Lupton 1999; Michael 2000, 2001; Beckmann 2004; Sheller 2004), so car accidents can be seen to weave together and implicate

a broad range of actors, from drivers, entire vehicles and vehicle components, to the road itself, vehicle manufacturers, mechanics, highway engineers and the weather (Merriman 2006b; cf. J. Law 2001). Nevertheless,
accident investigators, engineers, the police and judiciary do attempt to
unpick or dismantle such 'tightly-coupled' assemblages and systems, attributing or distributing blame.

After four major accidents in dense fog on 6 November 1959 – one of
which led to the death of two lorry drivers – journalists attempted to identify the cause of each accident and distribute blame. In his *Daily Express*
'investigation', Basil Cardew blamed the design of the narrow hard shoulder – which lorry drivers termed 'the strip of death' – for *causing* collisions,
as it prevented lorries from pulling off the carriageway in an emergency
(Cardew 1959b: 4). Other journalists blamed the drivers, suggesting that
they should have been driving with more care and less speed, while several
commentators blamed the fog itself:

> The first M1 casualty . . . seems to have been due to our old enemy, fog,
> which even the finest civil engineers cannot yet cope with. (*The Autocar*
> 1959e: 624)

> Fog causes the first fatal accidents on the M.1. (*The Sphere* 1959: 265)

Fog becomes bound into the social networks and spaces of motorway
driving, emerging as a major hazard, a force, a killer, when it engulfs the
motorway and menaces drivers travelling at high speeds. Fog emerges as a
mute target of blame, a disordering presence *responsible* for serious accidents and the deaths of innocent drivers (on the social and cultural significance of weather and climate, see Strauss and Orlove 2003; Boia 2005).
Motoring organizations focused their attention on the motorway designers
and drivers, arguing that the latter needed educating and warning about
the dangers of fog. In November 1961 the Automobile Association asked
the Ministry of Transport to install warning signs on M1 (*The Sunday
Telegraph* 1961), but it was only after the involvement of 200 vehicles in a
series of crashes in dense fog on the M1 on 21 January 1964 that politicians
and motoring organizations asserted greater pressure (*The Times* 1964a).
In the following month warning signs were trialled on the M5 motorway
(*The Times* 1964b). In December that year the trial was extended to the
M1,[61] while further 'pile-ups' on motorways throughout the mid- and
late 1960s resulted in the introduction of the 70 mph speed limit, and reinforced the association of motorway fog with accidents, serious injuries
and death.

Civil servants and Road Research Laboratory scientists paid close attention to the geographies of collisions and accidents, studying the sections of

motorway with the highest accident rates and examining the distinctive movements of vehicles during and after incidents. The government became particularly concerned by suggestions that the design of the motorway was implicated in certain kinds of accident. Engineers focused their attention on the function and effectiveness of many different design features, and in the remainder of this section I examine investigations into the effects of three motorway structures on accident rates: firstly, the open, unguarded central reservation; secondly, the terminal roundabouts; and, thirdly, Sir Owen Williams and Partners' two-span over-bridges.

When the motorway was opened there was no crash barrier along the length of the central reservation separating the two carriageways. In 1963 two Road Research Laboratory scientists, R. F. Newby and H. D. Johnson, set out to investigate the 'movements of vehicles involved in accidents', and the effect that a crash barrier *might have had* on the frequency and severity of collisions (Newby and Johnson 1963: 554). The 'facts' were to be studied and extrapolated to draw counter-factual conclusions. Newby and Johnson classified and mapped the irregular movements of 1155 vehicles involved in 741 accidents in 1960 and 1961 (Figure 5.14), and they showed that 'not more than 19 per cent' of the vehicles – those with movements C to K in Figure 5.14 – reached or crossed the central reservation, and would have been deflected by a crash barrier (Newby and Johnson 1963: 555). The statistics were seen to show that crash barriers might not

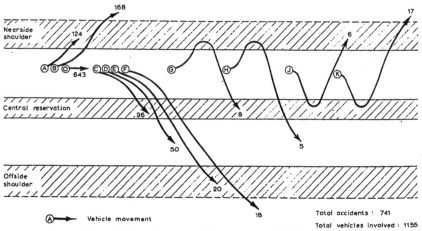

Fig 4. *Movements of vehicles involved in accidents (personal-injury + damage only) (1960 and 1961), excluding accidents at terminals, junctions and interchanges.*

Figure 5.14   The movements of vehicles involved in accidents on M1 in 1960 and 1961. Line drawing in *Traffic, Engineering and Control*, Volume 4, February 1963, p.554. Reproduced by permission of The Hemming Group.

prove that effective or cost-efficient in preventing collisions, while in 1961 W. F. Adams had suggested that crash barriers or fences 'might cause more accidents than they prevent' (W. F. Adams 1961: 180). Motoring organizations, the police and government engineers all emphasized the low probability of vehicles crossing the central reservation (*The Times* 1961a), but after 12 people died in crashes involving vehicles crossing the divide in just 18 days in August 1970 (*The Times* 1970b), Conservative Minister of Transport John Peyton announced that barriers would be installed on Britain's busiest motorways, including the M1, in 1971 (*The Times* 1970c).

Accident statistics were recorded for the different junctions on the motorway, and engineers focused their attention on the higher accident rates for the four terminal roundabouts at either end of the motorway. In July 1961 W. F. Adams mapped the specific locations of individual accidents during 1959–60, in an attempt to assess whether the design of the different roundabouts might be implicated in the accidents (Figure 5.15). Adams was quick to conclude that the drivers were largely to blame – in isolation from their vehicles, the light conditions or the design of the roundabouts: 'Nearly all the accidents at terminal roundabouts have been due to

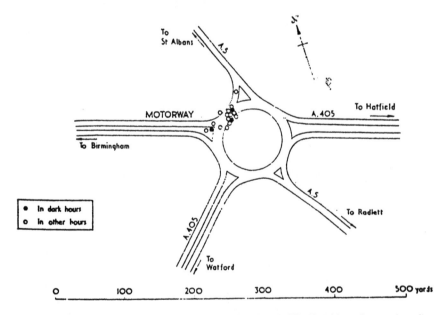

Figure 5.15 'Approximate position of accidents at Park St. Terminal (A5/A405) in twelve months ending Oct. 31, 1960'. Line drawing in *Traffic, Engineering and Control*, Volume 3, July 1961, p.179. Reproduced by permission of The Hemming Group.

drivers approaching too fast along the motorway, and over-running the island or striking vehicles already on the roundabout' (W. F. Adams 1961: 179). The solution was to educate and govern motorway drivers. Drivers must reduce their speed in advance of the junctions, and to ensure this the Road Research Laboratory placed 'reduce speed now' signs in advance of the roundabouts in February 1960, in an attempt to get drivers to regulate or govern their own conduct, movements and speed (W. F. Adams 1961).

My third and final example relates to attempts to attribute blame to the over-bridges designed by Sir Owen Williams and Partners. As I discussed in chapter 3, the appearance of the bridges was widely criticized by an array of architects, landscape architects and engineers, but a number of engineers also criticized their design on grounds of safety. Two features or effects of the bridges attracted attention. The first were the piers/columns on the central reservation (as visible on Figure 3.10). In the first three and a half months, six vehicles collided with the central piers/columns. Where the columns had been protected by a crash barrier, the accidents resulted in only minor damage and slight injuries, but where vehicles had hit columns protected with an unguarded concrete plinth (see Figure 3.10), the crashes resulted in two fatalities, two serious injuries and major vehicle damage (W. F. Adams 1961). When questioned about the function of the columns following the Institution of Civil Engineers' 1960 symposium on the London–Birmingham Motorway, Sir Owen Williams, Owen Tudor Williams and Ministry officials J. F. A. Baker and J. G. Smith all asserted that the central piers had proved necessary to lower the cost of motorway construction (*Proceedings* 1961). In February 1960, the Traffic Engineering Committee and Motorways Working Panel recommended that 'on future motorways piers should not generally be built on the central reserve',[62] and in designing the second, Crick to Doncaster, section of the motorway, Sir Owen Williams and Partners removed 'what was then considered to be the potential hazard of overbridge supports in the central reservation' (Sir OWP 1973: 46). Bridge piers on the central reservation might become *implicated* in serious accidents, *injuring* drivers; leading the designers and civil servants (as well as the bridges themselves) to become a target of blame.

The second aspect of the bridges which attracted attention and criticism was their shape and a series of shadows they appeared to cast. In June 1960 politicians and the press speculated about the role of a bridge and its shadows in a single fatal accident which had killed 28-year-old Valerie Hopkins in July 1959 (*Daily Express* 1960; *News Chronicle* 1960). At the inquest, her husband suggested that the car had overturned because Valerie swerved to avoid what appeared to be 'a furniture lorry with no lights' parked under one of Sir Owen Williams and Partners' bridges (*Daily*

Express 1960: 13). The police had failed to trace the van, and Mr Hopkins was unsure as to whether it had existed. One witness, a lorry driver, reinforced these doubts, arguing that he and his colleagues had regularly observed a shadowy 'phantom menace' under the bridges which assumed the shape of a parked vehicle (*Daily Express* 1960: 13). Journalists jumped on these reports of 'the phantom of the M1' (*Daily Express* 1960: 13), printing photographs of views under the bridges with captions such as: 'Is this the phantom van on the M1?' (*News Chronicle* 1960). Reporters started to attribute blame for the accident to the bridges, and it was not long before politicians questioned the Minister of Transport in the House of Commons. On 6 July 1960 Conservative MP Philip Hocking asked the Minister whether he would consider providing lighting under the bridges (*Parliamentary Debates* 1960b). On 20 July another MP, Mr A. Roberts, asked what style of bridges would be adopted on later sections of motorway 'to avoid the confusion and danger to motorists caused by the shadows cast by those bridges that have been constructed on the London–Birmingham section of the motorway' (*Parliamentary Debates* 1960c: 489).

The Ministry of Transport were concerned about the reports and they instructed the Road Research Laboratory to investigate the crash and the reported visual effects. In October 1960 the RRL's V. J. Jehu published his report on the 'Phantoms on the M1 motorway'.[63] Jehu attempted to conduct a scientific investigation which could explain away these disordering presences, proving that the bridges, their designers and the government were not to blame for the accident. He studied witness statements, police reports, and he photographed the accident site and bridges at different times of the day, in different light conditions, and in different densities of traffic. Jehu reported that his investigations were 'inconclusive', but he did conclude that the bridges and their engineers were not to blame. If there was a single contributing factor, it was likely to have been that the headlights of the family's Ford Popular had not been on at full beam:

> The photographs show that if the headlights of the accident car had been switched on then the nearside edge of the road would have been visible sufficiently far ahead to show that any obstruction was clear of the road. This is the only firm conclusion which can be arrived at from the known facts. It is significant that similar illusions are claimed to have been seen by lorry drivers, a large number of whom drive at night on unlighted roads, including the motorway, with only sidelights showing.[64]

The Road Research Laboratory's scientist makes his authoritative conclusions, highlighting decisions which the motorway driver *could* have made, which *would* have reduced the likelihood of such effects and accidents. The motorway's structures appear to be immutable, immobile, predictable and fairly standardized features of the landscape (Latour 2005). They appear to be easy to study, understand and grasp. Drivers, on the other hand, are presented as complex, mobile, variable actors. They make unpredictable and irrational decisions. Their actions are difficult to trace or reclaim. It is not so much that drivers are easy to blame, rather that the fixed structures of the motorway are easy to exonerate. If the motorway and its designers are not to blame, driver error emerges as the most probable cause. And so the story continues, as the M1 and other motorways continue to serve as spaces of experimentation and investigation into the twenty-first century, spaces for the enactments of both life and death.

Chapter Six

Motorways and Driving since the 1960s

In early and mid-twentieth-century Britain, motorways tended to be envisioned as distinctively modern spaces and forces. In chapters 3 to 5 I examined the debates about the planning, design, construction and use of Britain's first major motorway, the M1, and revealed how much of the excitement surrounding this experimental motorway reflected the novelty, uncertainty and mystery surrounding what was seen to be a distinctively modern British/English landscape. The M1 emerged as a modern spectacle, a modern landscape: a culturally significant site/sight; a well-travelled route; a popular destination; a site of post-war national reconstruction; an important space from which to view the landscape; a striking structure in the landscape; a space of scientific study; a presence which could reconfigure the economic geographies of regions; a route that would transform the travelling patterns of the public. The geographies of the M1, then, might be seen to be highly specific, located and distinctive, refracting the historical geographies of 1950s and 1960s Britain. As problems started to be identified with the design and construction of the M1, its structures started to age, and motorways became more familiar features of the English landscape, so the excitement surrounding the M1 and other motorways began to wane: 'It is now, astonishingly, nearly ten years since the M1 opened. The era of euphoria, when we drove up and down the M1 merely to savour the delights of driving up and down the M1, is past' (*The Architects' Journal* 1969: 1615). A trip on the motorway could still generate feelings of excitement, fear or adventure in drivers or passengers, but the M1 was no longer unique. In 1968, the M1, 'the great north–south motorway', was featured in volume five (1914–68) of S. E. Ellacott's *A History of Everyday Things in England* (Ellacott 1968: 179). The M1 and other motorways had become fairly familiar and ordinary features of our everyday landscape for an increasing number of motorists, and

I could follow a broad array of thinkers in arguing that, in the late twen-tieth century, such spaces were increasingly experienced and encountered as 'placeless' environments (Relph 1976), as 'non-places' (Augé 1995). As I discussed in chapter 1, and I have outlined in detail elsewhere (Mer-riman 2004b, 2005c, 2006b), I find such designations problematic and overly simplistic. Nevertheless, attitudes to driving and the spaces of the M1 and other motorways *have* changed over time, and in this chapter I explore how the geographies of the M1, other British motorways and driving were reworked and reinterpreted during the 1960s, 1970s and up to the present.

In section one I revisit ideas I traced out in chapter 5, exploring how the M1 was seen to carve a 'corridor' through the English landscape. In these accounts, the M1 is characterized as a modern force, reconfiguring the farming and hunting landscapes, and providing opportunities for urban, office and warehouse development. The M1 assumed a symbolic position in the geographies of the nation, and the motorway and its service areas emerged as significant passage points on north–south journeys. In section two I examine the criticisms which began to emerge in the mid-1960s and 1970s about the impact of motorways and motor vehicles on the urban and rural environment. Urban conservationists, environmental-ists, preservationists and local campaign groups started to challenge gov-ernment transport policies and oppose individual road building schemes, while cultural commentators criticized the consumption practices associ-ated with mass car ownership. In section three, I examine the futuristic, dystopian motorway landscapes presented in J. G. Ballard's novel *Concrete Island* (1974) and Peter Nichols's play *The Freeway* (1975). In their quite different visions of what might happen if motorists were brought to a halt and stranded on the spaces of the motorway verge, both authors explore the marginality and ambiguity of these landscapes. In section four, I examine the work of two artists who have attempted to articulate the ubiquitous, anonymous and placeless qualities of motorways in their work. Andrew Cross's exhibition *An English Journey* and film *3 Hours From Here* and the paintings of Julian Opie appear to refract the theoretical arguments of academics such as Marc Augé and Edward Relph, but I argue that their reflections upon the qualities of motorway landscapes, their awareness of the sensibilities they evoke, and their representation of these landscapes in paint and film contribute to the 'placing' of the motorway in distinctive and ongoing ways. In the final section, I trace how the M1 continues to be placed through all manner of movements, actions and events in the late twentieth and twenty-first centuries, being constructed, registered and at times celebrated as a significant, distinct and meaningful place – whether by ecologists, photographers, architectural historians, journalists, local residents or motorists.

The 'M1 Corridor'

During the 1960s commentators reflected on how the M1 *might* transform – or *was* shaping – the urban, rural, local, regional and national geographies of the London to Birmingham corridor. Writing in *Country Life* in October 1966, Brian Dunning remarked on how little attention had been paid to 'the growing tendency for motorways to act as local boundaries – artificial rivers, so to speak', changing 'the basic shape of the countryside' (Dunning 1966: 978). Whereas older roads, and their verges, could be crossed legally and fairly safely at ground level by pedestrians, farmers and animals, motorways were a space designed for unidirectional movement, with only a few multi-level crossing points. As Dunning observed, despite the presence of farm access bridges and tunnels, land owners were increasingly exchanging land, such that 'the number of farmers cultivating land on both the eastern and western margins is falling year by year' (Dunning 1966: 978). The M1, along with other motorways, appeared to be reshaping England's agricultural and hunting landscape. Hunt masters became 'unwilling to take their packs [of hounds] near to the heavy traffic', and foxes were observed to 'set up home alongside the motorways where they are assured, not only safety, but of a plentiful supply of dead carrion' (Dunning 1966: 979). In August 1959, this bisection, this severance, led the Oakley hunt to give up lands to the Whaddon Chase hunt (*The Times* 1959p: 15), while the Pytchley hunt and Grafton hunt discussed the possibility of redrawing their boundaries so that they could hunt to the West and East of the M1, respectively (Dunning 1966).

The M1, then, was seen to 'cut' the rural landscape in two, but as I discussed in chapter 5, commentators suggested that the motorway would also 'gather in' or 'open up' areas of rural Bedfordshire, Buckinghamshire and Northamptonshire. Writing in October 1959, conservationist Malcolm MacEwen suggested that the M1 might reinforce the regional inequality of the South-East and lead to the urbanization of this transport corridor:

> In Northampton . . . the Chamber of Commerce has suddenly awoken to the fact that the new London–Yorkshire motorway has created wonderful new opportunities for industry and commerce because Northampton lies only a mile or two to the east of the motorway, and about midway between Birmingham and London. Will the new road not also increase London's magnetism and the drift of population to the south-east? Could the new road not even become the central urban motor road of a new London–Birmingham conurbation? (MacEwen 1959: 270)

In *The South East Study 1961–1981*, the Ministry of Housing and Local Government presented the M1 as an important catalyst to urban expansion. Northampton was identified as an ideal candidate for major population expansion, while Bletchley (5 miles south of Newport Pagnell) – located adjacent to the M1 and on the main West Coast railway line – was styled as the ideal location for a new city (MHLG 1964). The Milton Keynes Development Corporation was subsequently established in May 1967, and in their city plan, published in March 1970, the M1 appears as one of two major motorways that would link the city to the surrounding region (the other being a new urban motorway along the route of the A5) (MKDC 1970). The proximity of the M1 and West Coast railway line helped influence the Roskill Commission's 1970 proposals to construct the Third London Airport at Cublington in Buckinghamshire (6 miles south of Bletchley, and 9 miles south-west of the M1), although the proposals were not endorsed by the government (Buchanan 1981).

The construction of the M1 sparked the imaginations of city planners and property developers, and with the extension of the motorway through Leicestershire in 1965, Nottinghamshire and Derbyshire in 1966–7, and the West Riding of Yorkshire in 1968, the 'M1 corridor' soon extended between London and Leeds (Charlesworth 1984). In the past few decades economists and real estate developers have paid increasing attention to the kinds of economic activity and urban development generated by the motorway. In the pages of *Chartered Surveyor Weekly* and *Estates Times*, property developers and surveyors regularly report on the changing fortunes of particular towns, office developments and industrial sites along the 'M1 corridor' (see, e.g., Sherman 1991; *Estates Times* 1993). The 'M1 corridor' is styled as a linear landscape of actual and potential development sites and more or less successful towns. The corridor is seen to be practised through the movements and actions of multi-national companies, property developers, investors, workers and their services and goods, as well as motorists, acting as both a barometer and shaper of fortunes in the region (on the practising of the company region, see Laurier and Philo 2003). Developers were particularly attracted to junction 18 at Crick, at the northern end of the first section of the motorway, where the M1, A5 and projected M6 came in close proximity. As Brian Turton explained in *The Geographical Magazine* in April 1978, the junction was styled as 'Britain's first motorway-orientated growth point', emerging as the location for several large depots and unsuccessful proposals to construct a £35 million 'privately developed new town' (Turton 1978: 453). In the 1990s, developers constructed DIRFT (the Daventry International Rail Freight Terminal) to the west of junction 18, at which Tesco, Royal Mail, Eddie Stobart, Wincanton, Exel and other large companies have all established distribution and warehouse centres, and which is actively promoted by its owners as a key site in England's important '"golden triangle" favoured for national distribution centres' (DIRFT 2006).[1]

The M1 opened up the landscapes of Bedfordshire, Buckinghamshire, Northamptonshire, Leicestershire, Nottinghamshire, Derbyshire and Yorkshire in other kinds of ways. Small villages and towns – including Toddington, Newport Pagnell, Trowell and Rothersthorpe – have gained renown due to the adoption of their names for motorway service areas.[2] In the case of Watford Gap, the service area and its location has come to assume an important symbolic position in the cultural geographies of the nation – serving as one location for a North/South divide. During the 1960s and 1970s, the M1 and M6 were envisioned as the new 'Great North Road[s]', the 'New Way North' (Banham 1972a: 241). The M1/M6 corridor became England's main North–South route, linking two worlds – North and South – which were and still are *mythologized* in popular culture.[3] As Lynne Pearce and Tim Edensor have shown, such north/south drives can evoke quite complex emotions and memories (Pearce 2000; Edensor 2003), and as many motorists use the M1 and M6 to travel north/south across England, it was perhaps inevitable that some motorists and cultural commentators would locate a national North/South divide on the M1. Watford Gap provided one such location and reference point: a resting stop just *beyond* South-East England, but *not quite of* the industrial landscapes of the Midlands and the North. A space beyond and between (Shields 1991; Jennings 1995).[4]

Motorways and 'the Environment'

During the 1960s and 1970s an increasing number of observers and commentators started to criticize Britain's reliance on the motor car and the government's expanding road programme. In the late 1950s and early 1960s amenity and farming groups had campaigned for alternative routes for specific sections of motorway, including the M1 through Charnwood Forest and the M4 through the Berkshire Downs, but with the emergence of proposals for urban motorways and inner ring roads in London and provincial towns and cities in the mid-1960s, a large number of amenity and local campaign groups formed to fight against such schemes (see Starkie 1982; Wall 1999). In 1973, after eight years of campaigns and criticism, the Labour-led Greater London Council dropped plans to build the controversial London 'Motorway Box' (Thomson 1969; Starkie 1982), but as schemes materialized across Britain, individuals and organizations in the increasingly active urban conservation movement expressed concern about the impact of motorways on the nation's urban fabric and urban life (Starkie 1982; Powers 2004).[5] In a 1975 essay, 'Roads, office blocks, and the new misery', cultural studies academic Fred Inglis described the appearance of a new urban motorway and urban redevelopment in Bristol:

At this point, a six-lane motorway in full play unrolls a thick carpet of noise and lead-poisonous fumes well beyond the point blocks whose vision of city life is now the roofs of the traffic thirty-five yards across the dreary brize-block [*sic*] patios which serve their hundreds of council tenants for playground and garden. If you walk across from the ground-floor flats to see the traffic, the road-planners' rationalization and flow-chart efficiencies are visible for what they are: a brutal and irrational effort to solve a trivial difficulty by destroying everything in the way. There was never a more telling occasion to accuse men of intellectual tunnel vision; the urban motorway is the symbol of their condition. It is as if, like American policy in Vietnam, you annihilate all natural and human life in an area, strip the trees and drive out the homesteaders in order to call the country pacified. (Inglis 1975: 174–5)

Urban motorways join high-rise housing and speculative office development as new evils threatening Britain's towns and cities. Two books published in European Architectural Heritage Year, 1975, articulated many of the fears of conservationists. In *The Rape of Britain* by Colin Amery and Dan Cruickshank, and *Goodbye Britain?* by Tony Aldous, the authors criticized the destruction brought by developers and planners to Britain's historic urban fabric and urban life, with 'comprehensive redevelopment' and planning for the motor car being identified as two key agents of destruction (Aldous 1975; Amery and Cruickshank 1975). Poet laureate and well-known conservationist Sir John Betjeman wrote forewords to both books, and in *Goodbye Britain?* he expressed his fear that the English landscape was being transformed into a mass-produced concrete environment occupied by a populace with somewhat suspect tastes: '. . . I write this at the thought of England becoming a few acres of preserved countryside between concrete fly-overs, spanned by cafeterias thrumming with canned music and reeking of grease' (Betjeman 1975: 7). Motorways become worked into different narratives of Englishness, as optimistic visions of modern motorways fitting snugly into and even enhancing the rural English landscape become accompanied and increasingly displaced by melancholic and dystopian visions of motorways destroying both rural and urban scenes.

The oil crisis of late 1973 provoked widespread concern about Britain's dependence on imported oil and petrol-driven transport, and while petrol rationing was narrowly avoided, the government introduced a temporary 50 mph speed limit on motorways between November 1973 and the spring of 1974 (Charlesworth 1984). Resource-depletion was not the only concern. As Britain's lorries grew in size and weight, so they emerged as a target for preservationists and conservationists (see, e.g., Inglis 1975). In the late 1960s and 1970s the Civic Trust campaigned against the impact of heavy lorries on urban and rural areas (Starkie 1982), and English artist Edward Burra – a critic of motorway building – expressed his concerns in two paintings of the Peak District from 1970 (Causey 1985). One of these, *An*

English Country Scene No. 1 – painted near Buxton during one of his many motoring tours – depicted what Burra felt to be an increasingly typical view of the English countryside, with a long line of lorries – replete with angry animalistic faces – snaking along a road, around hills and into the distance (see Causey 1985; also de Hamel 1976).

With the formation of the Department of the Environment in 1970, and the emergence of environmental lobby groups such as Friends of the Earth and Greenpeace in 1971, environmental concerns were increasingly being brought to the public's attention. In 1972 HMSO published *How Do You Want to Live?*, 'a study of public opinion' prepared by one of four working parties set up by Secretary of State for the Environment Peter Walker in advance of the United Nations' landmark Conference on the Human Environment at Stockholm in June 1972. The report confirmed a shift in public opinion around the car which 'amounted almost to a backlash', with 'the day of the supremacy of the motor car and the road-builder' having come to an end (Working Party 1972: 103). Nevertheless, the Working Party felt that modern planning and conservation could be combined, and they argued that urban motorways and bypasses represented 'the only hope of enabling a large proportion of the population to make use of private cars to get out into the country' (Working Party 1972: 105; see also Matless 1998). The Chairwoman of the Working Party, Greater London Councillor the Countess of Dartmouth, commissioned Philip Larkin to write a poem as a prologue to the report. The poem, 'Going, going', was printed over a photograph of the ICI chemical works in Teesside, and throughout it Larkin – in a not dissimilar vein to Betjeman – pondered on what would remain of England after the ever-increasing encroachment of urban and industrial development, tourism and pollution on the countryside. Larkin observed a younger generation, with questionable tastes, clamouring for consumer goods.[6] These were aspiring lower-class and middle-class consumers, who were increasingly mobile and visible, crowding the 'M1 café[s]' while escaping *en masse* to the countryside (Larkin 1972: x).

The M1 and its service areas become associated with mass leisure and a questionable consumerism, and commentators with diverse backgrounds directed increasingly strong criticisms at these spaces of consumption during the 1970s. Consumer groups, as well as restaurant critics such as Egon Ronay, published critical reviews of service area catering throughout the 1970s and 1980s (see, e.g., Ronay 1981), while record company EMI were successfully sued for defamation by service area operator Blue Boar over folk-rock musician Roy Harper's 'Watford Gap' on his 1977 album *Bullinamingvase* (*Daily Telegraph* 1979). In 'Watford Gap', Harper depicted the service area as a place of worship, 'the great plastic spectacular descendant of Stonehenge', amidst references to 'death defying meals', 'plastic cups of used bathwater', and a chorus rhyming 'Watford Gap' with 'a load of crap' (Roy Harper 1977). EMI withdrew the song from the album, but commentators continued to criticize the food and architecture of British service areas. In 1989 one journalist labelled Watford Gap 'one [of] the worst [service areas] in Britain': 'We'll gloss over the beans because I'm still feeling the after-effects. But if I'd thrown the toast out of the window it would probably still be bouncing up the M1' (Hurrell 1989).

In the 1970s environmental campaigners started to challenge the government's broader transport policy and their justification for the construction and location of individual road schemes. In 1973, John Tyme, a Senior Lecturer in Environmental Studies at Sheffield Polytechnic, joined the National Transportation Working Party of the Conservation Society, and he quickly gained a reputation for his legal challenges and vocal protests on behalf of campaigners at public inquiries across England. Tyme believed that the public inquiry process was undemocratic and that the motorway and trunk road programme posed 'a consummate evil, and constitutes the greatest threat to the interests of this nation in all its history' (Tyme 1978: 1). Tyme and his fellow campaigners succeeded in generating extensive publicity, delays, and in halting proposals to build the M27 at Chichester, M3 at Winchester, the new Archway Road in London, and the Aire Valley trunk road in Yorkshire (Tyme 1978). With the Aire Valley inquiry – held over nine days between November 1975 and February 1976 – widespread media coverage brought scenes of public anger, scuffles and subsequent arrests into the homes of the public, and the emerging tactics and the arguments of these and other campaigners were amusingly satirized in Tom Sharpe's 1975 book and the BBC's 1985 TV series *Blott on the Landscape* (Sharpe 1975).

Local campaigners continued to campaign against road building schemes throughout the 1980s and 1990s. In the early 1990s the vocal objections of environmental and local amenity groups were joined by the more direct-action and radical tactics of a younger generation of so-called 'eco-warriors', who succeeded in bringing extensive delays and escalating costs to the construction of such notorious schemes as the M3 at Twyford Down (Winchester), M11 link road (East London), M77 at Pollok (Glasgow), A34 Newbury Bypass and A30 at Fairmile near Exeter (see, e.g., McKay 1996; Routledge 1997; Wall 1999). Protestors justified their actions through both formal, mainstream understandings of the local and global environment, ecology, archaeology, heritage, landscape design, planning and the law, *and* more alternative, radical, 'New Age' understandings of the environment, political action and the history of protest in Britain. As revealed by Road Alert!'s 1997 handbook *Road Raging: Top Tips for Wrecking Road Building*, protestors with experience of direct-action protests on road schemes across Britain were sharing advice on organizing campaigns, recruitment, the media, the law and the most appropriate tactics and tools for occupying a route: from building tree camps, barricades and tunnels, to the use of ropes, lock-ons, scrap cars and smoke bombs (Road Alert! 1997).

Dystopian and Marginal Landscapes?

As the national network of motorways expanded, Britain's cultural commentators began to reflect upon the kinds of landscape, visuality, subjectivity and society which appeared to be ushered in by high-speed motorway driving. While an optimistic design commentator such as Reyner Banham could remark favourably on the new 'styles of seeing' necessitated by motorway driving (Banham 1972a: 243), Ian Nairn expressed regret that motorways created a less visually aware, automatized 'motorway person', who is 'unable to stop and stare' and is forced 'to slice across Britain' (Nairn 1975: 329). Motorway driving appeared to require more concentration and involve less communication than driving on ordinary roads, but what might happen if motorway drivers were brought to a halt in these spaces, pushed off the motorway and onto the verge, trapped in these landscapes of apparent freedom, and forced to reflect upon the geographies of such marginal landscapes? In 1973–4 – amidst the oil crisis and the growing furore about road building schemes – two well-known British writers independently explored such possibilities, tracing the futuristic, dystopian geographies of Britain's motorway landscapes.

In his 1974 play *The Freeway*, distinguished playwright Peter Nichols explored what could happen if hundreds of motorists became stranded in a three-day traffic jam, and motorists of different classes were forced to cooperate, pool resources and socialize on the grass verges. Nichols wanted to show how 'almost every civilized requirement in town and country has been sacrificed to keep the traffic moving', and the contradiction of the 'widely held conviction that the car is liberty incarnate' (Nichols 1975: 5, 6). The play is set in the future, when the A-roads and motorways have become dangerous, traversed by bandits, and are bypassed by a new network of high-security inter-city freeways. An accident blocks the Royal Freeway in central England, and the chaos is magnified when the crash site is occupied by 'the Scrubbers, the anti-motor group whose avowed aim is to paralyse the Freeway' Nichols 1975: 31). The Scrubbers are finally removed by the police, but the 'Wreckers' and Breakers' Union' then decide to engage in strike action, refusing to remove the crashed vehicles (Nichols 1975: 58). The play reflects key concerns from the early 1970s, when extensive strike action and work-to-rule policies were having a paralysing effect on the country. Class differences and relations underlie the entire play, which revolves around the pleasantries, deference and frank political discussions between the middle-aged occupants of two vehicles: a Lord and his mother (James and Nancy Rhyne) on their way to shoot in the Highlands, and the aspiring lower middle-class Lorimer family travelling in their American motor home to Yorkshire. The families spend much of their time on the ill-kept, litter-strewn verge, which Nancy Rhyne damns as 'the most emblematic feature of modern Britain' (Nichols 1975: 17). As supplies of food and water dwindle, motorists are forced to visit nearby villages and farms – purchasing, poaching and stealing – and a curfew is brought in as the freeway becomes increasingly dangerous. The play is comedic and satirical in tone, but while Nichols had emphasized that it was 'not an Orwellian nightmare' (Nichols 1975: 6), and that he had 'wanted a colourful landscape beyond the stranded cars – pastiche-Constable – with a bright blue cyclorama which only darkens at the end', the 'cut-price set had instead a black surround, a promise of gloom and futuristic horrors' (Nichols 2000: 377). The set appeared to have transformed this mildly satirical commentary on contemporary class relations and the landscapes of the modern motorway into a dark dystopian vision of a future England. Critics did not like *The Freeway* when it was first performed by the National Theatre at the Old Vic in October 1974, and Nichols remarked that this may have due to the fact that, 'by definition', a play about a traffic jam 'lacks movement' (Nichols 2000: 377).

The second fictional account I want to consider which explores the wastelands of the motorway verge is J. G. Ballard's 1974 novel *Concrete Island* (1974). As with Ballard's earlier novel *Crash* (1973), *Concrete Island* is a dystopian exploration of the landscapes of West London's urban motorways. Architect Roger Maitland swerves off the carriageway, crashing his car down an embankment and onto a large overgrown triangle of wasteland that is encased between three motorways – one of which is the Westway, the highly controversial urban motorway opened in 1970. Maitland becomes trapped on this 'forgotten island of rubble and weeds', this 'concrete wilderness', stranded like a late twentieth-century Robinson Crusoe (Ballard 1974: 5, 149). The wasteland has yet to be sanitized by 'compulsory landscaping', so Maitland is able to explore the remnants of buildings, cellars and air raid shelters destroyed to construct the surrounding motorways (Ballard 1974: 13). His injuries prevent him from scaling the embankments and escaping, and as time passes and he grows increasingly unwell, he begins to confuse the topography of his body with the landscapes of this wasteland:

> He surveyed the green triangle which had been his home for the past five days. Its dips and hollows, rises and hillocks he knew as intimately as his own body. Moving across it, he seemed to be following a contour line inside his head.

> The grass was quiet, barely moving around him. Standing there, like a shepherd with a silent flock, he thought of the strange phrase he had muttered to himself during his delirium: I am the island. (Ballard 1974: 131)

The boundaries of Maitland's body, car and the landscape become entwined and confused (Merriman 2004a). Just as the motorway vehicle driver may not reflect upon his or her presence, embodied actions and disposition at the wheel, or the appearance of the surrounding landscape, in his delirium and confusion Maitland becomes (at one with) 'the island'.[7]

In a series of books and essays, J. G. Ballard provided his readers with dystopian accounts of landscapes of modernity close to his home in Shepperton, West London, including the landscapes of Heathrow Airport, a tower block, and the carriageways and flyovers of the M4 and A40 (Sinclair 1999, 2002). While Peter Nichols was openly critical of Britain's increasing reliance on the motor car, Ballard took pleasure in exploring and exposing the darker desires and events which could emerge in the spaces of driving: the shadowy, seemingly marginal spaces we traverse and pass every day. *The Freeway* and *Concrete Island* reflected concerns about motorway construction and the aspirations of motorists that were widely expressed in the early 1970s. Nichols and Ballard presented visions of extraordinary, unexpected, 'abnormal' events – crashes and traffic jams – which were far removed from the expected and accepted statistician's 'normal accident' (cf. Perrow 1984). Here were nightmare scenarios, dystopian landscapes of confinement, and it was through these different processes of interruption, disordering and distraction that Maitland, the Lorimers and the Rhynes became all too conscious of their presence and confinement in spaces which may otherwise have passed them by or appeared rather generic, ubiquitous and placeless (Merriman 2006b). In *The Freeway* and *Concrete Island*, Nichols and Ballard *place* the marginal landscapes of Britain's motorway verges in distinctive ways, and their representations of these landscapes can usefully be considered alongside artistic engagements and cultural commentaries which appear to reflect upon the generic, seemingly placeless qualities of motorway landscapes.

Placeless Environments?

During the 1980s and 1990s, a significant number of academics and cultural commentators described motorways as generic, ubiquitous 'non-places', possessing similar qualities to other spaces of consumption, travel and exchange – including hotels, theme parks, airports, shopping centres, industrial estates, tourist sites and media spaces (Relph 1976; Augé 1995; cf. Merriman 2004b). In *Non-Places* Marc Augé (1995) suggested that such

spaces were characteristic of an era of 'supermodernity' – a term he preferred to postmodernity or late modernity – but while there was undoubt-
edly an expansion in the number of motorways, industrial estates, regional
shopping centres and suburban housing estates in the 1980s and 1990s, it
would be too sweeping to suggest that experiences of dislocation, boredom
and sameness were distinctively new *or* became increasingly prevalent in
the late twentieth century (Thrift 1995; Merriman 2004b). Augé, like
others before him, tends to overstate the novelty of such experiences and
environments, inferring that there was some kind of shift or break from an
earlier phase of modernity (for a critique of this tendency see Thrift 1995).
Motorways, industrial estates, shopping centres and airports have fairly
long, complex and specific histories, and while motorways may have functional designs, simplified and standardized aesthetic styles, and be uninspir-
ing to some observers, they are occupied, 'placed' and traversed through
the ongoing movements of an array of travellers, workers and consumers
on a continual basis in a variety of ways (Merriman 2004b). As I outlined
in chapter 1, a broad range of novelists, film-makers, artists and song-
writers have remarked upon or attempted to refract the distinctive qualities
of these (apparently placeless) environments, attempting to trace the
mundane geographies of the British motorway and motorway driving. Here
I examine the work of two British artists: Andrew Cross and Julian Opie.

Motorways form important routes in many long-distance English journeys, and when the photographer and film-maker Andrew Cross chose to
make a film and stage an exhibition marking the seventieth anniversary of
the first publication of J. B. Priestley's *English Journey* (1934), he focused
his attention on the motorways and dual carriageways which criss-cross
England (Cross 2004). As Steven Bode remarked in the catalogue of *An
English Journey*, Cross chose to focus on the contemporary, late twentieth-
century manifestations of Priestley's 'Third England' of arterial roads,
bypasses and petrol stations:

> Seventy years on, it is a manifestation of England that we see around us every
> day, however much the more familiar iconographies of countryside and city
> – mainstays of Priestley's two other Englands – tend to dominate the scene.
> It is an England, even now, that comparatively few people think to mention,
> and still less choose to highlight or celebrate. . . . (Bode 2004)

In the accompanying film, *3 Hours from Here*, Cross sets out from Priestley's departure point – Southampton – in a lorry heading north to DIRFT
(close to M1, junction 18), and then on to Manchester's Trafford Park
industrial estate.[8] The film intersperses camera shots looking straight
ahead through the lorry's windscreen, with shots out of the side window
which capture both the passing landscape and the rear view reflected in

the left wing-mirror. Captions state the location of the lorry and the number of miles from DIRFT. We hear faint sounds of passing traffic, and the occasional ticking sound of the indicator lights, but the cab is strangely quiet. The visual impressions of this 'English Journey' are quite complex. Without the captions, and occasional glimpses of traffic signs, viewers would probably find it difficult to judge their location, and the roads, roundabouts, motorways and industrial estates seen through the windscreen might be deemed to be somewhat dislocated, placeless, mundane and ubiquitous. The edited film contains few, if any, recognizable landmarks, and yet the lengthy shots through the side window of the lorry show how our experience of the surrounding countryside may not be that dissimilar to the motorway itself. Augé might argue that this is due to our separation from towns and villages, which can only be experienced through tourist signs or texts (Augé 1995: 67–8, 96–9); but I would argue that one could clearly see and experience the same ubiquitous qualities in the rural English landscape as one could in the landscapes of the motorway. The countryside can feel exciting, boring, solitary, placeless, distinctive, mediated, familiar or strange, as can a motorway, shopping centre or industrial estate. Landscape architects, again, have long recognized the significant but potentially boring and monotonous nature of such views of the road and landscape. As Brenda Colvin stressed in 1948:

> The great majority of people see the country mostly from roads, railways and footpaths. These points of view are those of the population in general, and should be regarded as being of national concern. . . . Good planting contributes interest and variety to the road which helps to keep the driver alert and vigilant. This interest and variety should be as great as possible, since it should serve to cancel the mechanical monotony of engine sound and road surface. From the motorist's point of view, this is one of the most important functions of roadside planting. (Colvin 1948: 243, 246)

Artists, as well as film-makers, novelists and landscape architects, have been struck by the somewhat detached yet engaging visual experience of motorway driving. In 1993, artist Julian Opie explored the landscapes and experiences of the British motorway in *Imagine You Are Driving*: a series of paintings, computer-generated films and sculptures based on scenes of the M40. The paintings and computer-generated films present the driver's view of a fairly anonymous simplified motorway carriageway, scenes which 'have no detail and seem to drag you physically into the painting' (Opie 2004: 84). In the opinion of Mary Horlock (2004: 118), Opie captures 'the anonymity and monotony of motorway travel', refracting contemporary experiences of, and Augé's thoughts on, the modern motorway:

Like the architecture of the post-war period, motorways encapsulated an ideal of efficiency, a promise of the future. But the initial excitement has long since waned. Motorways generate feelings of alienation: the road becomes a non-place where personal choices are abandoned in order to conform to certain rules, the monotony of driving subsuming the individual. (Horlock 2004: 58)

Of course, motorways *can* and *do* generate feelings of familiarity, homeliness and excitement, and drivers and passengers inhabit the spaces of the car and motorway in all manner of ways, making choices *and* breaking rules (Edensor 2003; Merriman 2004b). Opie was very familiar with the sections of the M40 he painted, just as he was with the M1,[9] which he depicted in a series of paintings in 2002 based on photographs taken (this time) from the hard shoulder rather than through the windscreen of a vehicle:

The difference in the resulting picture is greater than I would have thought. . . . These new images drawn from the kerb are much slower, calmer. . . . Instead of being physically drawn into the space, as you are in a computer game, you seem only to be drawn in visually, more like traditional landscape paintings which often use the device of a path leading into the picture. (Opie 2004: 84)

Opie's images of the M1, *I Dreamt I Was Driving My Car (Motorway)*, appeared on billboard installations in New York that were commissioned by the Museum of Modern Art, bringing a simplified, functional English motorway modernism, or 'the gentle, slightly dull, sweeping English landscape', into 'the claustrophobic, messy, dynamic American street' (Opie 2004: 84).

Julian Opie's paintings and Andrew Cross's film *3 Hours from Here* may appear to refract fairly common experiences of the contemporary motorway journey, but their focus on the visual qualities of driving overlooks or downplays the individualized, embodied, multi-sensory and kinaesthetic ways in which drivers and passengers inhabit the car and motorway, shifting their gaze, listening to music, singing, daydreaming, working or talking to passengers (e.g. Bull 2001, 2004; Edensor 2003; Laurier 2004). Travellers inhabit the landscapes of the M1 and other motorways in multiple ways, and their diverse movements and embodied engagements with the landscapes of driving are bound up with the ongoing placing of Britain's roads and motorways.

Placing the M1 in the Late Twentieth and Twenty-First Centuries

If British banality could be said to have a heart – an inner core of pure boredom – then the motorway and its environs might be expected to occupy a cosy place within it; alongside the carpet showrooms in out-of-town retail parks. . . . Increasingly, as the motorway features in the reclamation of shared and formative memory for successive generations, so its initial cultural status as a non-place is being exchanged for a new measure of significance.

You could say that the phenomenon of the 'non-place' is beginning to carry a romance – as a portal to nostalgia, and as a quasi-ironic metaphor of some lost state of innocence – that in Britain has been formerly ascribed to the Victorian suburb and the Edwardian seaside town. (Bracewell 2002a: 285–6)

During the twentieth century the landscapes of the M1 were practised and placed through the movements of an array of people and things. In contrast to Marc Augé, I believe that experiences of solitariness and boredom may emerge in all manner of spaces, and that academics do not need to invent new species of space and place – such as non-place – to comprehend such experiences (Merriman 2004b). The landscapes of the motorway are continually 'placed' through a broad range of actions and movements, and to conclude this chapter I want to examine a few of the ways in which the motorway continues to be practised, registered and at times celebrated as a significant, distinct, meaningful, located place.

Firstly, I want to start with the quotation above by Michael Bracewell. Bracewell suggests that as motorways have become worked into our everyday lives, acting as familiar backdrops for daily commutes and family holidays, so a new generation of nostalgists have begun to re-examine the modernist aesthetics of the motorway (Bracewell 2002a, 2002b). Accounts of motorway modernism vary widely. In his 1999 book *Boring Postcards*, photographer Martin Parr carefully arranged and presented a selection of mid- to late twentieth-century postcards of British motorway service areas, roads, bus stations, shopping centres, colleges, power stations, housing estates, holiday camps and airports (Parr 1999; Bracewell 2002a; Moran 2005). The book and collection reflects Parr's long-standing fascination with the mundane, the everyday, the kitsch and the boring, and specifically the ways in which, 'through the act of representation, the ordinary can become absurd and remarkable' (V. Williams 2002: 24). The first postcard in *Boring Postcards* is of the M1 motorway and Newport Pagnell service area; the same postcard which I have reproduced as Figure 5.11 and discussed in chapter 5. The excitement, novelty and modernity which became associated with the motorway, the service area and this postcard in the early 1960s, becomes displaced through the inclusion of this postcard in an amusing collection of 'boring postcards' in 1999 (Parr 1999). Parr's viewers may become nostalgic for these more optimistic and exciting times. They may remember when these places were not deemed to be boring. They may wonder why postcards were ever produced of these kitsch scenes and seemingly mundane places, or who actually bought such postcards. Parr explores the shifting geographies of taste and nostalgia, provoking a variety of responses through his witty and provocative selection of postcards from his personal collection, illustrating different dimensions of a vernacular modern British landscape (V. Williams 2002).

The modernist aesthetics of the M1 have also come to be examined and appreciated in more serious commentaries. With the introduction of the 'thirty year rule' in 1987, English Heritage launched its programme for the listing and preservation of exceptional post-World War II buildings. In 1992 the Royal Commission on the Historical Monuments of England and English Heritage outlined the way forward in an exhibition at the Royal College of Art entitled *A Change of Heart: English Architecture since the War. A Policy for Protection*. In the exhibition catalogue, Andrew Saint urged the British public and aesthetes 'to look again with an unprejudiced eye', placing buildings in their context, and putting aside 'stylistic prejudices' (Saint 1992: 29). Important buildings which some deemed ugly may need to be preserved, and in a subsequent interview with the architecture and motoring journalist Jonathan Glancey, Saint upheld Sir Owen's M1 bridges as 'an important part of our heritage' and as one possible target for listing (Saint, quoted in Glancey 1992: 30). Sir Owen Williams and Partners' much criticized concrete bridges had evidently aged, improved and become an important part of the nation's modernist architectural heritage, although (to date) their listing has not been formerly proposed or investigated.[10]

Secondly, in the early 1970s, amidst growing disquiet amongst environmentalists and conservationists, landscape architects and ecologists began promoting the ecological as well as the aesthetic importance of road and motorway verges. Let's start with the aesthetics of the M1's landscapes. In 1975 Ian Nairn reflected on the changes to the landscape between Southampton and Carlisle, which he had surveyed in 1955 for *The Architectural Review*'s 'Outrage' special issue. Nairn concluded that the 'only drastic change in 20 years' was the expansion of 'motorway land', and that strict planning controls ensured that 'motorway design and landscaping – after we got over the M1 – is one of the few genuinely collective and genuinely hopeful parts of design in Britain' (Nairn 1975: 329). Even W. G. Hoskins, a vehement critic of all post-1914 modern developments in *The Making of the English Landscape* (1955), felt able to accept and praise the landscaping of *well-designed* sections of motorway. In 1973 Hoskins criticized the first section of the M1 as a 'ghastly infliction on the English landscape', but the landscaping of the second section through Leicestershire was 'good', with 'lovely sweeping curves through Charnwood Forest in Leicestershire' (Hoskins 1973: 95): 'I do not think the landscape historian can resent what he sees now. It is a fine road, and already a bit of history' (Hoskins 1970: 14). In her 1970 book *New Lives, New Landscapes*, landscape architect Nan Fairbrother compared the excellent design, landscaping and planting of the new stretches of the M1 just north of London (opened in 1966) with the poor landscaping of the first section opened in 1959. Motorway design, landscaping and planting was improving, but Fairbrother joined ecologists and biologists in insisting that more attention needed to be paid to the conservation and sensitive management of road and motorway verges, which constituted 'important nature reserves' (Fairbrother 1970: 278). During the late 1960s, the Nature Conservancy's Dr J. Michael Way turned his attention to the conservation of roadside verges, and in European Conservation Year, 1970, he coordinated a botanical survey of the (then) 184 miles of M1 between London and Leeds (Way 1970). Nine students collected soil samples, calculated vegetation cover and recorded vegetation types and characteristics. The survey recorded 384 species of vegetation along the length of the motorway, including 54 species of grass, and a number of unusual and rare plants such as purging flax, yellow vetchling, pepper saxifrage and 'that alien stowaway, coriander' (Mabey 1974: 102; see also Way 1970; de Hamel 1976). Despite the effects of traffic fumes, salt spray and 'modern' management programmes, the M1's verges were supporting a wide range of vegetation, insects, birds and animals (Way 1970). Motorway verges were developing, ecologically as well as aesthetically, as pollen and plant seeds were blown in by air currents or carried in by insects, birds, animals and vehicles. Ecologists and biologists attached new meanings to discussions of the 'M1 corridor', which provided an important (if narrow) corridor of ecological diversity and conservation value, as well as a broader corridor for economic and urban development. The M1 emerges as a site of plant and animal movements and dwelling, a space of scientific study for ecologists as well as engineers. The verge is framed as a diverse and important, if still marginal, strip; a place with more to offer than the wastelands of confinement and danger explored by Peter Nichols and J. G. Ballard.

In my third example, I want to suggest that while the landscapes of the M1 form an important, meaningful and significant 'background' to an array of fairly everyday, routine movements and actions, the national importance of the M1 tends to be exposed and only discussed in the national media when something spectacular or unusual occurs and the traffic grinds to a halt.[11] The closure of the M1 following explosions at the nearby Buncefield oil depot on 11 December 2005 led to extensive delays for many motorists, while the strategic significance of the M1 and other motorways was recognized by the IRA during their terror campaigns of the 1970s and 1990s (Tendler et al. 1997). During the miners' strike in 1984–5, M1 junctions in Derbyshire and Nottinghamshire emerged as sites of strategic importance for both the police and miners. In March 1984, 200 protesting miners blocked the motorway after they abandoned their cars, while throughout the action, the police set up roadblocks at M1 junctions in Nottinghamshire and Derbyshire, arresting 'flying pickets' who were heading for collieries (Blomley 1994). When a British Midland Boeing 737–400 crashed on the M1 in January 1989 during its approach to East Midlands Airport, the M1 and the village of Kegworth became associated with a notable and controversial British air crash. In September 1997, Princess Diana's funeral cortège drove along the M1 on its way from Westminster Abbey to Althorp House in Northamptonshire. Maps of the route were printed in national newspapers, and many mourners lined the motorway, standing on bridges and on the edge of the deserted carriageways to watch the hearse pass (see Davies 1999).

Death and mourning, as well as celebration, become associated with this space of driving, but the M1 is not just 'placed' through the actions and attention surrounding such spectacular and extraordinary events. In my final point, I want to stress that the M1 and other motorways have been, and are, 'placed' through the fairly mundane and repeated practices, movements and actions of hundreds of workers, thousands of local residents and millions of travellers (Edensor 2003; Merriman 2004b). The motorway is a meaningful place of work for a large number of individuals. Company sales representatives and regional managers undertake work while on the move (Laurier 2004). Motorway service area staff, maintenance workers and breakdown patrols work the motorway on a daily basis. M1 police patrols have been the subject of a 'docu-soap' television series, with Carlton Television's *Motorway* following police officers patrolling the M1, M40, M6 and M42. In the wake of a new generation of 'docu-soaps' came a 'docu-soap' parody set on a motorway. In 1998 comedian Peter Kay played five different characters in *The Services*, a comic Channel 4 mock-documentary about workers in a motorway service area on the M61 near Kay's home town of Bolton.

The M1 has become a meaningful site for local residents living near or travelling on the motorway. In Leicester, the nightclub 'Junction 21' appears to have been named after the city's main M1 junction, while the website accompanying BBC Northampton's *Sense of Place* radio documentary presented the M1 not as a placeless motorway, but as 'the backbone of Britain', a significant landscape feature contributing to the economic success of the county (BBC 2002). In Milton Keynes, the Living Archive and the Open University's Institute of Educational Technology have piloted an online local history project entitled 'The coming of the M1 motorway to Newport Pagnell' to demonstrate how local schools can undertake local history projects. Researchers collected film footage of the motorway from local residents as well as interviewing local farmers and others who were affected by construction. The project aimed to demonstrate how a local feature such as the M1 became incorporated into the lives and memories of local residents (Barrett 1999).

Individual motorists and families experience and narrate the spaces of the motorway in a plethora of ways (Pearce 2000; Edensor 2003; Merriman 2004b). Families may have favourite (or preferred) service areas to stop at, landmarks they point out, or signs or bridges they anticipate in order to break up a journey and keep children entertained. Car games may involve features of the motorway and landscape, and booklets like Margaret Baker's *Discovering M1* (1968) instruct passengers on how to read the landscape and anticipate features on the motorway. I remember my own childhood journeys from Kent, along the M2, around London, and up the M1 and M6, to visit relatives in North-West England, stopping at particular service areas, and attempting to earn points by spying motorway features detailed in my copy of *I-Spy on the Motorway* (Big Chief I-Spy 1976).

The practice of driving or being a passenger in particular cars, travelling to and from particular places, along particular stretches of motorway, provokes a range of emotions, thoughts and sensations: from feelings of anxiety or excitement about being in motorway traffic, to emotions surrounding one's departure or arrival at another place. We may get bored, feel strangely alone, or feel quite excited or relaxed on the motorway, but our movements and actions are still implicated in the working and performance of the motorway landscape, the ongoing 'placing' of these driving environments in a myriad of different ways.

Appendix: Archival Sources

The majority of the archive work was undertaken between 1998 and 2000. Since this time, a number of records have been moved (see below).

AUTOMOBILE ASSOCIATION ARCHIVE (AAA), FANUM HOUSE, BASINGSTOKE

Unnumbered files entitled: motorways; motorway signs.
   Miscellaneous photographs of AA activities on the M1

BBC WRITTEN ARCHIVE CENTRE, CAVERSHAM PARK, READING

Microfilms and microfiches of BBC radio scripts.
   M5/155/1, production file of 'Song of a Road'.

BIRMINGHAM CENTRAL LIBRARY, LOCAL STUDIES COLLECTION

Press cuttings books on the London to Birmingham Motorway/M1.

BIRMINGHAM CITY ARCHIVES (BCA), BIRMINGHAM CENTRAL LIBRARY

Charles Parker Archive (MS4000):

Numerous files consulted, particularly:

   MS4000/2/74, series of files on 'Song of a Road';
   MS4000/LC83–88, audio cassettes relating to 'Song of a Road';

MS4000/2/85, file on 'People Today: Owen Williams';
MS4000/2/107, file on 'The Crack'/'The Irishmen';
Video of the 1965 film 'The Irishmen' by Philip Donnellan.

BRITISH LIBRARY OF POLITICAL AND ECONOMIC SCIENCE (BLPES), LONDON

Rees Jeffreys Archive (RJ):

Numerous files consulted, particularly those in Section 16: Motorways.

Pamphlet Collection: Microfilms of 1930s publications by the British Road
Federation.

BRITISH ROAD FEDERATION (BRF), LONDON

Boxes of British Road Federation publications, 1940s–present day.
    Since the winding-up of the British Road Federation in 2005, these
publications have been deposited in the archives of the Institution of Civil
Engineers, London.

CIVIC TRUST LIBRARY, LONDON

Roads Beautifying Association (RBA) Archive (boxes 110–17):

Publications and press cuttings books relating to the work of the
Association.

CTIS LIBRARY, LAING TECHNOLOGY GROUP, HEMEL HEMPSTEAD

Miscellaneous uncatalogued publications, unpublished reports and materials
relating to the opening of the M1; albums of M1 photographs; *Team Spirit*
newsletter, 1946–present day (including card index of contents); unpublished manuscript of 'A Laing company history' by L. T. C. Rolt; Video of
Laing's film *Motorway* (1959); 16 mm films of *Major Road Ahead* (1958) and
Motorway (1959).
    I understand that this library/archive has now been disbanded. Many of
the materials I consulted now form part of the Laing Motorway Archive in
the Northamptonshire County Record Office.

HERTFORDSHIRE COUNTY RECORD OFFICE, HERTFORD

Records relating to the St Albans Bypass motorway (M1, junctions 5 to 10, and M10) are held in the Hertfordshire County Record Office. This includes contracts, some correspondence and photographs.

These records are fairly limited and I do not refer to them in this book due to my focus on the sections of the M1 which were designed by Sir Owen Williams and constructed by John Laing and Son Limited.

INSTITUTION OF CIVIL ENGINEERS (ICE) ARCHIVES, LONDON

Plan chest A, drawer 4: London–Yorkshire Motorway Contract LY/CA drawings. Engineering drawings of the M1 submitted to ICE by Sir Owen Williams and Partners.

Box of miscellaneous reports on the M1 by Sir Owen Williams and Partners.

MUSEUM OF ENGLISH RURAL LIFE, THE UNIVERSITY OF READING

Council for the Preservation of Rural England (CPRE) Archive:

Files in the following series were consulted:

    241   Roads
    27/21   Leicestershire roads 1957–1969
    87/36   Special Roads Act 1949

THE NATIONAL ARCHIVES OF THE UK (TNA), KEW

During the research I consulted over one hundred files in the National Archives. These included files on early motorway proposals, signage, service areas, the landscaping of roads and motorways, and extensive records relating to the planning, design, construction, landscaping, opening, use and regulation of the M1. The exact file and document references are listed in the endnotes, but the majority of the files were in these series:

BP1   Royal Fine Art Commission, minutes of meetings, 1924–99
DSIR12   Department of Scientific and Industrial Research, Road Research Laboratory, 1931–82

MT39   Highways, correspondence and papers 1862–1973

MT95   Ministry of Transport, Highways Engineering, registered files, 1929–81

MT105   Ministry of Transport, Highways Lands Division, registered files, 1930–81

MT113   Ministry of Transport, Road Traffic Division, registered files, 1931–68

MT117   Ministry of Transport, Highways (Trunk Roads) Division, registered files, 1925–72

MT120   Ministry of Transport, Highways General Planning Division, registered files, 1925–80

MT121   Ministry of Transport, Highways Special Roads Division 1942–77. The majority of the files I consulted were in this series.

MT123   Ministry of Transport, Contracts Division, registered files 1937–85

NATIONAL MOTOR MUSEUM TRUST, BEAULIEU, HAMPSHIRE

Papers and publications relating to life and work of John Montagu of Beaulieu.

OWEN WILLIAMS, COMPANY HEADQUARTERS, BIRMINGHAM

Correspondence about the design and construction of the M1 motorway.
Unpublished reports produced by Sir Owen Williams and Partners.
Photographs and slides of the construction and use of the M1 motorway.
Press cuttings books.
'London to Birmingham Motorway during construction at Milton near Northampton', 1958 painting by Terence Cuneo.

ROYAL AUTOMOBILE CLUB, LONDON

Miscellaneous motoring books and journals, including runs of *The Autocar, The Car (Illustrated), The Motor, The Motor Car Magazine* and *The Automotor and Horseless Vehicle Journal.*

A commemorative album of the 1937 visit of the German Roads Delegation.

RUSKIN COLLEGE LIBRARY, OXFORD

Ewan MacColl and Peggy Seeger Archive:

Includes written material and sound recordings relating to the BBC Radio Ballads, including the script and reviews of *Song of a Road*. Much of this material duplicates that held in the Birmingham City Archives.

Notes

1 Geographers have explored a vast array of topics and drawn upon a range of philosophical approaches, from Marxism and phenomenology, to feminism and post-structuralism. Of particular interest, here, is the extensive body of work which has attempted to examine the dynamic and non-representational qualities of particular movements, landscapes, places, buildings, etc. (see, e.g., Thrift 1999; Cloke and Jones 2001; Lees 2001; Cresswell 2002, 2003, 2006; M. Rose 2002; Wylie 2002, 2005, 2006; Hinchliffe 2003; Massey and Thrift 2003; Merriman 2004b, 2005a, 2006a; J. M. Jacobs 2006; M. Rose and Wylie 2006). This is not to mention work by anthropologists, architects, art historians and others which expresses similar intentions.

2 As Nigel Whiteley has pointed out, Banham generalizes the experience of Los Angeles drivers and makes no attempt to highlight the multiple and varied embodied experiences of different kinds of vehicle driver (Whiteley 2002: 234).

3 Cresswell has shown how this 'nomadic metaphysics' was embraced by a range of postmodern, post-structuralist and feminist theorists, from Marc Augé, James Clifford, Edward Said, Iain Chambers and Rosi Braidotti, to Michel de Certeau and Deleuze and Guattari (Cresswell 1997, 2001, 2006). Feminist scholars, in particular, have argued that (gender) differences tend to get eroded in such generalized and frequently uncritical celebrations of mobility and nomadism (see Wolff 1993; Kaplan 1996; Cresswell 2001).

4 Tim Cresswell draws a distinction between 'mobility' and 'movement'. Movement is defined as 'abstracted mobility', while 'mobility is not a simple function in abstract space but a meaningful and power-laden geographical phenomenon' (Cresswell 2001: 20; 2006).

5 My concerns about the use of complexity theories are twofold. Firstly, theorists like Ilya Prigogine (1976) were frequently attempting to develop sophisticated

statistical techniques and provide more 'accurate' and authoritative calculations or approximations of social processes and human behaviour, and in his writings on traffic the qualitative and affective dimensions of driving and decision-making in traffic get lost (see Prigogine and Herman 1971). Secondly, as Böhm et al. (2006) point out, a focus on self-organization and 'autopoiesis' can draw our attention away from the power relations and political and economic interests of the different actors involved with automobility.

6 The majority of literatures on automobility focus on car cultures and car driving, although there is a significant literature on cycling, and there are some notable publications on lorry driving, and motorbike and scooter riding (e.g. Hollowell 1968; Hebdige 1988; McDonald-Walker 2000).

7 Cars, motorcycles, lorries and other vehicles (along with the acts of driving) have long been entwined in particular narratives of masculinity, femininity, domesticity and (largely heterosexual) sexuality. The marketing of four-wheel-drive off-road vehicles and SUVs is particularly instructive in this respect, being embroiled in complex over-lapping discourses of masculinity, family life, exploration, adventure, femininity and safety, as these vehicles have become popular amongst urban professionals doing the school run, as well as off-road enthusiasts (Bishop 1996; Sheller 2004; Böhm et al. 2006).

8 There is a long history of government legislation, educational programmes, governmental techniques and material technologies designed to affect and shape the conduct and movements of vehicle drivers, pedestrians and other road users (see, e.g., O'Connell 1998; Emsley 1993; Joyce 2003; Thrift 2004; Merriman 2005b; Moran 2006).

9 It is not just academics who argue that streets and roads have become placeless non-places. In a 1996 poster the British anti-car group Reclaim the Streets reproduced a quote (without acknowledging the source) from Jane Jacobs' *The Death and Life of Great American Cities* (1961: 352) to emphasize how the spaces and activities of car-dominated roads blur the character of cities, 'until every place becomes more like every other place, all adding up to Noplace' (reproduced in Reclaim the Streets poster, issued at RTS street party, M41 motorway, London, 13 July 1996; author's collection).

10 Early cinematographers developed novel techniques and technologies for invoking the visual experiences, excitement and sensations of movement and speed associated with modes of transport such as the railway. Prominent examples include the 'phantom ride' – where a film was shot using a camera mounted on the front of a railway locomotive – and 'the Hale's Tour' – where viewers watched phantom ride films while seated in a mock railway carriage subjected to jolts, sound effects and even simulated wind (Nead 2004; D. B. Clarke and Doel 2005).

11 Richard Hamilton's motion studies were exhibited at London's Hanover Gallery in 1955. They included a series of four 'Trainsition' paintings – inspired by the writings of James Gibson on visual perception – in which Hamilton painted views from trains travelling between London and Newcastle. His 1954 painting of the view through a car windscreen, entitled *Carapace*, is much less well known (Tate Gallery 1992).

12 In addition to producing experimental photographic motion studies, László
    Moholy-Nagy wrote *Vision in Motion* (1947), in which he discussed changing
    attempts to represent and understand motion and speed from the fourteenth
    century onwards.

13 Motorists may choose a particular route because it is scenic, involves/avoids
    motorways, or they believe it to be the shortest, quickest or least congested
    route. They may be more familiar with one route, or have been advised by a
    friend or satellite navigation system to take that route. The chosen route may
    avoid speed cameras, toll booths or be the only route to the chosen
    destination.

14 Developments in road surfacing, lane markings, lighting, pedestrian crossings,
    traffic islands, bollards, traffic lights, one-way systems, parking meters, crash
    barriers, signage, roundabouts, multi-level interchanges, multi-storey car
    parks, speed cameras and a host of other technologies have all helped to shape
    the spaces and practices of driving (McShane 1999; Merriman 2005b, 2006b;
    Moran 2005, 2006; see also Thrift and French 2002; Joyce 2003; Thrift
    2004).

15 Reclaim the Streets poster, issued at RTS street party, M41 motorway,
    London, 13 July 1996 (author's collection).

16 The last two of these volumes (Bridle and Porter 2002; P. Baldwin and
    Baldwin 2004) were products of the UK's Motorway Archive Trust funded
    by the Institution of Highways and Transportation, Institution of Civil Engi-
    neers, Rees Jeffreys Road Fund, Transport Research Laboratory and corpo-
    rate sponsors from the construction industry. The Motorway Archive Trust
    has deposited materials in archives and county record offices, recorded the
    locations of existing records, and set up a website on the history of Britain's
    motorways (see http://www.iht.org/motorway/).

17 J. B. Jackson pioneered much of this work in his journal *Landscape*
    in the 1950s and 1960s (see Jackson 1997). More recently John Jakle,
    Keith Sculle and the Society for Commercial Archaeology (founded
    in 1977) have continued to focus on the vernacular American roadside land-
    scape, with studies of gas stations, roadside restaurants, parking lots and
    motels.

18 British architects, architectural historians and design historians have been less
    conservative than British landscape historians and industrial archaeologists,
    demonstrating a long-standing interest in the nation's vernacular modern
    landscapes. This is evident in the writings of Reyner Banham. In the past few
    decades architectural historians and preservationists associated with organiza-
    tions such as English Heritage and the Twentieth-Century Society (formerly
    the Thirties Society) have actively worked to record and in some cases list and
    preserve important mid-twentieth-century vernacular modern structures and
    landscapes.

19 Iain Sinclair and Chris Petit also made a film entitled *London Orbital* (2002).
    While Sinclair walked the route of the motorway, Petit chose to drive the
    circuit and film the view through the windscreen. Footage of Petit's drive is
    combined with interviews, shots of Sinclair's 'collaborators', and archive film
    of the opening of the motorway. Scenes switch between single frames and a

split-screen format in which shots relating to Sinclair's encounters and explorations are juxtaposed with film of the unceasingly hypnotic view through Petit's windscreen. The film's soundtrack is comprised of both music and extracts of a radio phone-in about the M25.

20 The 72 miles of motorway were designed and built in two parts. The largest stretch was the 55 mile-long, tarmac-surfaced London to Birmingham section of the London–Yorkshire Motorway (between Luton and Rugby, M1 junctions 10 and 18, and incorporating the M45 Birmingham Spur). Officials variously referred to this as the London to Birmingham Motorway or London to Yorkshire Motorway, although the latter was the more common official name. The motorway was designed by Sir Owen Williams and Partners and constructed by John Laing and Son Limited. In this book I focus on the design and construction of this section of the M1. At the southern end of this section was the concrete-surfaced 17 mile-long St Albans Bypass Motorway (stretching between Watford and Luton, M1 junctions 5 to 10, and incorporating the M10). The St Albans Bypass was first proposed in 1936, and it was designed by Hertfordshire County Council (although Sir Owen Williams and Partners designed a number of the bridges). The Bypass was constructed in two contracts – one by Tarmac Civil Engineering Limited, the other by a consortium of Cubitts, Fitzpatrick and Shand – and it was constructed at the same time as Laing's contracts on the London Yorkshire Motorway.

CHAPTER 2 ENVISIONING BRITISH MOTORWAYS

1 John (later Lord) Douglas-Scott-Montagu of Beaulieu (1866–1929) served as a member of the government's Road Board and wrote regular articles on motoring and roads in *The Times, Country Life* and *Spectator* (see Troubridge and Marshall 1930; Tritton 1985). The Automobile Club, founded in 1897, was the most prominent of a number of 'gentleman's clubs' established by motorists in the 1890s. In 1907 it became the Royal Automobile Club (see Brendon 1997).

2 On attitudes to 'distressed' and 'derelict' areas in the 1920s and 1930s, see Linehan (2000).

3 Abercrombie had expressed concern at the increasing levels of 'ribboning' across the countryside, which was 'a result of the new motor-omnibus services and use of private motor-cars' (Abercrombie 1925: 15).

4 William Rees Jeffreys (1871–1954) was Administrative Secretary to the Royal Automobile Club (RAC), Secretary of the Motor Union (from 1903) and the Road Board (1910–18), and later Chairman of the Roads Improvement Association (*The Times* 1954).

5 *Germany Speaks* is a quite remarkable book which Thornton Butterworth Limited published to provide 'the British public' with 'reliable information on Germany's political and economic aims and aspirations', with the hope of increasing understanding and promoting long-term peace (Thornton Butterworth Ltd 1938: 11). Amongst other topics, the book contains chapters

by German ministers on 'population policy', 'National Socialist racial thought', 'the place of women in the New Germany', 'the essence of "propaganda" in Germany', 'German culture and landscape', 'the colonial problem' and 'Germany and England. What has been: what is: what ought to be'.

6   *Autobahn* construction was only one of the National Socialist strategies to aid the 'motorization' of the German people. In 1934 Hitler indicated his intention to remove the privilege of car ownership by encouraging the production of a popular automobile or *Volkswagen* (Sachs 1992). On British attitudes to German cars and car manufacturing in the 1920s and 1930s, see Koshar (2004).

7   The National Archives (TNA) MT 39/96, Report of German Roads Delegation (1937) by Malcolm Heddle, 5 November 1937, 6.

8   Germany was not alone is seeing road construction as an important step in the unification of nations and empires. Civil and military engineers, colonial administrators and academics had long stressed the importance of modern transport infrastructure (especially roads) in the development and control of Britain's colonies and dominions. As the prominent politician, writer and colonial administrator Lord Hailey wrote in his preface to the third edition of *Road Making and Road Using* by the City Engineer of Delhi, Major T. Salkield: 'The expansion of road communications is the pre-condition of development in most of our colonies; its results alike in the social and economic field have had a significance that can well be described as dramatic' (Hailey 1947: viii). In a letter to the journal *Roads and Road Construction* in March 1942, Rees Jeffreys lamented the British government's neglect of the Empire's roads, and their failure to recognzse 'the importance of roads for defence as well as for development'. Jeffreys wanted a British organization similar to the *Reichsautobahnen*, and the journal titled his letter 'Roads of empire. Wanted – a British "Todt" organisation' (Jeffreys 1942: 57).

9   TNA MT 39/96, 'The German motor roads', undated personal note by Leslie Burgin, c.January 1938. See also the account of Burgin's visit in *The Times* (1938e).

10   TNA MT 39/145, Minute, E. Bull to Chief Engineer, 'London–Birmingham Motorway', 25 October 1937.

11   TNA MT 39/96, Memorandum, F. C. Cook to The Minister, 3 November 1937. Preliminary reconnaissance of the route had been completed by early February 1938. See TNA MT 39/145, Minute, Edward Bull to Chief Engineer, 'Investigation into the alignment of a motorway from London to Preston', 4 February 1938. Details of the route appeared to have been leaked to the public by a member of Buckinghamshire County Council, who addressed a protest meeting at Chalfont St Giles in May 1938 (see *The Times* 1938d).

12   TNA MT 39/556, 'Post war planning and motorways', by F. C. Cook, 14 August 1942. An edited version of Cook's memorandum has been published by the Motorway Archive Trust (Cook 1942).

13   British Library of Political and Economic Science, Rees Jeffreys Archive [hereafter BLPES RJ] 16/3, 'Motorways: statement prepared by

the County Surveyors' Society in connection with their scheme for motorways', by Arthur Floyd (Secretary of the CSS), with letter dated 30 July 1938.

14 BLPES RJ 16/3, 'Motorways: statement prepared by the County Surveyors' Society in connection with their scheme for motorways', by Arthur Floyd (Secretary of the CSS), with letter dated 30 July 1938. See also *The Times* (1938b).

15 Roger Gresham-Cooke was Secretary to the British Road Federation, the German Roads Delegation and the Modern Roads Movement. The Movement planned to take a delegation to the USA in September and October 1939 to tour the roads of New York and Washington DC, but with the outbreak of war in September I doubt whether the tour proceeded (see *The Times* 1939).

16 The Ministry of Transport published their statement in May 1939, but it drew heavily upon a speech by the Minister of Transport at the annual dinner of the Society of Motor Manufacturers and Traders in October 1938. See TNA MT 39/146, 'Extract from the Minister's speech at the annual dinner of the SMMT on October 12th 1938'.

17 In 1934 the Minister of Transport appointed government engineer Sir Charles Bressey and architect Sir Edwin Lutyens to conduct a comprehensive survey and plan for the remodelling of London's roads. The survey took three years to complete, and their *Highway Development Survey 1937 (Greater London)* was eventually published on 16 May 1938. Bressey and Lutyens were in favour of constructing a number of new roads (possibly as motorways) radiating from London: north-east towards Birmingham, north towards Nottingham, south-west towards Basingstoke and south towards Brighton (Bressey and Lutyens 1938).

18 The Institution of Highway Engineers proposed the construction of a national network of 51 motorways, 2826 miles in length. This plan had originally been drafted by the Institution, and forwarded to the Ministry of Transport, in May 1936 (IHE 1943).

19 TNA MT 39/556, 'Post war planning and motorways', by F. C. Cook, 14 August 1942.

20 Ibid.

21 TNA MT 39/556, Map no. 5, Ministry of War Transport, 1942. For a black and white reproduction of the map, see Cook (1942: 143).

22 TNA MT 39/556, Minute [with illegible signature], 3 September 1942.

23 See Council for the Preservation of Rural England Archives, Museum of English Rural Life, University of Reading (hereafter CPRE), 241/16/I, letters from H. G. Griffin to Robert Moses, the Department of Metropolitan Parkways and Herr Dr Schoenichen, 1 January 1937.

24 See 'Road architecture. List of exhibits and detailed instructions', D.522/9, unpublished document bound into the front of a copy of RIBA (1939), held in the Royal Institute of British Architects' Library, London.

25 Rotha had 'long been a co-opted member of the R.I.B.A. Publicity Film Sub-Committee' (*JRIBA* 1939: 401). On Rotha's work see Boon (2000).

26   Dr Fox was 'Consulting Physician for Diseases of the Skin at St. George's
      Hospital and St. John's Hospital, Leicester Square' (*The Times* 1962: 25).
      Later in life he ran his father's firm of general merchants, Duncan, Fox &
      Company, which had eleven offices in Chile, five in Peru, and one each in
      London, Liverpool and New York. Fox also established an arboretum at
      Winkworth Farm in Surrey, which he later gifted to the National Trust.

27   On the history of the Roads Beautifying Association see Spitta (1952); Wells
      (1970); Bassett (1980); Sheail (1981); E. Ford (1994); Merriman (2001,
      2007).

28   TNA MT 39/63, Note to Chief Engineer, 26 June 1935.

29   Lionel Nathan de Rothschild (1882–1942) served as a Unionist Member of
      Parliament between 1910 and 1923 and was a partner in N. M. Rothschild
      and Sons, the well-known banking firm established by his great grandfather
      (de Rothschild 1996). He was a key figure in the Roads Beautifying Associa-
      tion, chairing the Technical Sub-Committee, and co-authoring the RBA's
      170-page book *Roadside Planting* (RBA 1930b).

30   The RBA were paid to act as official advisers to the Ministry of Transport
      between 1938 and 1947, and 1954 and 1956. See TNA MT 121/73, Letter
      from P. Faulkner to W. Fox, 12 May 1947; TNA MT 121/575, Letter from
      A. Lennox-Boyd to W. Fox, 28 July 1954. The AA funded and co-authored
      several of the RBA's early pamphlets, including *Roadside Trees* (AA and RBA
      1930), *The Roadside Halt* (AA and RBA 1935) and *The Highway Beautiful* (AA
      and RBA 1937).

31   CPRE, 241/16/I, Letter, H. G. Griffin to Lord Crawford, 9 July 1928. See
      also Matless (1990).

32   CPRE, 241/16/I, Letter, W. Fox to H. G. Griffin, 8 December 1936.

33   See CPRE, 241/16/I, Letter, H. G. Griffin to Lord Crawford, 15 December
      1936.

34   The RBA had one representative on the Joint Committee and the CPRE had
      12, although the CPRE's representatives were drawn from many of its con-
      stituent organizations. The membership included E. H. Fryer (AA), Guy
      Dawber (RIBA), Patrick Abercrombie (Town Planning Institute), Lawrence
      Chubb (Commons and Footpaths Preservation Society), A. D. C. Le Sueur
      and W. D. Dallimore (Royal English Forestry Association), F. R. Yerbury
      (Architectural Association), George Langley-Taylor and H. G. Griffin (CPRE),
      Professor E. J. Salisbury (University of London) and W. Harding Thompson
      (unaffiliated) (see Trunk Roads Joint Committee 1937).

35   Several paragraphs in the Joint Committee's final report were extracted from
      a letter by E. J. Salisbury to the Committee's chairman, E. H. Fryer, 13 July
      1937 (CPRE 241/16/I).

36   The Institute of Landscape Architects were formed in 1929 after it was decided
      to change the name of the British Association of Garden Architects. The latter
      had been formed in 1928 at a meeting of 30–40 interested people at the
      Chelsea Flower Show (Fricker 1969; Colvin 1979).

37   A number of influential architects, geographers and planner-preservationists
      became members or honorary members of the Institute of Landscape Archi-
      tects (ILA) during World War II, including Patrick Abercrombie, Edwin

Lutyens, Lord Reith, Clough Williams-Ellis, Dudley Stamp and Thomas Sharp. In 1947 there were 170 members of the ILA, with 65 being categorized as horticulturists and 81 as architects or town planners (Fricker 1969).

CHAPTER 3 DESIGNING AND LANDSCAPING THE M1

1   The gendering of the language is striking. One can only assume that women's blood flows were seen to be unimportant or less important to the strength, defence and life of the nation.
2   In the same year as Lloyd's proposals, the British Transport Commission set out its own plans to modernize Britain's railways through an extensive programme of electrification and a focus on inter-city and commuter travel (BTC 1955).
3   TNA MT 39/558, Memorandum, Divisional Road Engineer (Eastern) to Ministry of Transport, 18 April 1945.
4   TNA MT 39/558, 'Quarterly report to Minister – Construction Section', 30 December 1950, and 'Quarterly report to the Minister – Construction Section', 4 April 1951.
5   Sir Owen Williams's partners were Thomas S. Vandy and his son, Owen Tudor Williams (1916–96).
6   TNA MT 123/59, Letter, J. D. W. Jeffery (Ministry of Transport) to Sir Owen Williams, 16 September 1955.
7   TNA MT 117/28, Minute 12, by J. A. Bedford, 23 August 1955.
8   The Special Roads Act (1949: 187) states that 'the Minister shall give due consideration to the requirements of local and national planning, including the requirements of agriculture', when routing a motorway, and the Ministry of Transport agreed that they would consult the Ministry of Agriculture, Fisheries and Food, National Farmers' Union and Country Landowners' Association at an early stage. See TNA MT 121/576, Minute 32 by R. S. Chettoe, 5 May 1955.
9   TNA MT 117/28, 'Objections received during the week ending 24 December 1955, List 9'.
10   TNA MT 117/29, Doc. 41, by T. R. Newman, 22 August 1956.
11   TNA MT 117/29, Minute 8, 5 March 1956. The signature on this minute is difficult to read, although it may be by the Minister, Harold Watkinson.
12   See discussions in TNA MT 117/28.
13   TNA MT 95/503, Letter, T. R. Newman to Messrs Rumball & Edwards, 8 May 1956.
14   On the routing of the Leicestershire sections of the M1, see CPRE 27/21; TNA MT 121/84.
15   Nairn published his own positive response to 'Outrage' in the 'Counter-attack' special issue of *The Architectural Review* in 1956 (Nairn 1956).
16   The choice of trees was decided by the Association of Planning and Regional Reconstruction, the Royal Botanic Gardens at Kew, the Institute of Landscape Architects and the Roads Beautifying Association.

17   Government policy on the landscaping and planting of roads changed
      frequently between 1946 and 1956. Up until 1947 the Roads Beautifying
      Association were paid £200 a year to act as official advisers to the
      Ministry, but in 1946–7 the CPRE succeeded in persuading the Minister
      of Transport to replace them with a Ministry horticulturist (with effect
      from May 1947). The appointment of a young New Zealander as
      horticulturist incensed both the RBA and preservationists in the CPRE. The
      CPRE set up its own committee in 1949, whose main task was
      to prepare a revised version of the Institute of Landscape Architect's
      1946 report, *Roads in the Landscape*, for republication as *The Landscape Treat-
      ment of Roads* (see ILA 1946; Joint Committee 1954). In 1954
      the President of the RBA, the Marquess of Salisbury, succeeded in getting
      the Ministry of Transport to reinstate the RBA as their official advisers,
      and the Association fulfilled this role between July 1954 and December 1956.
      See TNA MT 121/73, Letter from P. Faulkner to W. Fox, 12
      May 1947; TNA MT 121/575, Letter from A. Lennox-Boyd to W. Fox, 28
      July 1954; TNA MT 121/575, 'History of the Roads Beautifying
      Association from formation in 1928 to dismissal by the Minister of
      Transport in 1947', note prepared for Lord Salisbury by the Roads
      Beautifying Association and forwarded to the Ministry of Transport, 4
      December 1951.
18   TNA MT 121/74, 'Advisory Committee on the Landscape Treatment of
      Trunk Roads. Minutes of First Meeting held in Room 6042, Berkeley Square
      House, at 2.30p.m. on 30 April 1956', 17 May 1956.
19   TNA MT 121/81, Letter from W. Fox to Mr Watkinson, 6 September 1956.
      Fox was replaced on the Committee by the RBA's secretary, Madeleine Spitta,
      whose views were more in line with the rest of the Committee.
20   Of course, there were a significant number of other British thinkers and organi-
      sations during the 1930s, 1940s and 1950s who *did* draw links between race,
      landscape, ecology, health and the nation (see Matless 1998).
21   TNA MT 121/576, Minute 15 by Mr Haynes, 23 December 1954.
22   TNA MT 121/577, Letter from Godfrey Samuel to Ministry of Transport and
      Civil Aviation, 20 April 1956; TNA MT 121/182, 'Royal Fine Art Commis-
      sion. Landscaping of Motorways. Note of a meeting held at 5 Old Palace Yard,
      at 12 noon, 5th November 1959'; TNA MT 121/77, Minute 70 by L. S. Mills,
      26 November 1959.
23   TNA MT 121/77, 'Note of a meeting held on the 31st July to discuss the
      appointment of a landscape architect for the London–Yorkshire Motorway',
      10 August 1956.
24   TNA MT 121/77, Letter, A. E. Dale (ILA) to A. H. M. Irwin, 14 August
      1956.
25   Owen Williams Archive, Correspondence file, Letter, A. H. M. Irwin to A. E.
      Dale, 5 November 1956; TNA MT 95/503, Loose minute by T. R. Newman,
      29 August 1956. However, cf. TNA MT 121/74, 'Institute of Landscape
      Architects. Statement by Mr James W. R. Adams as to professional services
      on Landscaping of Roads', 12 November 1956.

26 Long was employed as Consultant on Forestry and Landscape and Clay as a junior forestry officer. See TNA MT 95/503, Letter from Sir Owen Williams to J. D. W. Jeffery, MTCA, 13 February 1957.

27 TNA MT 121/77, 'Motorways: Statement by the Royal Fine Art Commission', HT.33/2/06, 15 April 1959.

28 TNA MT 121/77, Letter, Godfrey Samuel (Sec, RFAC) to MTCA, 18 December 1958.

29 TNA MT 121/77, Minute 56, by C. P. Scott-Malden, 7 April 1961. After consulting with Sylvia Crowe, Jellicoe had suggested that Mr Holiday, Mr Porter, Mr Booth and Mr Edwards should all be approached and interviewed for the job. No forenames are given.

30 TNA MT 121/577, Letter, Godfrey Samuel (Sec, RFAC) to Assistant Chief Engineer (Bridges), MTCA, 24 May 1956; TNA BP 1/11, Minutes of the 358th meeting of the Royal Fine Art Commission, 12 February 1958.

31 The central supports helped to minimize construction costs and enabled the bridges to be built in two separate halves, but a series of fatal collisions with the unprotected columns in the early 1960s led a number of engineers to criticize their design (see *Proceedings* 1961; Sir OWP 1973; chapter 5, below).

32 See TNA MT 121/182, 'Royal Fine Art Commission. Landscaping of Motorways. Note of a meeting held at 5 Old Palace Yard, at 12 noon, 5th November 1959'.

33 In his *Guide to Modern Architecture*, Reyner Banham chose his words more carefully, stating that Sir Owen Williams's early work 'helped to build him an impressive reputation' which was 'now somewhat diminished by his rather unimpressive work on the M1 motorway' (Banham 1962: 62).

34 As Gavin Stamp has argued, Owen Williams 'was, in many ways, a man of the Establishment' and was never a 'conventional avant-garde architect', although his pioneering use of reinforced concrete and glass in the 1920s and 1930s brought him great respect from British and European proponents of the Modern Movement (Stamp 1986: 7). Williams turned down two separate invitations to join the Modern Architectural Research Group (the British group affiliated to the Congrès Internationaux d'Architecture Moderne) (Stamp 1986; Yeomans and Cottam 2001).

35 *The Architects' Journal*'s 'Astragal' was particularly critical of the bridges which were constructed where the London to Birmingham Railway runs alongside the M1 (near Watford Gap). Here, the existing brick-arched bridges passing over the railway were 'cobbled with brickwork' onto the newly built 'standard *in situ* concrete bridges' designed by Sir Owen Williams and Partners (Astragal 1959: 412). Architectural critic John Gloag added further criticisms in a letter published in *The Architects' Journal* three weeks later, stating that 'the English genius for compromise' typified by these 'entertaining' structures had 'produced a new form of Sharawadgi' (Gloag 1959: 523). The term Sharawadgi (or Sharawaggi) was first used in English by Sir William Temple in 1685 to refer to irregular or informal gardens. During the 1940s the editors of *The Architectural Review* used the term 'Sharawaggi' interchangeably with 'the

picturesque' in their calls for town planners to accept compromise and inconsistency, and adopt 'an aesthetic method which is designed to reconcile by various means – contrast, concealment, surprise, balance – the surface antagonisms of shape which a vital democracy is liable to go on pushing up in its architecture in token of its own liveliness' (The Editor 1943: 8). For the antipicturesque and Brutalist critics of *The Architectural Review*'s town planning theories, Sharawaggi became a term of abuse (see Banham 1968), a shorthand for chaotic and ugly design compromises and juxtapositions of the kind witnessed by Gloag and *The Architects' Journal*.

36   TNA MT 123/59, 'Advisory Committee on the Landscape Treatment of Trunk Roads. Minutes of 15th meeting held in Room 6042 Berkeley Square House at 3p.m. on Wednesday 17th July, 1957', LT/M 15, 9 August 1957.

37   TNA MT 123/59, 'Advisory Committee on the Landscape Treatment of Trunk Roads. Note of a meeting of sub-committee on 28th January, 1958', LT/61; 'Advisory Committee on the Landscape Treatment of Trunk Roads. Minutes of 21st meeting held in Room 6042 in Berkeley Square House at 3p.m. on Wednesday 12th February, 1958', LT/M 21; TNA MT 121/78, L. E. Morgan, 'London–Yorkshire Motorway: landscaping proposals. Meeting between representatives of Sir Owen Williams & Partners and Ministry of Transport and Civil Aviation, 27 Feb 58', 7 March 1958.

38   TNA MT 123/59, 'London–Birmingham Motorway. Mr. Williams-Ellis' observations arising from the committee's inspection of the motorway on 21st May, 1959', LT/113. In his autobiography, Williams-Ellis described the 'exciting and stimulating' experience of flying in a helicopter on his first visit to the motorway in 1958 (Williams-Ellis 1978: 19).

39   TNA MT 123/59, 'London–Birmingham Motorway. Mr. Williams-Ellis' observations arising from the committee's inspection of the motorway on 21st May, 1959', LT/113.

40   Sir William Ling Taylor took over as Sir Owen's consultant following the death of Archibald Long on 7 March 1959.

41   See TNA MT 121/34, MT 121/37, MT 121/38, MT 121/39, MT 121/164, MT 121/165, MT 121/166.

42   TNA MT 121/34, Note by Sir Owen Williams and Partners on services at service areas, 10 February 1956.

43   TNA MT 121/33, Minute 1, P. A. Robinson to Mr Gillender, 14 April 1957. The intention was that the major service areas would cover 10 acres and the minor service areas 6½ acres, but the Ministry eventually decided to acquire 10 acres for all sites, unless 'agricultural or other opposition' should arise.

44   TNA MT 121/36, Note on London–Birmingham Motorway service areas, 23 February 1959.

45   TNA MT 121/33, 'Motorways – facilities at service areas. Note of a meeting held by the Ministry on 13th March, 1958', 17 March 1958.

46   TNA MT 121/359, 'Advisory Committee on the Landscape Treatment of Trunk Roads. Meeting of sub-committee held in Berkeley Square House on Wednesday, 24th June 1959. Service areas on London–Birmingham Motorway', LT/123.

47   Ibid. One civil servant, E. Y. Bannard, commented that, 'of our two developers, Blue Boar are the more impressive and enjoy our full confidence. They are a compact, go ahead, rough and ready organisation who regard their concession as a privilege, and the deadline as a challenge. Motorway Services, the Forte-Blue Star combination at Newport Pagnell will probably produce the more polished article in the long-run, but they seem less effective on a rush job'. TNA MT 121/183, 'Service area development', note by E. Y. Bannard, 16 September 1959.

48   TNA MT 121/182, 'Royal Fine Art Commission. Landscaping of Motorways. Note of a meeting held at 5 Old Palace Yard, at 12 noon, 5th November 1959'.

49   TNA MT 121/359, letter by L. S. Mills, 4 April 1960. The President of the RIBA had suggested that Lionel Brett, Sir Hugh Casson, J. W. M. Dudding, Frederick Gibberd, Geoffrey Jellicoe, Sir Leslie Martin, Peter Shepheard and Ralph Tubbs would all be suitable candidates to act as coordinating architects. See TNA MT 121/359, Letter from Godfrey Samuel, 24 March 1960.

50   TNA MT 121/150, 'Advisory Committee on the Landscape Treatment of Trunk Roads. Minutes of 34th meeting held in Room 6042 Berkeley Square House at 3p.m. on Wednesday 8th July, 1959', LT/M 34.

51   Quotation in ibid. Williams-Ellis's unsigned sketch is bound into TNA MT 121/359, after doc.10.

52   TNA MT 121/355, 'Advisory Committee on the Landscape Treatment of Trunk Roads. Minutes of 38th meeting held in Room 6042 Berkeley Square House at 3p.m. on Wednesday 9th December, 1959', LT/M 38.

53   TNA MT 121/182, Letter from G. Langley-Taylor to T. R. Newman, Ministry of Transport, 26 January 1960.

54   TNA MT 121/355, 'ACLTTR. Minutes of 38th meeting', LT/M 38.

55   TNA MT 123/59, 'Advisory Committee on the Landscape Treatment of Trunk Roads. Minutes of 21st meeting held in Room 6042 in Berkeley Square House at 3p.m. on Wednesday 12th February, 1958', LT/M 21.

56   TNA MT 123/59, Letter from E. P. King to J. D. W. Jeffery, 15 July 1959.

57   TNA MT 121/360, 'Advisory Committee on the Landscape Treatment of Trunk Roads. Landscaping of service areas on M1', LT/167, undated, c.July 1960.

58   Ibid.

59   TNA MT 121/355, 'Advisory Committee on the Landscape Treatment of Trunk Roads. Minutes of 45th meeting held in Room 6042, Berkeley Square House at 3p.m. on Wednesday 10th August, 1960', LT/M 45.

60   Sir Colin Anderson (1904–80) served as director, president, chairman or on the committees of a large number of companies and organizations associated with shipping, banking and the arts. Anderson was a partner in the Orient Line and he became well known for commissioning prominent modern designers to style the company's new ships of the 1930s. In later years he acted as Director of Midland Bank, P&O, the Royal Opera House, Chairman of the Trustees of the Tate Gallery, Chairman of the Council of the Royal College

of Art, President of the Design and Industries Association, and Chairman of the Royal Fine Art Commission.

61   TNA MT 121/72, 'Advisory Committee on Traffic Signs for Motorways. Agenda of the first meeting', item 10.

62   TNA MT 121/72, 'Advisory Committee on Traffic Signs for Motorways. Minutes of the 4th meeting held on 27 March 1958'.

63   TNA MT 121/72, 'Advisory Committee on Traffic Signs for Motorways. Minutes of the 7th meeting held on 25 June 1958'.

64   TNA MT 121/72, Press notice issued by the Ministry of Transport and Civil Aviation, 1 December 1958.

65   Serifs are the stylistic projections which appear at the end of the strokes of letters in many fonts. Sans serif type has no stylistic projections.

66   TNA MT 113/46, Letter, B. Crutchley to P. D. Proctor, 4 December 1958.

67   TNA MT 113/46, Letter, D. Kindersley to D. Bowes-Lyon, 17 January 1959.

68   TNA MT 113/46, Letter, C. Anderson to G. Jenkins, 2 February 1959.

69   TNA MT 113/46, Letter, C. Anderson to D. Bowes-Lyon, 2 February 1959.

70   See also TNA MT 113/46, 'Transcript of meeting held at the Design Centre on 6th May 1959'.

71   TNA MT 121/72, 'Advisory Committee on Traffic Signs for Motorways. Minutes of the 13th meeting held on 25 May 1959'.

72   TNA MT 113/46, Letter, D. Bowes-Lyon to C. Anderson, 13 February 1959.

73   TNA MT 113/46, Road Research Laboratory, 'Proposals for an experiment to compare the relative effectiveness of upper and lower case scripts for use on traffic signs', RN/3562/RLM.AWC, by R. L. Moore and A. W. Christie, August 1959.

CHAPTER 4 CONSTRUCTING THE M1

1   On the inaugural ceremony, see TNA MT 121/23.

2   TNA MT 121/23, J. D. W. Jeffery, 'London–Birmingham Motorway and St. Albans By-pass. Notes of meeting held at Hereford Road 27th January, 1958 to discuss proposed Inaugural Ceremony'.

3   Ibid.

4   *Hell Drivers* (1957), Rank/Aqua, was directed by C. Raker Endfield and starred Stanley Baker, Herbert Lom, Patrick McGoohan, Sid James and Sean Connery as lorry drivers paid by the quantity of ballast they transport. The drivers take short cuts, overtake on bends, drive on verges and travel at breakneck speeds along country lanes.

5   On the role of other communications technologies, see *Team Spirit* 1958b; Rolt 1959.

6   Military metaphors and analogies were commonly used in a positive way, but this was also a time when campaigners and commentators were remarking on the negative impact of military activities in the British landscape (see Brown 1955; Gruffudd 1995a; Wright 1995; Woodward 2004).

7   This information is based on interviews the author conducted with Douglas Elbourne (Project Manager, Contract D) on 11 February 2000, and Michael May (Chief Engineer) on 18 February 2000.

8   A copy of the article is held in the press cuttings book in the Owen Williams archives. While the title of the newspaper and date were not noted, it appears alongside other articles from August 1958.

9   The authorship of the Radio Ballads was originally attributed to MacColl and Parker. The production script of *Song of a Road* lists MacColl and Parker as authors, and states that MacColl wrote the lyrics and music, Seeger was responsible for the arrangements and musical direction, and Parker was the producer (see MacColl and Parker 1959). In reality, Seeger had been involved in recording actuality, she had transcribed all of the tapes, arranged the music, directed the performance, and suggested possible songs and actuality sequences (MacColl 1990). In September 1959, Parker wrote to his seniors suggesting that 'it may be that in these circumstances she should be acknowledged as co-author and composer of the final work' (BBC Written Archives Centre M5/155/1, Memo, C. Parker to MR Programming, 3 September 1959). However, it was only with the release of the Radio Ballads on compact disc by Topic Records in 1999 that all three were listed as authors of the eight ballads.

10   BBC Written Archives Centre M5/155/1, Memo, D. Morris to C. Parker, 1 December 1958.

11   Birmingham City Archives [BCA] MS4000/2/74/2/1, Letter, C. Parker to Mr Bird, 15 October 1959.

12   BCA MS4000/2/74/1/1, C. Parker, '"Song of a Road" – Treatment'.

13   BCA MS4000/2/74/1/2, Letter, K. G. Gerrard to C. Parker, 2 December 1958.

14   BCA MS4000/2/74/2/3, C. Parker, '"Song of a Road" – Analysis I: Professional difficulties encountered in collecting material for the programme'. The subject matter of these programmes was indicated in an editorial in the July 1958 issue of *Team Spirit*: 'The B.B.C. has twice recently drawn attention to the long hours that are being worked on the motorway' (*Team Spirit* 1958g: 2).

15   BCA MS4000/2/74/2/3, C. Parker, '"Song of a Road" – Analysis I: Professional difficulties encountered in collecting material for the programme'.

16   Ruskin College, Oxford, Ewan MacColl and Peggy Seeger Archive, Ruskin College, Oxford [EMPSA] file: 'Broadcasts: Radio/Radio Ballads/ MacColl, Ewan: "The Radio Ballads"', Ewan MacColl, 'Song of a Road'.

17   Ibid.

18   BCA MS4000/LC83, Cassette of sound effects.

19   BCA MS4000/LC84, LC85 and LC86, Cassettes of interviews; interview transcripts in BCA MS4000/2/74/1/3. Parker recorded several hours of discussion with Sir Owen Williams on a range of issues including engineering, traffic, architecture, aesthetics and the motorway itself. Parker planned to use the recordings to produce a programme on Sir Owen Williams for the radio series *People Today*, but the poor quality of the recordings – coupled with a lack of

opportunity to re-record – meant that the programme was removed from the broadcasting schedule. See BCA MS4000/2/85.

20   BCA MS4000/LC86, Cassette.

21   BCA MS4000/2/74/1/1, Letter, C. Parker to E. MacColl, 27 July 1959. A copy of Morgan's (1958) article is in BCA MS4000/2/74/1/4.

22   EMPSA file: 'Broadcasts: Radio/Radio Ballads/MacColl, Ewan: "The Radio Ballads"', Ewan MacColl, 'Song of a Road'.

23   See BCA MS4000/2/74/1/4.

24   BCA MS4000/2/74/2/3, C. Parker, '"Song of a Road" – Analysis I: Professional difficulties encountered in collecting material for the programme'.

25   Transcribed from *Song of a Road*, Topic Records CD TSCD 802.

26   BCA MS4000/2/74/1/1, Letter, C. Parker to Joyce, 15 October 1959.

27   BCA MS4000/2/74/1/1, Memo, C. Parker to Art Editor, Radio Times, 15 October 1959.

28   BCA MS4000/2/74/2/1, Memo, C. Parker to S.D. Usherwood, 26 November 1959.

29   BCA MS4000/2/74/2/1, Memo, C. Parker to S.D. Usherwood, 22 December 1959.

30   BCA MS4000/2/74/2/3, C. Parker, '"Song of a Road" – Analysis I: Professional difficulties encountered in collecting material for the programme'.

31   Ibid.

32   MacColl remained a member of the Communist Party at a time when many supporters had left following Khrushchev's 'crushing of the Hungarian Revolt' in 1956 (Mulhern 1996: 30).

33   Working titles for *The Irishmen* included *The Crack* and *The Irish Navvy*. Draft shooting scripts for *The Irishmen* are held in BCA MS 4000/2/107. The film was never shown on television and I viewed a copy held on VHS video in the Birmingham City Archive. Donnellan joined the BBC in 1948, and he met Parker when Parker transferred to the Midland Region in 1954. Donnellan transferred to BBC television in 1958, but the two remained friends, and worked together on a number of films: *The Irishmen* (1965), BD8 (1967), *Shoals of Herring* (1973) (based on the Radio Ballad *Singing the Fishing*), *The Fight Game* (1973) (based on the Radio Ballad), *The Big Hewer* (1972) (based on the Radio Ballad) and *Passage West* (1975). Donnellan and Parker were both leading figures in the West Midlands Gypsy Liaison Group, and after Parker's death Donnellan became a trustee of the Charles Parker Archive Trust. *The Irishmen* was the first of a series of six films which Donnellan made on Irish subjects, all of which exposed the complex relations between Britain and Ireland. Donnellan was born in England and attended public school, but his surname led him to experience anti-Irish prejudice (his grandfather had migrated from Ireland in 1847) (see Pettitt 2000a, 2000b).

34   *Song of a Road* and *The Irishmen* provide complex (and quite different) representations of Irishness, masculinity and Irish men's experiences of migration. In some senses they engage with and reinforce, and in other senses they cut against, contemporary stereotypes.

CHAPTER 5 DRIVING, CONSUMING AND GOVERNING THE M1

1   Automobile Association Archive, Basingstoke [AAA], Motorways file, 'Guide to the motorway' by Automobile Association, 1959. As the Royal Automobile Club's records were not available for consultation when I undertook the research, I focus my attention on the activities of the Automobile Association.

2   AAA, Motorways file, 'Guide to the motorway' by Automobile Association, 1959.

3   The excitement surrounding the new motorway and the symbolism of the opening ceremony was clearly felt by a group of thirty students from Cranfield College of Aeronautics, who broke through the barriers on Sunday, 1 November with eight 'gaily-decorated' cars and four motorcycles. The students had formed a 'Motorway Society' and intended to hold their own opening ceremony, but they were escorted off the carriageways by Laing officials before they could stop and stretch their ceremonial 'tape' of toilet paper across the motorway (*The Guardian* 1959a).

4   Marples served as Minister of Transport during a period of extensive and often controversial change in the British transport system (1959–64). He instructed Colin Buchanan's influential committee to report on the problem of *Traffic in Towns* (Steering Group 1963), and he brought Dr Richard Beeching to the British Transport Commission from ICI (who as chairman of the British Railways Board proposed the closure of 2000 stations and 5000 miles of track nationwide) (BRB 1963; Dutton 2004).

5   AAA, Motorways file, 'Speech of the Rt. Hon. Ernest Marples, Minister of Transport at the opening ceremony of the London-Birmingham Motorway on Monday, November 2nd, at 9.30 am', issued with a Ministry of Transport press release, 2 November 1959.

6   Ibid.

7   Ibid.

8   I have reconstructed these extracts of Ernest Marples's speech using two contemporary newspaper reports (*The Guardian* 1959b; Mennem 1959: 5).

9   The 400 guests included engineers, journalists and civil servants. Ernest Marples, Harold Watkinson, Sir Owen Williams, J. M. Laing, C. H. ffolliott (County Surveyor, Hertfordshire County Council) and J. C. Burman (Chairman, Tarmac Civil Engineering) all spoke at the lunch. See Laing Technology Group, CTIS Library, M1 file, 'Savoy Hotel. John Laing and Son Limited and Tarmac Civil Engineering Limited. Luncheon on the opening of the London–Birmingham Motorway. Lancaster Room. Monday, 2nd November, 1959'.

10 The film *Motorway* was produced by the BBC Midland Region Film Unit and was shown at 9.15 pm on Thursday, 29 October 1959. The film is not on open access to researchers.

11 AAA, Motorways file, Ministry of Transport press release 330A on speech by Harold Watkinson, 2 November 1959. Watkinson had invited Stanley Roberts,

founder of the British School of Motoring and College of Automobile Engineering, to establish the Institute of Advanced Motorists (IAM) in June 1956, and by May 1959 membership had reached 10,000 (Noble 1969).

12 Chief Superintendent John Gott also had the distinction of being captain of the British Motor Corporation rally driving team (*The Motor* 1959f).

13 TNA MT 121/225, 'Police report on the London to Birmingham Motorway', c.18 March 1960.

14 A number of daily motorway situation reports are held in TNA MT 121/225. See Merriman (2005b).

15 AAA, Motorways file, Automobile Association Memo, 'Motorway Services', undated.

16 AAA, Motorways file, Memo, Patrol Services Manager to All Area Secretaries, 'Opening of the London/Birmingham Motorway 2 November, 1959', 29 October 1959.

17 AAA, Motorways file, Memo, Patrol Services Manager to All Area Secretaries, 'Opening of the London/Birmingham Motorway 2 November, 1959', 29 October 1959.

18 AAA, Motorways file, Automobile Association Memo, 'Motorway Services', undated.

19 AAA, Motorways file, 'Motorway course' syllabus, with letter by John Kipps, Public Relations Officer, 8 October 1959.

20 AAA, Motorways file, 'AA records 13,500 M1 breakdowns', Press Release, 2 November 1960. Including drawing of breakdown rates shown in Figure 5.6.

21 The public's fascination with watching motorway traffic continued well into 1960. When *The Times* reported record traffic levels on Easter Sunday 1960, it remarked: 'On the London–Birmingham motorway flyovers and bridges were often crowded with sightseers watching the traffic' (*The Times* 1960b: 4).

22 *Semi-detached* was commissioned in 1961 for a 1962 festival celebrating the opening of Basil Spence's modernist Coventry Cathedral. The play ran for a week. It was directed by Tony Richardson, and its cast included Leonard Rossiter, Gillian Raine and Ian McKellen. Dowlihull appears to be based on Solihull, an affluent town forming part of the Birmingham conurbation (D. Turner 1962).

23 At the time 'M1' was released, in 1960, the Ted Taylor Four were Ted Taylor on clavioline, Bob Rogers on guitar, Teddy Wadmore on bass and Bobby Wilkinson on drums. The 7 inch single of 'M1' was released on Oriole Records, backed with a cover version of Jimmie Davis and Charles Mitchell's 1940 hit 'You are My Sunshine' (Ted Taylor Four 1960).

24 'M1' is a rather unusual, almost comic track. After listening to it I learned that Ted Taylor had gone on to act as musical adviser to *The Benny Hill Show* (Burke 1992). I was not surprised.

25 My descriptions of these toys are based on internet research. Tri-Ang's Minic Motorways sets are catalogued on the website *www.tri-ang.co.uk*. Details of the Marx sets were gleaned from images and descriptions of items for sale at *www.ebay.co.uk* in October 2003.

26 BBC Written Archives Centre (WAC), Schools Microfilm 153/154, 23 June 1959–7 December 1959, 'Current affairs. No.8. The new motorway' by Wynford Vaughan-Thomas, Transmitted 11 November 1959.

27 Ibid.

28 BBC WAC, Schools Microfilm 169, 1 May 1968–28 June 1968, 'Exploration Earth, Unit II: Around Britain. 1. Along the M1', by Paddy Feeny, Transmitted 29 May 1968. The other programmes in the *Around Britain* unit were: 'Euston to Liverpool by Rail' by Judith Chalmers; 'Hovercraft' by Paddy Feeny; and 'London to Birmingham by Canal' by Roger Pilkington.

29 In looking through the national press, motoring press and articles aimed specifically at women motorists in *Vogue*, I could find no explicit discussion of the gendering of motorway driving which paralleled the kinds of discussions printed in the 1910s, 1920s and 1930s (cf. O'Connell 1998).

30 In the book Mike is constructed as a product of the late 1950s, a relatively affluent teenager with a car, who one middle-aged man in the story sees as somewhat unruly: '. . . you teenage yobs have it good these days, don't you?' (Martin 1961: 5).

31 Bruce Carter was the pseudonym of Richard Alexander Hough, a highly successful writer, biographer and publisher who was Director of Hamish Hamilton's Children's Books Limited between 1955 and 1970.

32 Midland Red's coaches were the first coaches in Britain to be licensed to travel at over 30 mph. It was not surprising, then, that the existing record of 5 hours 20 minutes for travelling by coach between London and Birmingham was almost cut in half to 2 hours 51 minutes (*The Guardian* 1959d).

33 The death of two of the Duke of Bedford's predecessors in fairly close succession (1940 and 1953), left the family estate with substantial death duties.

34 The literature on motorway service areas and petrol stations is somewhat limited. In the past decade, design and architectural historians have begun to pay increasing attention to the history of British motorway service areas and petrol stations (Jones 1998; Croft 1999; D. Lawrence 1999). Sociologists have also started to examine the social dimensions of petrol station use (Laegran 2002; Normark 2006), but, as yet, no in-depth studies have emerged of consumption practices in service areas.

35 TNA MT 121/183, 'Motorway service areas', note by E.Y. Bannard, 30 October 1959.

36 TNA MT 121/182, 'Draft code of conduct for service area users', Appendix I attached to letter from T. R. Newman, 'London–Birmingham Motorway. Newport Pagnell and Watford Gap Service Areas', undated but late August 1959.

37 See TNA MT 121/353.

38 TNA MT 121/33, 'Note for the files', by P.A. Robinson, H.M. Division, 5 March 1957.

39 Throughout the twentieth century petrol was commodified, fetishized and entwined with particular narratives of national identity.

40 TNA MT 121/33, 'Motorways – facilities at service areas. Note of a meeting held by the Minister on 13th March 1958', 17 March 1958.

41   When the Ministry of Transport and Civil Aviation consulted organizations about future service area facilities in 1956, opinions varied about the sale of alcohol. The British Road Federation, Caterers' Association of Great Britain and the Director of the Ace Café in North London all felt that licensed premises should be provided for non-commercial drivers. The British Holidays and Restaurants Association felt that no alcohol should be sold, while there was a general consensus that as lorry and coach drivers did not drink and drive, alcohol should not be made available in transport cafés. See TNA MT 121/34.

42   At Trowell service area (situated west of Nottingham on the second section of M1) the operator Mecca created a 'Robin Hood'-style atmosphere, with a 'Sheriff's Restaurant', 'colourful mosaic sculptures, wall decorations, ornaments, shields and even a replica of the Major Oak' (*Evening Post & News* 1967: 13). This medieval English decorative interior was in stark contrast with the modernist exterior designed by Kett and Neve (D. Lawrence 1999).

43   On the wall of Giles's transport café is a poster advertising *Headlight*, a monthly magazine published by the Headlight Drivers' Association which contained news relating to lorry driving, motoring law and other matters. The magazine was read by many lorry drivers in the 1950s and 1960s (see, e.g., Hollowell 1968).

44   TNA MT 121/353, Minute by T. R. Newman, 28 February 1964.

45   TNA MT 120/13, Minute 3, by S. R. Walton, 23 February 1955; Minute 7, 'Research into the Economics of Motorways', by A. E. M. Walter, 18 March 1955.

46   TNA MT 120/13, Minute 3, by S. R. Walton, 23 February 1955.

47   TNA MT 120/13, Letter, J. Boyd-Carpenter to Prof. Gilbert Walker, 8 July 1955.

48   TNA MT 120/13, Minute 65, by A. W. Clarke, 24 August 1959.

49   TNA MT 120/13, Minute 68, by L. S. Mills, 12 October 1959.

50   The named authors of the final report were T. M. Coburn, Michael Beesley and D. J. Reynolds.

51   TNA MT 95/501, Loose minute, by J. G. Smith, 16 November 1959.

52   TNA MT 95/501, 'London–Birmingham Motorway. First report by the Traffic Engineering Committee', by W. F. Adams, 4 February 1960.

53   TNA MT 95/501, 'London–Birmingham Motorway. Second report by the Traffic Engineering Committee', by W. F. Adams, 25 February 1960.

54   See the reports in TNA DSIR 12/152 on the reconstruction of the hard shoulders and the work of the Sub-Committee on Hard Shoulders.

55   See the minutes and reports of both committees in TNA MT 95/501 and TNA MT 95/642.

56   See TNA DSIR 12/474, especially: Letter, W. H. Glanville (RRL) to Secretary of the Ministry of Transport, 10 December 1957; 'Consultation between the Ministry of Transport and Civil Aviation and the Road Research Laboratory on the London–Birmingham Motorway', by the Road Research Laboratory, 30 January 1959; and 'Note of a meeting between W. H. Glanville and Harold Watkinson, 4 February 1959'.

57 With the launch of the new E-type Jaguar in March 1961 British Movietone News released a short film entitled 'Flat out for dollars' in which an E-type is shown driving along the M1, the camera capturing the driver's view through the windscreen, as the speedometer edges upwards from 130 mph towards 140 mph (see British Movietone News, 10/4/1961, story: 82220A, issue number: 1662, *www.movietone.com*). The title of the film reflected the fact that the E-type was aimed at the North American market, and had attracted nearly £11 million of orders from the USA (*The Times* 1961b).

58 Ballard first explored these themes in an exhibition of crashed cars, accompanied by semi-naked female models, at the New Arts Laboratory in London in April 1970 (*New Society* 1970). He subsequently developed these ideas in *The Atrocity Exhibition* (Ballard 1970).

59 The numbers of fatalities increased slightly from 29 in 1958–9 to 39 in 1959–60. The statistics were measured from 2 November to 31 October (*Parliamentary Debates* 1961).

60 There are, of course, complex ethical and political implications and problems which result from both the 'normalization' of accidents within society *and* the labelling of these events as 'normal' or 'inevitable' by academics. For example, this can lead observers to naturalize the interests of powerful social and political actors, drawing our attention away from questions of corporate and governmental social and environmental responsibility.

61 AAA, Motorway Signs file, Press Release issued by the Ministry of Transport, 'Warning signs on M.1 motorway', 21 December 1964.

62 TNA MT 95/501, 'London–Birmingham Motorway. Second report by the Traffic Engineering Committee', by W. F. Adams, 25 February 1960; reprinted in MT 95/501, 'London–Birmingham Motorway. Second report by the Working Party', by J. G. Smith, 25 April 1960.

63 TNA DSIR 12/200, 'Phantoms on the M1 motorway', DSIR/RRL Research note no.RN/3860/VJJ, October 1960, BRL263, by V. J. Jehu.

64 Ibid.

CHAPTER 6 MOTORWAYS AND DRIVING SINCE THE 1960S

1 The 'golden triangle' is formed in the space between motorways M1, M6 and M69, although here, as elsewhere, it is used to refer to surrounding areas.

2 In the first four years the names of service areas were not featured on approach signs, but in November 1961 one civil servant argued that this might enable them to serve as 'landmarks on a fairly monotonous road. . . . As the service areas are intended to replace towns by catering for the needs of travellers so they can replace them as landmarks.' TNA MT 121/72, Minute 15, by Mr Pander, 2 November 1961. The names of service areas (though not the operating companies) were added to approach signs in 1964.

3 Motorway signs are, of course, a key medium/space in which we see references to 'the North'. Despite early criticisms of the size and layout of motorway signs, they have come to be accepted as fairly familiar features of our everyday

landscapes. In 1972, an experimental English rock band were so struck by the aesthetics and poetics of these signs that they named their band after a sign on the A1(M) seen when heading north out of London. The band were called 'Hatfield and the North'.

4   References to 'north of Watford Gap' rarely mention the service area, and commentators more frequently refer to being 'north of Watford', i.e. north of the town of Watford (Hertfordshire) located on the northern edge of London, 50 miles south of Watford Gap (and the village of Watford, Northamptonshire). The town of Watford was frequently referred to as its position on the northern edge of London suggested a certain snobbery, narrow-mindedness and ignorance among metropolitan cultural commentators who refused to acknowledge England's and Britain's cultural life outside of (and particularly north of) London. As I suggested in chapter 3, the geographical significance of Watford Gap has long been recognized (Appleton 1960), but it is difficult to be clear about the longer history of these North/South references or divides. One interesting representation is a map produced by the Doncaster and District Development Council that was reproduced in Gould and White's 1974 book *Mental Maps*. The Council produced 'Ye newe map of Britain' to illustrate 'how Londoners see the North'. On this map, the M1 is portrayed as the only transport link running up the centre of England: from London, through 'Potters Bar (End of Civilisation)' (which is actually located near the A1, 8 miles east of the M1), to Birmingham (Gould and White 1974: 40).

5   The construction of the A40(M) Westway in London between 1966 and 1970, the protests at its opening, and ongoing media coverage of the plight of residents living by the road made the public increasingly aware of the stakes involved with such schemes. Houses alongside the Westway were soon deemed uninhabitable and were demolished. Public opposition to urban road construction quickly grew (Starkie 1982; Duncan 1992; Robertson 2007).

6   The Countess of Dartmouth asked Larkin to cut the fifth stanza of the poem, which criticized the economic motives and polluting practices of companies, and the industrial relocation strategies of government. The stanza was restored in the version of 'Going, going' printed in Larkin's book *High Windows* (Larkin 1974; see Matless 1998).

7   Maitland encounters two people who live in ruined buildings on the island – Jane, a prostitute, and Proctor, a mentally ill vagrant. Maitland tries to get them to help him off the island, but following several days of struggles, the vagrant is killed, Jane leaves, and at the end of the book Maitland remains on the island, planning his next escape attempt.

8   Cross journeys along a succession of dual-carriageway roads and motorways, including the M271, M27, M3, A34, M40, A43, M1, M6 and M62.

9   Julian Opie's comments suggest that it was his familiarity with these motorways, and his repeated journeys along them, that led him to reflect upon and explore particular aesthetic experiences and effects. Opie made regular trips along the M40 from London to his parents' house in Oxfordshire, while he got to know the M1 during regular drives to Luton Airport to catch flights to see his future wife in Geneva (Opie 2002). It is interesting to contrast the relative anonymity of Opie's M40 paintings and computer animations with the

intricately detailed and localized landscapes and surreal events set on and around the M40 in Will Self's short story 'Scale' (Self 1994).

10   When I telephoned English Heritage in October 2000 they stated that listing of the M1's bridges and structures had never been investigated in depth, and I understand this still to be the case at the time of writing.

11   One exception are the articles which have appeared in the press marking the M1's significant 'birthdays' or 'anniversaries': for example, at 20, 21, 25, 30 and 40 after opening. Journalists have tended to construct a golden age of British motorways, a time when motorways were exciting and congestion was rare (see, e.g., Glancey 1999). I suspect that similarly nostalgic and amusing articles will be penned for the fiftieth anniversary of the opening of the M1 in November 2009.

References

Where a reference to a newspaper article does not contain an exact page number, the article was included in a cuttings book in one of three archives. National and local newspaper articles relating to the construction of the M1 are contained in two cuttings books in the offices of Owen Williams. Articles relating to the opening and use of the M1 are in press cuttings books in Birmingham Central Library. Articles relating to the Roads Beautifying Association and roadside planting are held in press cuttings books in the RBA Archive, Civic Trust Library. On these records, see Appendix: Archival Sources.

[AA] The Automobile Association 1959: *Members Handbook 1959–60*. London: AA.

[AA and RBA] The Automobile Association and The Roads Beautifying Association 1930: *Roadside Trees: Their Care and Treatment in the Interests of Safety and Beauty*. London: AA and RBA.

[AA and RBA] The Automobile Association and The Roads Beautifying Association 1935: *The Roadside Halt: A Joint Plea by The Automobile Association and The Roads Beautifying Association*. London: AA and RBA.

[AA and RBA] The Automobile Association and The Roads Beautifying Association 1937: *The Highway Beautiful*. London: AA and RBA.

Abbs, P., Inglis, F., Bantock, G. H., Morris, B., Calthorp, K., Thompson, D., Holbrook, D., Whitehead, F., Ewart Evans, G. and Hall, S. M. 1972: Leaving the BBC [letter]. *The Times*, 15 November, 17.

Abercrombie, P. 1925: Town-planning in rural England (letter). *The Times*, 6 April, 15.

Abercrombie, P. 1945: *Greater London Plan 1944*. London: HMSO.

Abrams, M. 1959: *The Teenage Consumer*. London: The London Press Exchange.

[ACTSM] Advisory Committee on Traffic Signs for Motorways [Ministry of Transport] 1962: *Motorway Signs: Final Report of Advisory Committee on Traffic Signs for Motorways*. London: HMSO.

Adams, J. W. R. 1962: Trunk roads: the work of the Landscape Advisory Committee. *Journal of the Institute of Landscape Architects*, 60, 3–4.

Adams, W. F. 1961: Safety aspects of motorway design. *Traffic Engineering and Control*, 3, 178–81.

Adshead, S. D. 1925: Motor tracks to south coast (letter). *The Times*, 21 April, 15.

Adshead, S. D. 1929: A motor road to Brighton (letter). *The Times*, 11 January, 8.

Aldous, T. 1975: *Goodbye Britain?* London: Book Club Associates.

Allen of Hurtwood, Lady 1943: A review of policy. *Wartime Journal of the Institute of Landscape Architects*, 3, 4–12.

Amery, C. and Cruickshank, D. 1975: *The Rape of Britain*. London: Paul Elek.

Amin, A. and Thrift, N. 2002: *Cities*. Cambridge: Polity.

Andrews, T. 1958: I join Britain's long-distance men [newspaper and exact date unknown, copy held in the Owen Williams archive].

Appleton, J. H. 1960: Communications in Watford Gap, Northamptonshire. *The Institute of British Geographers Publication No. 28: Transactions and Papers 1960*, 215–24.

Appleton, J. H. 1962: *The Geography of Communications in Great Britain*. London: Oxford University Press.

Appleton, J. H. 1975: *The Experience of Landscape*. London: Wiley.

Appleyard, D., Lynch K. and Myer, J. R. 1964: *The View from the Road*. Cambridge MA: The MIT Press.

The Architects' Journal 1959: RIBA and motorways. 129, 613–14.

The Architects' Journal 1969: Keeping the motorway away: Scratchwood service area. 149, 1614–16.

The Architectural Review 1937: Roads supplement. 81, 155–78.

The Architectural Review 1951: Foreword. 110, 73–9.

Armitage, J. 1999a: Paul Virilio: an introduction. *Theory, Culture and Society*, 16(5–6), 1–23.

Armitage, J. 1999b: From modernism to hypermodernism and beyond: an interview with Paul Virilio. *Theory, Culture and Society*, 16(5–6), 25–55.

Arthurs, J. and Grant, I. (eds) 2003: *Crash Cultures*. Bristol: Intellect Books.

Astragal 1959: Not half a good design. *The Architects' Journal*, 130, 412–13.

Astragal 1960: Pull-up for socks? *The Architects' Journal*, 132, 417.

Aubury, N. 1959: Whoosh! 80 mph in a motorway bus. *Daily Sketch*, 3 November.

Augé, M. 1995: *Non-Places: Introduction to an Anthropology of Supermodernity*. London: Verso.

Augé, M. 1996: Paris and the ethnography of the contemporary world. In M. Sherringham (ed.), *Parisian Fields*. London: Reaktion, 175–81.

Augé, M. 1999: *An Anthropology for Contemporaneous Worlds*. Stanford, CA: Stanford University Press.

Augé, M. 2004: An itinerary. *Ethnos*, 69, 534–51.

The Autocar 1959a: Motorway one: new way from the Midlands to London. 111, 30 October, 525–8.

The Autocar 1959b: M1 – the great occasion. 111, 6 November, 582.

The Autocar 1959c: Is your driving advancing? Part 5: on the motorway. 111, 6 November, 560–1.

The Autocar 1959d: A series of posters . . . 111, 20 November, 671.

The Autocar 1959e: Fog accidents. 111, 13 November, 624.

Automotive Products Associated Limited 1959: [Advertisement]. *The Times*, 2 November, 9.

The Automotor and Horseless Vehicle Journal 1902: Motor roads. 6, January, 138.

Baker, M. 1968: *Discovering M1*. Tring, Hertfordshire: Shire Publications.

Baldwin, M. (ed.) 1994: *L. T. C. Rolt: A Bibliography* (Second Edition). Cleobury Mortimer, Shropshire: M & M Baldwin.

Baldwin, P. and Baldwin, R. (eds) 2004: *The Motorway Achievement, Volume 1. The British Motorway System: Visualisation, Policy and Administration*. London: Thomas Telford.

Ballard, J. G. 1970 [1993]: *The Atrocity Exhibition*. London: Flamingo.

Ballard, J. G. 1973: *Crash*. London: Jonathan Cape.

Ballard, J. G. 1974 [1994]: *Concrete Island*. London: Vintage.

Bamford, C. G. and Robinson, H. 1978: *Geography of Transport*. London: MacDonald and Evans.

Banham, R. 1960: The road to ubiquopolis. *New Statesman*, 59, 784 and 786.

Banham, R. 1962: *Guide to Modern Architecture*. London: The Architectural Press.

Banham, R. 1966: *The New Brutalism: Ethic or Aesthetic?* London: The Architectural Press.

Banham, R. 1968: Revenge of the picturesque: English architectural polemics, 1945–1965. In J. Summerson (ed.), *Concerning Architecture*. Harmondsworth: Allen Lane The Penguin Press, 265–273.

Banham, R. 1971 [1973]: *Los Angeles: The Architecture of Four Ecologies*. Harmondsworth: Penguin.

Banham, R. 1972a: New way north. *New Society*, 20, 4 May, 241–3.

Banham, R. 1972b: Big Brum artwork. *New Society*, 21, 13 July, 84–5.

Barrett, C. 1999: The Open University and Living Archive Project Millennium awards scheme. Demonstration web design project: 'The coming of the M1 motorway to Newport Pagnell'. Brief overview. Unpublished document available from the Living Archive, Milton Keynes.

Barry, A. 2001: *Political Machines*. London: Athlone.

Bartram, R. and Shobrook, S. 2001: Body beautiful: medical aesthetics and the reconstruction of urban Britain in the 1940s. *Landscape Research*, 26, 119–35.

Barty-King, H. 1980: *The AA: A History of the First 75 Years of The Automobile Association, 1905–1980*. Basingstoke: The Automobile Association.

Bassett, P. 1980: *A List of the Historical Records of The Roads Beautifying Association*. Birmingham and Reading: Centre for Urban and Regional Studies, University of Birmingham and Institute of Agricultural History, University of Reading.

Baudrillard, J. 1988: *America*. London: Verso.

Baxendale, J. 2001: 'I had seen a lot of Englands': J. B. Priestley, Englishness and the people. *History Workshop Journal*, 51, 87–111.

BBC 2002: The backbone of Britain (accessed 8 March 2007): http://www.bbc.co.uk/northamptonshire/asop/northampton/m1.shtml

BBC Radio for Schools 1968: *Exploration Earth (Summer 1968)*. London: British Broadcasting Corporation.

Beane, G. 1959: Train. *Birmingham Mail*, 2 November.

Beaumont, C. 1903: The roads of the future: motor car ways, and how they will influence trade. *The Car Magazine*, 1(2), 174–83.

Beckmann, J. 2001: Automobility – a social problem and theoretical concept. *Environment and Planning D: Society and Space*, 19, 593–607.

Beckmann, J. 2004: Mobility and safety. *Theory, Culture and Society*, 21(4/5), 81–100.

Belloc, H. 1923: *The Road*. Manchester: The British Reinforced Concrete Engineering Co. Ltd.

Bender, B. 2001: Introduction. In B. Bender and M. Winer (eds), *Contested Landscapes: Movement, Exile and Place*. Oxford: Berg, 1–18.

Berman, M. 1983: *All That is Solid Melts Into Air*. London: Verso.

Betjeman, J. 1960: Men and buildings: style on road and rail. *Daily Telegraph and Morning Post*, 27 June, 15.

Betjeman, J. 1975: Foreword. In T. Aldous, *Goodbye Britain?* London: Book Club Associates, 6–7.

Beynon, H. 1973: *Working for Ford*. Harmondsworth: Allen Lane.

Big Chief I-Spy 1976: *I-Spy on the Motorway*. London: Polystyle Publications.

BIP [British Industrial Plastics] 1959: [Advertisement]. *The Guardian*, 2 November, 9.

Birmingham Mail 1959a: Breakdown rush on M1. 2 November.

Birmingham Mail 1959b: It's like driving in a dream world. 22 October.

Bishop, P. 1996: Off road: four-wheel drive and the sense of place. *Environment and Planning D: Society and Space*, 14, 257–71.

Bishop, P. 2002: Gathering the land: the Alice Springs to Darwin rail corridor. *Environment and Planning D: Society and Space*, 20, 295–317.

Blackwood Hodge 1959: Advertisement. *The Times*, 2 November, 8.

Blizard, G. P. 1926: Midlands to the Mersey. Big motorway scheme. Route via the Potteries. *Cox's Pottery and Glass Trade Year Book* [reprint held in the BLPES HE/31(SPEC)].

Blizard, G. P. 1928: Roads specially reserved for motor cars. *Transport Management*, January [copy held in BLPES Rees Jeffreys Archive 16/4].

Blomley, N. 1994: *Law, Space, and the Geographies of Power*. London: Guilford Press.

Blue Boar 1959: Petrol on the motorway [advertisement]. *Birmingham Post*, 2 November, motorway supplement, iii.

Bode, S. 2004: An English journey. In A. Cross, *An English Journey*. London and Southampton: Film and Video Umbrella and John Hansard Gallery, no pagination.

Böhm, S., Jones, C., Land, C. and Paterson, M. 2006: Introduction: Impossibilities of automobility. In S. Böhm, C. Jones, C. Land and M. Paterson (eds), *Against Automobility*. Oxford: Blackwell Publishing/Sociological Review, 3–16.

Boia, L. 2005: *The Weather in the Imagination*. London: Reaktion.

Boogaart, P. 2000: *A272 – An Ode to a Road*. London: Pallas Athene.

Boon, T. 2000: 'The shell of a prosperous age': history, landscape and the modern in Paul Rotha's *The Face of Britain* (1935). In C. Lawrence and A. K. Mayer (eds), *Regenerating England*. Amsterdam: Rodopi, 107–48.

Boorman, A. G. 1966: The economic justification of motorways – some second thoughts. *Roads and Road Construction*, March, 74–76.

Boumphrey, G. 1939: *British Roads*. London: Thomas Nelson and Sons Ltd.

Boyd Whyte, I. 1995: National socialism and modernism. In D. Britt (ed.), *Art and Power: Europe Under the Dictators 1930–45*. London: Hayward Gallery, 258–69.

Boyes, G. 1993: *The Imagined Village: Culture, Ideology and the English Folk Revival*. Manchester: Manchester University Press.

Bracewell, M. 2002a: *The Nineties: When Surface Was Depth*. London: Flamingo.

Bracewell, M. 2002b: Fade to grey: motorways and monotony. In P. Wollen and J. Kerr (eds), *Autopia: Cars and Culture*. London: Reaktion, 288–92.

[BRB] British Railways Board 1963: *The Reshaping of British Railways. Part 1: Report*. London: HMSO.

Brendon, P. 1997: *The Motoring Century: The Story of the Royal Automobile Club*. London: Bloomsbury.

Bressey, C. 1944: Roads. In British Road Federation, *Roads and Road Transport: Reconstruction of Roads and Development of Road Transport*. London: British Road Federation, 30–47.

Bressey, C. and Lutyens, E. 1938: *Highway Development Survey 1937 (Greater London)*. London: HMSO.

[BRF] British Road Federation 1933: *A Statement by the British Road Federation Supplemental to the Case of Trade and Industry Against What is Known as 'The Salter Report'*. London: British Road Federation.

[BRF] British Road Federation 1937a: *National Motor Roads*. London: British Road Federation.

[BRF] British Road Federation 1937b: *New Roads or Old?* London: British Road Federation.

[BRF] British Road Federation 1937c: *Get a Move On*. London: British Road Federation.

[BRF] British Road Federation 1938: *The Crisis and the Roads*. London: British Road Federation.

[BRF] British Road Federation 1939: *From Here to There*. London: British Road Federation.

[BRF] British Road Federation 1943: Motorways for Britain [exhibition advertisement]. *The Times*, 7 December, 3.

[BRF] British Road Federation 1944: *Roads and Road Transport: Reconstruction of Roads and Development of Road Transport*. London: British Road Federation.

[BRF] British Road Federation 1946: *Roads Exhibition: Organised by the British Road Federation*. London: British Road Federation.

[BRF] British Road Federation 1948: *The Case for Motorways*. London: British Road Federation.

[BRF] British Road Federation 1949: *The Case for Motorways* (Second Edition). London: British Road Federation.

[BRF] British Road Federation 1951a: *Memorandum on Roads and National Defence*. London: British Road Federation.

[BRF] British Road Federation 1951b: *South Wales Roads*. London: British Road Federation.

[BRF] British Road Federation 1952: *New Roads for Lancashire*. London: British Road Federation.

Bridle, R. and Porter, J. (eds) 2002: *The Motorway Achievement: Frontiers of Knowledge and Practice*. London: Thomas Telford.

The British Council 2000: *Landscape*. London: The British Council.

Brodrick, A. H. 1938: The new German motor-roads. *The Geographical Magazine*, 6, 193–210.

Brodsly, D. 1981: *L. A. Freeway: An Appreciative Essay*. London: University of California Press.

Brooks, T. 1959: The hazards of M1. *The Observer*, 8 November, 5.

Brottman, M. (ed.) 2001: *Car Crash Culture*. New York: Palgrave.

Brown, R. D. 1955: *The Battle of Crichel Down*. London: The Bodley Head.

Brunner, C. T. 1948: *The Road Way to Recovery*. London: British Road Federation.

[BTC] British Transport Commission 1955: *Modernisation and Re-equipment of British Railways*. London: British Transport Commission.

Buchanan, C. 1981: *No Way to the Airport*. Harlow: Longman.

The Bucks Standard 1958a: Motorway contractors 'very helpful'. 18 October.

The Bucks Standard 1958b: Motorway film show. 20 December.

The Bucks Standard 1958c: It's like a military operation on . . . the helicopter highway. 19 July.

Bugge, C. 2004: 'Selling youth in the age of affluence': marketing to youth in Britain since 1959. In L. Black and H. Pemberton (eds), *An Affluent Society? Britain's Post-War 'Golden Age' Revisited*. Aldershot: Ashgate, 185–202.

Bugler, J. 1966: The lorry men. *New Society*, 8, 4 August, 181–4.

Bull, M. 2001: Soundscapes of the car: a critical ethnography of automobile inhabitation. In D. Miller (ed.), *Car Cultures*. Oxford: Berg, 185–202.

Bull, M. 2004: Automobility and the power of sound. *Theory, Culture and Society*, 21(4–5), 243–59.

Bullock, N. 2002: *Building the Post-War World*. London: Routledge.

Burke, D. 1992: The Ted Taylor Four story: an interview with Ted Taylor and Bob Rogers. *Pipeline: Instrumental Review*, 12, January, 1–10.

Burns, M. (ed.) 1999: *George Thomson in Ireland and Birmingham: Exhibition on Floor 6, Birmingham Central Library Until Thursday 30th September* (Exhibition Booklet). Birmingham: Birmingham City Council, Department of Leisure and Community Services.

Buxton, Lord 1929: London to Brighton motorway: the suspended bill. *The Times*, 15 February, 15.

Calder, A. 1969 [1992]: *The People's War: Britain 1939–1945*. London: Pimlico.

Callon, M. 1998a: Introduction. In M. Callon (ed.), *The Laws of the Market*. Oxford: Blackwell/The Sociological Review, 1–57.

Callon, M. 1998b: An essay on framing and overflowing: economic externalities revisited by sociology. In M. Callon (ed.), *The Laws of the Market*. Oxford: Blackwell/The Sociological Review, 244–69.

Campbell, E. 1928: Roadside tree planting: art *versus* utility. *The Queen*, 19 December.

Campbell, L. 1996: *Coventry Cathedral: Art and Architecture in Post-War Britain*. Oxford: Clarendon Press.

Cardew, B. 1959a: Marples gets a scare on the M1. *Daily Express*, 3 November.

Cardew, B. 1959b: How did it all happen? *Daily Express*, 7 November, 4.

Carkeet, J. and Cooper, D. 1929: London to Brighton motorway: the engineers' reply (letter). *The Times*, 20 February, 10.

Carrington, N. 1939: *The Shape of Things*. London: Nicholson and Watson Limited.

Carrington, N. 1959: Signs for Preston motorway. *The Times*, 20 March, 13.

Carrington, N. 1976: *Industrial Design in Britain*. London: George Allen and Unwin.

Carter, B. 1961: *The Motorway Chase*. London: Hamish Hamilton Ltd.

Castle, B. 1990: *The Castle Diaries 1964–1975*. London: Papermac.

Caterpillar 1959: [Advertisement]. *The Times*, 2 November, 5.

Causey, A. 1985: The late landscapes. In *Edward Burra, Hayward Gallery, 1 August–29 September 1985*. London: Arts Council of Great Britain, 52–7.

Charlesworth, G. 1984: *A History of British Motorways*. London: Thomas Telford.

Christie, A. W. and Rutley, K. S. 1961: Lettering and legibility: research on road signs. *Design*, 152, 59–60.

Chronicle and Echo [Northampton] 1958a: Motorway men's pay grabbed. 5 December.

Chronicle and Echo 1958b: Motorway firm's appeal allowed. 11 September.

Chronicle and Echo 1958c: Lorries 'speed' in narrow lanes. 18 September.

Chronicle and Echo 1958d: Motorway rumours scotched. 26 November.

Chronicle and Echo 1958e: Motorway code. 26 November.

Chubb, L. 1925: Motorways to the coast [letter]. *The Times*, 22 April, 15.

Clark, H. F. 1951: Landscape architecture in the Festival of Britain. *Journal of the Institute of Landscape Architects*, 21, 2–4.

Clarke, D. B. and Doel, M. A. 2005: Engineering space and time: moving pictures and motionless trips. *Journal of Historical Geography*, 31, 41–60.

Clarke, F. 1937: Germany as I saw it. *Anglo-German Review*, December, 9–10.

Clements, R. G. H. 1937: *The System of Motor Roads in Germany: A Record of Facts and of Technical Details*. London: German Roads Delegation.

Cloke, P. and Jones, O. 2001: Dwelling, place, and landscape: an orchard in Somerset. *Environment and Planning A*, 33, 649–66.

Cohan, S. and Hark, I. R. (eds) 1997: *The Road Movie Book*. London: Routledge.

Collens, G. and Powell, W. (eds) 1999: *LDT Monographs No. 2: Sylvia Crowe*. Reigate: Landscape Design Trust.

Colvin, B. 1939: Roadside planting in country districts. *Landscape and Garden*, 6(2), 86–8.

Colvin, B. 1947: Introduction. In The Association for Planning and Regional Reconstruction, *Trees for Town and Country*. London: Lund Humphries, 5–7.

Colvin, B. 1948: *Land and Landscape*. London: John Murray.

Colvin, B. 1959: The London–Birmingham motorway: a new look at the English landscape. *The Geographical Magazine*, 32, 239–46.

Colvin, B. 1979: 1929–1979 ILA–LI golden jubilee: beginnings. *Landscape Design*, 125, 8.

Conekin, B. 2003: *The Autobiography of a Nation: The 1951 Festival of Britain*. Manchester: Manchester University Press.

Conekin, B., Mort, F. and Waters, C. (eds) 1999: *Moments of Modernity: Reconstructing Britain 1945–1964*. London: Rivers Oram Press.

Cook, F. 1942 [2004]: Post war planning and motorways. In P. Baldwin and R. Baldwin (eds), *The Motorway Achievement, Volume 1. The British Motorway System: Visualisation, Policy and Administration*. London: Thomas Telford, 107–44.

Cooter, R. and Luckin, B. 1997: Accidents in history: an introduction. In R. Cooter and B. Luckin (eds), *Accidents in History*. Amsterdam: Rodopi, 1–16.

Cottam, D. 1986a: Selected projects. In G. Stamp (ed.), *Sir Owen Williams 1890–1969*. London: The Architectural Association, 33–143.

Cottam, D. 1986b: Chronology. In G. Stamp (ed.), *Sir Owen Williams 1890–1969*. London: The Architectural Association, 149–74.

Crang, M. 2002a: Between places: producing hubs, flows and networks. *Environment and Planning A*, 34, 569–74.

Crang, M. 2002b: Rethinking the observer: film, mobility, and the construction of the subject. In T. Cresswell and D. Dixon (eds), *Engaging Film: Geographies of Mobility and Identity*. London: Rowman and Littlefield, 13–31.

Crang, P. 2000: Organisational geographies: surveillance, display and the spaces of power in business organisation. In J. Sharp, P. Routledge, C. Philo and R. Paddison (eds), *Entanglements of Power*. London: Routledge, 204–218.

Crary, J. 1999: *Suspensions of Perception: Attention, Spectacle and Modern Culture*. London: The MIT Press.

Crawford and Balcarres, Earl of 1936: Roads and rural beauty – work of CPRE – a federation of societies. *The Times*, 18 December, 15.

Cresswell, T. 1993: Mobility as resistance: a geographical reading of Kerouac's *On the Road*. *Transactions of the Institute of British Geographers*, 18, 249–62.

Cresswell, T. 1997: Imagining the nomad: mobility and the postmodern primitive. In G. Benko and U. Strohmayer (eds), *Space and Social Theory: Interpreting Modernity and Postmodernity*. Oxford: Blackwell, 360–79.

Cresswell, T. 2001: The production of mobilities. *New Formations*, 43, 11–25.

Cresswell, T. 2002: Introduction: Theorizing place. In G. Verstraete and T. Cresswell (eds), *Mobilizing Place, Placing Mobility*. Amsterdam: Rodopi, 11–32.

Cresswell, T. 2003: Landscape and the obliteration of practice. In K. Anderson, M. Domosh, S. Pile and N. Thrift (eds), *Handbook of Cultural Geography*. London: Sage, 269–81.

Cresswell, T. 2006: *On the Move: Mobility in the Modern Western World*. London: Routledge.

Croft, C. (ed.) 1999: *On the Road: The Art of Engineering in the Car Age*. London: The Architectural Foundation.

Cross, A. 2004: *An English Journey* [includes the DVD, *3 Hours From Here*]. London and Southampton: Film and Video Umbrella and John Hansard Gallery.

Crow, T. 1990: Saturday disasters: trace and reference in early Warhol. In S. Guilbaut (ed.), *Reconstructing Modernism: Art in New York, Paris and Montreal 1945–1964*. London: The MIT Press, 311–31.

Crowe, S. 1956: *Tomorrow's Landscape*. London: The Architectural Press.

Crowe, S. 1958: *The Landscape of Power*. London: The Architectural Press.

Crowe, S. 1959a: Roads through the landscape. *The Times*, 20 May, 11.

Crowe, S. 1959b: The London/York motorway: a landscape architect's view. *The Architects' Journal*, 130, 10 September, 156–61.

Crowe, S. 1960: *The Landscape of Roads*. London: The Architectural Press.

Crowe, S. 1962: The landscape of roads. *The Journal of the Institution of Highway Engineers*, 9, 217–26.

Crowe, S. 1966: *Forestry in the Landscape* (Forestry Commission booklet no. 18). London: HMSO.

Crowe, S., Nairn, I., Boynton, J. K. and Adams, J. 1956: Symposium on subtopia. *Journal of the Institute of Landscape Architects*, 35, 2–9.

Crutchley, B. 1959: Signs on Preston motorway. *The Times*, 17 March, 11.

Cullen, G. 1956: Alphabet or image. *The Architectural Review*, 120, 240–7.

Cullen, G. 1961: *Townscape*. London: The Architectural Press.

Cuneo, T. 1977: *The Mouse and His Master: The Life and Work of Terence Cuneo*. London: New Cavendish Books.

Cuneo, T. 1984: *The Railway Painting of Terence Cuneo*. London: New Cavendish Books.

Curnock, G. C. 1944: *New Roads for Britain: A Practical Plan for the Solution of a Great Post-War Problem*. London: British Road Federation.

Daily Express 1928: Ugly arterial roads, and a move to beautify them. 26 July.

Daily Express 1959a: The hovering eye above the M1. 20 October, 5.

Daily Express 1959b: When a tyre bursts on a coach at 70 mph. 3 November.

Daily Express 1960: The phantom of the M1. 21 June, 13.

Daily News 1928: Mrs Ashley and the trees. 29 December.

Daily Telegraph 1958: Lorry short cuts. 12 September.

Daily Telegraph 1959a: 100 breakdowns on first motorway day – Mr Marples 'frightened' by early users. 3 November, 1.

Daily Telegraph 1959b: 75 mph coach keeps same time as train. 3 November.

Daily Telegraph 1979: EMI apologises for food song. 25 January, 11.

Dallas, K. 1989: MacColl – the man, the myth, the music. *English Dance and Song*, 51(4), 11–14.

Daniels, S. 1985: Images of the railway in nineteenth century paintings and prints. In Nottingham Castle Museum, *Train Spotting: Images of the Railway in Art*. Nottingham: Nottingham Castle Museum, 5–19.

Daniels, S. 1993: *Fields of Vision*. Cambridge: Polity.

Daniels, S. 1996: On the road with Humphry Repton. *Journal of Garden History*, 16, 170–91.

Daniels, S. 1999: *Humphry Repton: Landscape Gardening and the Geography of Georgian England*. London: Yale University Press.

Daniels, S. and Rycroft, S. 1993: Mapping the modern city: Alan Sillitoe's Nottingham novels. *Transactions of the Institute of British Geographers*, 18, 460–80.

Dannefer, W. D. 1977: Driving and symbolic interaction. *Sociological Inquiry*, 47(1), 33–8.

Dant, T. 2004: The driver-car. *Theory, Culture, and Society*, 21(4–5), 61–79.

Dant, T. and Martin, P. J. 2001: By car: carrying modern society. In J. Gronow and A. Warde (eds), *Ordinary Consumption*. London: Routledge, 143–57.

Davies, D. 1999: The week of mourning. In T. Walter (ed.), *The Mourning for Diana*. Oxford: Berg, 1–18.

Dawson, R. F. F. 1961: The effect of the M1 on journey time and fuel consumption. *Traffic Engineering and Control*, 3, 406–8 and 411.

de Certeau, M. 1984: *The Practice of Everyday Life*. London: University of California Press.

de Hamel, B. [Department of the Environment] 1976: *Roads and the Environment*. London: HMSO.

de Maré, E. 1956: Notes on books: *To-morrow's Landscape*. By Sylvia Crowe. *Journal of the Royal Society of Arts*, 105, 121–2.

de Rothschild, L. 1996: Exbury. In M. Rothschild, K. Garton and L. de Rothschild, *The Rothschild Gardens*. London: Gaia Books Limited, 48–68.

de Rothschild, L. N. 1936a: Roads and rural beauty – risk of overlapping societies – the Minister's advisers. *The Times*, 17 December, 17.

de Rothschild, L. N. 1936b: The defence of amenities – overlapping of societies. *The Times*, 30 December, 6.

de Wolfe, I. 1949: Townscape. *The Architectural Review*, 106, 354–62.

Dean, M. 1999: *Governmentality*. London: Sage.

Denselow, R. 1996: MacColl, Ewan (1915–1989). In C. S. Nicholls (ed.), *The Dictionary of National Biography, 1986–1990*. Oxford: Oxford University Press, 270–1.

Dery, M. 2006: 'Always crashing in the same car': a head-on collision with the technosphere. In S. Böhm, C. Jones, C. Land, and M. Paterson (eds), *Against Automobility*. Oxford: Blackwell Publishing/Sociological Review, 223–39.

Design 1959: Which signs for motorways? 129, 28–32.

Development and Road Improvement Funds Act 1909. In *The Public General Acts, 9 Edward VII*. London: HMSO, Chapter 47, 215–26.

DIA [Design and Industries Association] 1930: *The Village Pump: A Guide to Better Garages*. London: Sidgwick & Jackson Ltd.

Dimendberg, E. 1995: The will to motorization: cinema, highways, and modernity. *October*, 73, 91–137.

[DIRFT] Daventry International Rail Freight Terminal 2006: DIRFT Logistics 2. Location. Website (accessed 8 March 2007), http://www.dirft.com/location.asp

Divall C. and Revill G. 2005: Cultures of transport: representation, practice and technology. *Journal of Transport History*, 26, 99–111.

Dixon, D. 1961: M1 squad: progress on the Bedfordshire motorway experiment. *The Motor Cycle*, 106, 2 February, 132–4.

Doubleday, E. H. [County Planning Officer] 1951: *Hertfordshire: Survey Report and Analysis of County Development Plan 1951*. Hertford: Hertfordshire County Council.

Drake, J., Yeadon, H. L. and Evans, D. I. 1969: *Motorways*. London: Faber and Faber.

Duncan, A. 1992: *Taking on the Motorway: North Kensington Amenity Trust 21 Years*. London: Kensington and Chelsea Community History Group.

Dunning, B. 1966: The impact of the motorways. *Country Life*, 20 October, 978–9.

Dutton, D. J. 2004: Marples, (Alfred) Ernest, Baron Marples (1907–1978). *Oxford Dictionary of National Biography*. Oxford: Oxford University Press (accessed online, 8 March 2007), http://www.oxforddnb.com/view/article/31411

Eason Gibson, J. 1959: The pros and cons of M1. *Country Life*, 3 December, 1089.

Eason Gibson, J. 1961: Problems of motorway cruising. *Country Life*, 22 June, 1490.

The Economist 1960: The motorway: first lessons. 195, 14 May, 665.

Edensor, T. 2003: M6 – Junction 19–16: Refamiliarizing the mundane roadscape. *Space and Culture*, 6(2), 151–68.

Edensor, T. 2004: Automobility and national identity: representation, geography and driving practice. *Theory, Culture and Society*, 21, 101–20.

The Editor 1943: Exterior furnishing or Sharawaggi: the art of making urban landscape. *The Architectural Review*, 95, 3–8.

Ellacott, S. E. 1968: *A History of Everyday Things in England, Volume V: 1914–1968*. London: B. T. Batsford Ltd.

Ellis, H. F. 1959: M1 for murder. *Punch*, 28 October, 362–3.

Emsley, C. 1993: 'Mother, what *did* policemen do when there weren't any motors?' The law, the police and the regulation of motor traffic in England, 1900–1939. *The Historical Journal*, 36, 357–81.

The Engineer 1959: Motorways and vehicle design. 208, 13 November, 583–4.

Enright, T. 1988: George Thomson: a memoir. In G. Thomson, *Island Home: The Blasket Heritage*. Dingle, Co. Kerry: Brandon Book Publishers.

Erasmus, N. J. de W. 1958: Distant eye on the motorway. *Team Spirit: The Monthly News Sheet Issued by John Laing and Son Limited*, 145, 7.

Estates Times 1993: Survey: M1 corridor. 3 December, 10–34.

Evans, H. 1981: *Downing Street Diary: The Macmillan Years 1957–1963*. London: Hodder and Stoughton.

Evening Post & News [Nottingham] 1967: Robin Hood flavour at M1 service area. 23 March, 13.

Eyerman, R. and Löfgren, O. 1995: Romancing the road: road movies and images of mobility. *Theory, Culture and Society*, 12, 53–79.

Fairbrother, N. 1970 [1972]: *New Lives, New Landscapes*. Harmondsworth: Penguin.

Fairlie, H. 1959 [1962]: The B.B.C. In H. Thomas (ed.), *The Establishment: A Symposium*. London: The New English Library Limited, 175–92.

Featherstone, M. 1998: The *Flâneur*, the city and virtual public life. *Urban Studies*, 35, 909–25.

Featherstone, M. 2004: Automobilities: an introduction. *Theory, Culture, and Society*, 21(4–5), 1–24.

Field, R. 1958: Rain stops the motorway machines – view from air of brown scar shows idleness. *Chronicle and Echo*, 4 June.

Fisher, T. 1986: *Charles Parker: Aspects of a Pioneer*. Birmingham: The Charles Parker Archive.

Ford 1959: [Advertisement for Thames Articulated Trader]. *The Times*, 3 November, 8.

Ford, E. 1994: Byways revisited. *Landscape Design*, 234, 34–8.

Forshaw, J. H. and Abercrombie, P. 1943: *County of London Plan*. London: Macmillan and Company.

Forte, C. 1986: *Forte: The Autobiography of Charles Forte*. London: Sidgwick and Jackson.

Foster, C. D. 1999: Michael Beesley [obituary]. *The Guardian*, 8 October, 24.

Foster, H. 1996: Death in America. *October*, 75, 37–59.

Fothergill, C. 1959: I'm so scared! Marples threatens speed limit on the super-road. *Daily Sketch*, 3 November.

Foucault, M. 1985: *The History of Sexuality, Volume 2: The Uses of Pleasure*. Harmondsworth: Penguin.

Foucault, M. 1986a: Space, knowledge, and power. In P. Rabinow (ed.), *The Foucault Reader*. Harmondsworth: Penguin, 239–56.

Foucault, M. 1986b: *The History of Sexuality, Volume 3: The Care of the Self*. Harmondsworth: Penguin.

Foucault, M. 1988: Technologies of the self. In L. H. Martin, H. Gutman and P. H. Hutton (eds), *Technologies of the Self: A Seminar With Michel Foucault*. London: Tavistock, 16–49.

Foucault, M. 1991: Governmentality. In G. Burchell, C. Gordon and P. Miller (eds), *The Foucault Effect: Studies in Governmentality*. Hemel Hempstead: Harvester Wheatsheaf, 87–104.

Foucault, M. 1997: Subjectivity and truth. In P. Rabinow (ed.), *Ethics, Subjectivity and Truth: The Essential Works of Michel Foucault 1954–1984, Volume One*. London: Allen Lane/The Penguin Press, 87–92.

Fox, W. 1944: Roadside planting (including post-war suggestions). *Journal of the Royal Horticultural Society*, 69, 231–9.

Freeman, M. 2001: Tracks to a new world: railway excavation and the extension of geological knowledge in mid-nineteenth-century Britain. *British Journal for the History of Science*, 34, 51–65.

Freund. P. and Martin, G. 1993: *The Ecology of the Automobile*. Montréal: Black Rose Books.

Fricker, L. J. 1969: Forty years a growing. *Journal of the Institute of Landscape Architects*, 86, 8–15.

Friedberg, A. 1993: *Window Shopping: Cinema and the Postmodern*. London: University of California Press.

Friedberg, A. 2002: Urban mobilities and cinematic visuality: the screens of Los Angeles – endless cinema or private telematics. *Journal of Visual Culture*, 1(2), 183–204.

Fyfe, N. (ed.) 1998: *Images of the Street*, London: Routledge.

Gandy, M. 2002: *Concrete and Clay: Reworking Nature in New York City*. London: The MIT Press.

Gardiner, J. 1999: *From the Bomb to the Beatles*. London: Collins and Brown.

The Geographical Journal 1986: The M25 – a new geography of development [special theme section]. 152(2), 155–75.

The Geographical Magazine 1938: [Untitled editor's summary of Brodrick's 'The new German motor roads']. 6, 193.

Geological Survey of Great Britain 1959: *Summary of Progress of the Geological Survey of Great Britain and the Museum of Practical Geology for the Year 1958*. London: HMSO.

Giddens, A. 1990: *The Consequences of Modernity*. Cambridge: Polity.

Gilbert, D., Matless, D. and Short, B. (eds) 2003: *Geographies of British Modernity*. Oxford: Blackwell.

Giles 1959: Cartoon. *Daily Express*, 5 November.

Gilroy, P. 2001: Driving while black. In D. Miller (ed.), *Car Cultures*. Oxford: Berg, 81–104.

Glancey, J. 1992: A bridge too far? *The Independent Magazine*, 18 July, 24–31.

Glancey, J. 1999: Motorway madness. *The Guardian*, 19 October, G2 section, 2–3.

Glendinning, M. and Muthesius, S. 1994: *Tower Block: Modern Public Housing in England, Scotland, Wales and Northern Ireland*. London: Yale University Press.

Gloag, J. 1959: Two bridges. *The Architects' Journal*, 130, 523.

Goffman, E. 1971: *Relations in Public*. Harmondsworth: Allen Lane/The Penguin Press.

Gold, J. R. 1997: *The Experience of Modernism: Modern Architects and the Future City 1928–1953*. London: E. & F. N. Spon.

Goldberg, N. L. 1978: *John Crome the Elder: I – Text and Critical Catalogue*. Oxford: Phaidon Press.

Goldthorpe, J. H., Lockwood, D., Bechhofer, F. and Platt, J. 1968–9: *The Affluent Worker* (3 Volumes). Cambridge: Cambridge University Press.

Gordon, C. 1991: Governmental rationality: an introduction. In G. Burchell, C. Gordon and P. Miller (eds), *The Foucault Effect: Studies in Governmentality*. London: Harvester Wheatsheaf, 1–51.

Gott, J. 1971: The police and operational systems. In J. S. Davis (ed.), *Motorways in Britain: Today and Tomorrow*. London: The Institution of Civil Engineers, 91–6.

Gould, P. and White, R. 1974: *Mental Maps*. Harmondsworth: Penguin Books.

Graham, P. 1983: *A1 The Great North Road*. Bristol: Grey Editions.

Graham, S. and Marvin, S. 2001: *Splintering Urbanism*. London: Routledge.

Graves-Brown, P. 1997: From highway to superhighway: the sustainability, symbolism and situated practices of car culture. *Social Analysis*, 41(1), 64–75.

[GRD] German Roads Delegation 1938: *Report upon the Visit of Inspection and Its Conclusions*. London: German Roads Delegation.

Greaves, S. 1985: Motorway nights with the stars. *The Times*, 14 August, 8.

Green, G. R. 1961: First year's traffic on M1. *Traffic Engineering and Control*, 2, 710–11.

Green, R. 2004: *Destination Nowhere: A South Mimms Motorway Service Station Diary*. London: Athena Press.

Griffiths, R. 1983: *Fellow Travellers of the Right: British Enthusiasts for Nazi Germany 1933–9*. Oxford: Oxford University Press.

Gröning, G. 1992: The feeling of landscape – a German example. *Landscape Research*, 17, 108–15.

Gruffudd, P. 1995a: Remaking Wales: nation-building and the geographical imagination, 1925–50. *Political Geography*, 14, 219–39.

Gruffudd, P. 1995b: 'Propaganda for seemliness': Clough Williams-Ellis and Portmeirion, 1918–1950. *Ecumene*, 2, 399–422.

Gruffudd, P., Herbert, D. T. and Piccini, A. 2000: In search of Wales: travel writing and narratives of difference, 1918–50. *Journal of Historical Geography*, 26, 589–604.

The Guardian 1959a: Students in the way. 2 November.

The Guardian 1959b: Mr Marples appalled by motorway speeds. 3 November.

The Guardian 1959c: Mad rush on motorway. 3 November.

The Guardian 1959d: Buses break a speed barrier. 3 November.

[GWR] Great Western Railway 1924: *Through the Window. Number One: Paddington to Penzance*. London: The Great Western Railway.

Hailey, Lord 1947: An appreciation. In T. Salkield, *Road Making and Road Using* (Third Edition). London: Sir Isaac Pitman and Sons, vii–viii.

Halprin, L. 1966: *Freeways*. New York: Reinhold Publishing Corporation.

Halprin, L. 1986: *Lawrence Halprin: Changing Places*. Exhibition Catalogue ed. L. Creighton Neall. San Francisco: San Francisco Museum of Modern Art.

Hannam, K., Sheller, M. and Urry, J. 2006: Editorial: Mobilities, immobilities and moorings. *Mobilities*, 1, 1–22.

Harker, D. 1980: *One for the Money: Politics and Popular Song*. London: Hutchinson.

Harker, D. 1985: *Fakesong: The Manufacture of British 'Folksong' 1700 to the Present Day*. Milton Keynes: Open University Press.

Harley, J. B. 1988: Maps, knowledge, and power. In D. Cosgrove and S. Daniels (eds), *The Iconography of Landscape*. Cambridge: Cambridge University Press, 277–312.

Harris, A. J. 1959: The London/York motorway: an engineer's view. *The Architects' Journal*, 130, 162–5.

Harrison, M. 1998: *Young Meteors: British Photojournalism 1957–1965*. London: Jonathan Cape.

Harrison, S., Pile, S. and Thrift, N. (eds) 2004: *Patterned Ground: Entanglements of Nature and Culture*. London: Reaktion.

Hartwell, E. 1959: Provision of catering facilities. *The Guardian*, 2 November, 13.

Harvey, D. 1990: *The Condition of Postmodernity*. Oxford: Blackwell.

Harvey, S. (ed.) 1998: *LDT Monographs No. 1: Geoffrey Jellicoe*. Reigate: Landscape Design Trust.

Hawkins, R. 1986: A road not taken: sociology and the neglect of the automobile. *California Sociologist*, 8, 61–79.

Hay, J. 1959: Day of breakdowns on motorway. *Birmingham Post*, 3 November.

Hebbert, M. 2000: Transpennine: imaginative geographies of an interregional corridor. *Transactions of the Institute of British Geographers*, 25, 379–92.

Hebdige, D. 1988: *Hiding in the Light: On Images and Things*. London: Routledge.

Hemingway, A. 1979: *The Norwich School of Painters 1803–1833*. Oxford: Phaidon.

The Herts Advertiser & St Albans Times 1959a: No time to wave back! 6 November, 3.

The Herts Advertiser & St Albans Times 1959b: Comparing problems. 30 October, motorway supplement, iv.

The Herts Advertiser & St. Albans Times 1959c: Drivers are mastering the motorway code. 6 November, 13.

Hewitt, J. 1992: The 'nature' and 'art' of Shell advertising in the early 1930s. *Journal of Design History*, 5(2), 121–39.

Hinchliffe, S. 2003: 'Inhabiting' – landscapes and natures. In K. Anderson, M. Domosh, S. Pile and N. Thrift (eds), *Handbook of Cultural Geography*. London: Sage, 207–25.

Hoggart, R. 1957 [1958]: *The Uses of Literacy*. Harmondsworth: Pelican.

Hollowell, P. G. 1968: *The Lorry Driver*. London: Routledge and Kegan Paul.

Horlock, M. 2004: *Julian Opie*. London: Tate Publishing.

Horvath, R. J. 1974: Machine space. *The Geographical Review*, 64(2), 167–88.

Hoskins, W. G. 1955 [1983]: *The Making of the English Landscape*. Harmondsworth: Penguin.

Hoskins, W. G. 1970: *Leicestershire: A Shell Guide*. London: Faber and Faber.

Hoskins, W. G. 1973: *English Landscapes*. London: British Broadcasting Corporation.

Howkins, A. 1989: Greensleeves and the idea of national music. In R. Samuel (ed.), *Patriotism: The Making and Unmaking of British National Identity, Volume III: National Fictions*. London: Routledge, 89–98.

Hoyle, T. 1979: *The Man Who Travelled on Motorways*. London: John Calder.

Hudson, R. and Schamp, E. (eds) 1995: *Towards a New Map of Automobile Manufacturing in Europe?* Berlin: Springer.

Hughes, J. 2002: Modernism, medicine and movement in 1960s Britain. *Twentieth-Century Architecture*, 6, 91–104.

Hurrell, I. 1989: It's all in the line of duty . . . *Chronicle and Echo* [Northampton], 17 April.

[ICE] The Institution of Civil Engineers 1942: *Post-War National Development Report No. III: Public Works*. London: The Institution of Civil Engineers.

[IHE] Institution of Highway Engineers 1943: *The Post-War Development of Highways*. London: Institution of Highway Engineers.

[ILA] Institute of Landscape Architects 1946: *Roads in the Landscape: A Report Prepared by the ILA, 21 March 1946*. London: Institute of Landscape Architects.

[IMCE] The Institution of Municipal and County Engineers 1942: *Post-War Planning and Reconstruction*. London: The Institution of Municipal and County Engineers.

India Tyres 1959: [Advertisement]. *The Times*, 3 November, 5.

Inglis, F. 1975: Roads, office blocks, and the new misery. In P. Abbs (ed.), *The Black Rainbow*. London: Heinemann, 168–88.

Ingold, T. 1993: The temporality of landscape. *World Archaeology*, 25, 152–74.

Interior Design 1970: Motorway restaurant on the M1. June, 358.

Jackson, J. B. 1997: *Landscape in Sight: Looking at America*, ed. H. Lefkowitz Horowitz. London: Yale University Press.

Jacobs, J. 1961 [1965]: *The Death and Life of Great American Cities*. Harmondsworth: Pelican.

Jacobs, J. M. 2006: A geography of big things. *Cultural Geographies*, 13, 1–27.

Jakle, J. A. and Sculle, K. A. 1994: *The Gas Station in America*. London: Johns Hopkins University Press.

Jakle, J. A. and Sculle, K. A. 1999: *Fast Food: Roadside Restaurants in the Automobile Age*. Baltimore, MD: Johns Hopkins University Press.

Jakle, J. A. and Sculle, K. A. 2004: *Lots of Parking: Land Use in a Car Culture*. London: University of Virginia Press.

Jakle, J. A., Sculle, K. A. and Rogers, J. S. 1996: *The Motel in America*. Baltimore, MD: Johns Hopkins University Press.

Jeffery, J. D. N. 1959: Motorways and trunk roads in Great Britain: some technical and administrative problems. *Roads and Road Construction*, May, 130–1.

Jeffreys, W. R. 1927a: New arterial roads. *The Times*, British Motor Number, 5 April, xvii.

Jeffreys, W. R. 1927b: Special motor roads. *The Times*, British Motor Number, 5 April, xvi.

Jeffreys, W. R. 1942: Roads of Empire. Wanted – a British 'Todt' organisation [letter]. *Roads and Road Construction*, 20, 1 April, 57.

Jeffreys, W. R. 1949: *The King's Highway: An Historical and Autobiographical Record of the Developments of the Past Sixty Years*. London: Batchwood Press.

Jeffries. S. 1998: Romance roadblock. *The Guardian*, G2 section, 13 July, 14–15.

Jellicoe, G. A. 1945: The aesthetic aspect of civil engineering design (fifth lecture). In The Institution of Civil Engineers, *The Aesthetic Aspect of Civil Engineering Design*. London: The Institution of Civil Engineers, 83–95.

Jellicoe, G. A. 1958: Motorways – their landscaping, design and appearance. *Journal of the Town Planning Institute*, 44, 274–83.

Jellicoe, G. A. 1980: *Blue Circle Cement Hope Works Derbyshire: A Progress Report on a Landscape Plan 1943–93*. London: Blue Circle Industries Limited.

Jellicoe, G. A. 1985: The Wartime Journal of the Institute of Landscape Architects. In S. Harvey and S. Rettig (eds), *Fifty Years of Landscape Design 1934–84*. London: The Landscape Press, 9–25.

Jellicoe, G. A. 1987: Sir Geoffrey Jellicoe. In S. Harvey (ed.), *Reflections On Landscape: The Lives and Works of Six Landscape Architects*. Aldershot: Gower Technical Press, 1–29.

Jennings, C. 1995: *Up North: Travels Beyond the Watford Gap*. London: Abacus.

[*JILA*] *Journal of the Institute of Landscape Architects* 1959: Motorways. 47, 13 and 16.

Joad, C. E. M. 1934: *A Charter for Ramblers, or the Future of the Countryside*. London: Hutchinson.

Joad, C. E. M. 1946: *The Untutored Townsman's Invasion of the Country*. London: Faber and Faber.

Joint Committee including representatives of The Council for the Preservation of Rural England, The Institute of Landscape Architects, The Roads Beautifying Association, The Royal Forestry Society of England and Wales, and The Standing Joint Committee of the Royal Automobile Club, The Automobile Association and The Royal Scottish Automobile Club 1954: *The Landscape Treatment of Roads*. London: The Council for the Preservation of Rural England.

Joint Committee of the British Road Federation, The Institution of Highway Engineers and The Society of Motor Manufacturers & Traders 1948: *Economics of Motorways (Report presented to the Minister of Transport on 18 October 1948)*. London: Joint Committee of the British Road Federation, The Institution of Highway Engineers and The Society of Motor Manufacturers & Traders.

Jones, H. 1998: Buildings designed to advertise fuel. *British Archaeology*, 38, October, 6–7.

Joyce, P. 2003: *The Rule of Freedom: Liberalism and the Modern City*. London: Verso.

[*JRIBA*] *Journal of the Royal Institute of British Architects* 1939: Road Architecture: the need for a plan. 46, 400–1.

Kaplan, C. 1996: *Questions of Travel*. Durham, NC: Duke University Press.

Katz, J. 1999: *How Emotions Work*. London: The University of Chicago Press.

Kern, S. 1983: *The Culture of Time and Space 1880–1918*. Cambridge, MA: Harvard University Press.

Kindersley, D. 1960: Motorway sign lettering. *Traffic Engineering and Control*, December, 463–5.

King-Salter, E. J. 1958: Excavated top soil [letter]. *The Times*, 23 January, 11.

Kirsch, S. and Mitchell, D. 1998: Earth-moving as the 'measure of man': Edward Teller, geographical engineering, and the matter of progress. *Social Text*, 54, 101–34.

Koshar, R. 2004: Cars and nations: Anglo-German perspectives on automobility between the world wars. *Theory, Culture and Society*, 21(4–5), 121–44.

Krim, A. 2005: *Route 66: Iconography of the American Highway*. Santa Fe, NM: Centre for American Places.

Kunstler, J. H. 1994: *The Geography of Nowhere*. New York: Touchstone.

Lackey, K. 1997: *Roadframes: The American Highway Narrative*. London: University of Nebraska Press.

Laegran, A. S. 2002: The petrol station and the internet café: rural technospaces for youth. *Journal of Rural Studies*, 18, 157–68.

Laing, W. K., Broadbent, B. H. and Fisher, J. M. 1960: The London–Birmingham motorway: Luton–Dunchurch: construction. *Proceedings of the Institution of Civil Engineers*, 15, Jan.–April, 387–400.

Lane, L. 1959: That 'M for murder' road. *Punch*, 25 November, 504.

Larkin, P. 1972: Prologue. In *How Do You Want to Live? A Report on the Human Habitat*. London: HMSO, x–xi.

Larkin, P. 1974: *High Windows*. London: Faber and Faber.

Larsen, J. 2001: Tourism mobilities and the travel glance: experiences of being on the move. *Scandinavian Journal of Hospitality and Tourism*, 1, 80–98.

Latour, B. 1992: Where are the missing masses? The sociology of a few mundane artefacts. In W. E. Bijker and J. Law (eds), *Shaping Technology, Building Society: Studies in Sociotechnical Change*. London: MIT Press, 225–58.

Latour, B. 1993: *We Have Never Been Modern*. Hemel Hempstead: Harvester Wheatsheaf.

Latour, B. 1996: *Aramis, or the Love of Technology*. London: Harvard University Press.

Latour, B. 2005: *Re-assembling the Social*. Oxford: Oxford University Press.

Laurier, E. 2004: Doing office work on the motorway. *Theory, Culture and Society*, 21(4–5), 261–77.

Laurier, E. and Philo, C. 2003: The region in the boot: mobilising lone subjects and multiple objects. *Environment and Planning D: Society and Space*, 21, 85–106.

Law, J. 1994: *Organizing Modernity*. Oxford: Blackwell.

Law, J. 2001: Ladbroke Grove, or how to think about failing systems. Published by the Centre for Science Studies and the Department of Sociology, Lancaster University at http://www.lancaster.ac.uk/fss/sociology/papers/law-ladbroke-grove-failing-systems.pdf (version: paddington5.doc, August 2000, accessed 12 March 2007).

Law, R. 1999: Beyond 'women and transport': towards new geographies of gender and daily mobility. *Progress in Human Geography*, 23, 567–88.

Lawrence, D. 1999: *Always a Welcome: The Glove Compartment History of the Motorway Service Area*. Twickenham: Between Books.

Lawrence, W. 1929: London-to-Brighton motorway: Surrey residents' opposition [letter]. *The Times*, 25 January, 12.

Lees, L. 2001: Towards a critical geography of architecture: the case of an Ersatz colosseum. *Ecumene*, 8, 51–86.

Leonard Manasseh and Partners 1960: A model service area on M1. *The Architectural Review*, 128, 417–19.

Linehan, D. 2000: An archaeology of dereliction: poetics and policy in the governing of depressed industrial districts in interwar England and Wales. *Journal of Historical Geography*, 26, 99–113.

Linehan, D. 2003: A new England. In D. Gilbert, D. Matless and B. Short (eds), *Geographies of British modernity*. Oxford: Blackwell, 132–50.

Liniado, M. 1996: *Car Culture and Countryside Change*. Cirencester: The National Trust.

Livingstone, D. 1992: *The Geographical tradition*. Oxford: Blackwell.

Lloyd, A. L. (ed.) 1952: *Come All Ye Bold Miners: Ballads and Songs of the Coalfields*. London: Lawrence and Wishart.

Lloyd, A. L. 1961: The folk-song revival. *Marxism Today*, 5, 170–3.

Lloyd, T. I. 1955: Potentialities of the British Railways' system as a reserved roadway system. *Proceedings of The Institution of Civil Engineers*, 4(1), 732–88.

Lloyd, T. I. 1957: *Twilight of the Railways – What Roads They'll Make! Railway Conversion as the Cure for Britain's Transport Troubles*. London: Forster Groom & Co. Ltd.

Lowerson, J. 1980: Battles for the countryside. In F. Gloversmith (ed.), *Class, Culture and Social Change*. Brighton: Harvester Wheatsheaf, 258–80.

[LRRC] London Regional Reconstruction Committee of the Royal Institute of British Architects 1943: *Greater London: Towards a Master Plan. Second Interim Report – May 1943*. London: Royal Institute of British Architects.

Luckin, B. 1997: War on the roads: traffic accidents and social tension in Britain, 1939–45. In R. Cooter and B. Luckin (eds), *Accidents in History: Injuries, Fatalities and Social Relations*. Amsterdam: Rodopi, 234–54.

Lupton, D. 1999: Monsters in metal cocoons: 'road rage' and cyborg bodies. *Body and Society*, 5, 57–72.

Luton News 1958a: Race against time won on motorway bridge. 15 May.

Luton News 1958b: Motorway death. 25 July.

Luton News 1958c: Pitch is 'reprieved' – roadmakers aid sport. 27 April.

Luton News 1958d: Where they dare not wear shoes: M-U-D spells trouble for Leagrave mothers. 18 December.

Luton News 1958e: Duke doesn't mind a land 'grab'. 27 June.

Lynch, M. 1993: *Scientific Practice and Ordinary Action*. Cambridge: Cambridge University Press.

Lyons, G. and Urry, J. 2005: Travel time use in the information age. *Transportation Research Part A*, 39, 257–76.

Mabey, R. 1974: *The Roadside Wildlife Book*. Newton Abbot: Readers Union.

McClintock, A. 1995: *Imperial Leather: Race, Gender and Sexuality in the Colonial Contest*. London: Routledge.

MacColl, E. 1954: Preface. In E. MacColl (ed.), *The Shuttle and Cage: Industrial Folk Ballads*. London: Workers' Music Association.

MacColl, E. 1961: Going American? [text of BBC interview with Ewan MacColl and Peggy Seeger]. *English Dance and Song*, New Year special number, 19–20.

MacColl, E. 1985: Theatre of action, Manchester. In R. Samuel, E. MacColl and S. Cosgrove, *Theatres of the Left 1880–1935: Workers' Theatre Movements in Britain and America*. London: Routledge, 205–55.

MacColl, E. 1990: *Journeyman: An Autobiography*. London: Sidgwick and Jackson.

MacColl, E. and Goorney, H. (eds) 1986: *Agit-Prop to Theatre Workshop: Political Playscripts, 1930–50*. Manchester: Manchester University Press.

MacColl, E. and Parker, C. 1959: 'Song of a Road', unpublished production script TBM 21189, Birmingham City Archives MS4000/2/74/1/6.

MacColl, E. and Seeger, P. 1963: *Songbook*. New York: Oak Publications.

McCreery, S. 1996: Westway: Caught in the speed trap. In I. Borden, J. Kerr, A. Pivaro and J. Rendell (eds), *Strangely Familiar: Narratives of Architecture in the City*. London: Routledge, 37–41.

McDonald-Walker, S. 2000: *Bikers: Culture, Politics and Power*. Oxford: Berg.

McDowell, L. 1996: Off the road: alternative views of rebellion, resistance, and 'the beats'. *Transactions of the Institute of British Geographers*, 21, 412–19.

MacEwen, M. 1959: Motropolis: a study of the traffic problem. Can we get out of the jam? *The Architects' Journal*, 130, 254–75.

MacInnes, C. 1959 [1964]: *Absolute Beginners*. Harmondsworth: Penguin.

MacInnes, C. 1961: *English, Half English*. London: MacGibbon and Kee.

McKay, G. 1996: *Senseless Acts of Beauty*. London: Verso.

Mackenzie, A. 2002: *Transductions: Bodies and Machines at Speed*. London: Continuum.

McKenzie, W. A. 1959a: More lane discipline by M1 drivers. *Daily Telegraph*, 4 November.

McKenzie, W. A. 1959b: Motoring: shedding light all along the motorways. *Daily Telegraph and Morning Post*, 25 November, 13.

McKenzie, W. A. 1959c: Motoring: much to learn on motorways. *Daily Telegraph and Morning Post*, 11 November, 13.

McKenzie, W. A. 1962: Motoring: Hard facts about M1 standards. *Daily Telegraph*, 7 November.

McLintock, J. D. 1959: Motorway technique. *Road Way*, December, 22.

Macmillan, H. 1972: *Point the Way, 1959–1961*. London: Macmillan.

McShane, C. 1994: *Down the Asphalt Path: The Automobile and the American City*. New York: Columbia University Press.

McShane, C. 1999: The origins and globalization of traffic control systems. *Journal of Urban History*, 25, 379–404.

Maddox, A. 2003: *Classic Cafés*. London: Black Dog Publishing.

The Manchester Guardian 1929: From Midlands to the Mersey. 23 October, 4.

The Manchester Guardian 1958: No 'pubs' on new motorways. 1 November.

Manzoni, H. 1958: Closing session: review of proceedings. In The Institution of Civil Engineers, *Conference on the Highway Needs of Great Britain, at the Institution 13–15 November 1957: Proceedings.* London: The Institution of Civil Engineers, 299–311.

Martin, D. 1959: Britain's first motorway. *Radio Times*, 23 October.

Martin, R. 1961: *The Mystery of the Motorway.* London: Thomas Nelson and Sons Ltd.

Massey, D. 1984: *Spatial Divisions of Labour.* London: Macmillan Education.

Massey, D. 1991: A global sense of place. *Marxism Today*, June, 24–9.

Massey, D. 2000: Travelling thoughts. In P. Gilroy, L. Grossberg and A. McRobbie (eds), *Without Guarantees: In Honour of Stuart Hall.* London: Verso, 225–32.

Massey, D. 2004: The A34. In A. Cross, *An English Journey.* London and Southampton: Film and Video Umbrella and John Hansard Gallery, no pagination.

Massey, D. 2005: *For Space.* London: Sage.

Massey, D. and Thrift, N. 2003: The passion of place. In R. Johnston and M. Williams (eds), *A Century of British Geography.* Oxford: Oxford University Press, 275–99.

Matless, D. 1990: Ordering the land: the 'preservation' of the English countryside, 1918–1939. Unpublished Ph.D. thesis, University of Nottingham.

Matless, D. 1996: New material? Work in cultural and social geography, 1995. *Progress in Human Geography*, 20, 379–91.

Matless, D. 1998: *Landscape and Englishness.* London: Reaktion.

Matless, D. 2002: Topographic culture: Nikolaus Pevsner and the buildings of England. *History Workshop Journal*, 54, 73–99.

Mauger, P. 1959: *Buildings in the Country.* London: B. T. Batsford.

Mennem, P. 1959: Motorway 1 opens – and Mr. Marples says: 'I was appalled'. *Daily Mirror*, 3 November, 5.

Mercury and Herald [Northampton] 1958a: Camera record of pre-autoway Northants – club presents 92 prints to community centre. 12 July.

Mercury and Herald 1958b: Hartwell men stole tools from motorway. 19 December.

Merriman, P. 2001: M1: A cultural geography of an English motorway, 1946–1965. Unpublished Ph.D. thesis, University of Nottingham.

Merriman, P. 2003: 'A power for good or evil': geographies of the M1 in late-fifties Britain. In D. Gilbert, D. Matless and B. Short (eds), *Geographies of British modernity.* Oxford: Blackwell, 115–31.

Merriman, P. 2004a: Freeways. In S. Harrison, S. Pile and N. Thrift (eds), *Patterned Ground: Entanglements of Nature and Culture.* London: Reaktion, 86–8.

Merriman, P. 2004b: Driving places: Marc Augé, non-places and the geographies of England's M1 motorway. *Theory, Culture, and Society*, 21(4–5), 145–67.

Merriman, P. 2005a: 'Operation motorway': landscapes of construction on England's M1 motorway. *Journal of Historical Geography*, 31, 113–33.

Merriman, P. 2005b: Materiality, subjectification and government: the geographies of Britain's Motorway Code. *Environment and Planning D: Society and Space*, 23, 235–50.

Merriman, P. 2005c: Driving places: Marc Augé, non-places and the geographies of England's M1 motorway. In M. Featherstone, N. Thrift and J. Urry (eds), *Automobilities*. London: Sage, 145–67.

Merriman, P. 2005d: 'Respect the life of the countryside': the Country Code, government, and the conduct of visitors to the countryside in post-war England and Wales. *Transactions of the Institute of British Geographers*, 30, 336–50.

Merriman, P. 2006a: 'A new look at the English landscape': landscape architecture, movement and the aesthetics of motorways in early post-war Britain. *Cultural Geographies*, 13, 78–105.

Merriman, P. 2006b: 'Mirror, signal, manoeuvre': assembling and governing the motorway driver in late fifties Britain. In S. Böhm, C. Jones, C. Land and M. Paterson (eds), *Against Automobility*. Oxford: Blackwell Publishing/Sociological Review, 75–92.

Merriman, P. 2007: ' "Beautified" is a vile phrase': the politics and aesthetics of landscaping roads in postwar Britain. In C. Mauch and T. Zeller (eds), *The World Beyond the Windshield: Driving and the Experience of Landscape in 20th-Century Europe and America*. Athens, OH: Ohio University Press.

[MHLG] Ministry of Housing and Local Government 1964: *The South East Study 1961–1981*. London: HMSO.

Michael, M. 1998: Co(a)gency and the car: attributing agency in the case of the 'road rage'. In B. Brenna, J. Law and I. Moser (eds), *Machines, Agency and Desire*. Oslo: Centre for Technology and Culture, 125–41.

Michael, M. 2000: *Reconnecting Culture, Technology and Nature*. London: Routledge.

Michael, M. 2001: The invisible car: the cultural purification of road rage. In D. Miller (ed.), *Car Cultures*. Oxford: Berg, 59–80.

Millar, J. and Schwartz, M. 1998: Introduction – Speed is a vehicle. In J. Millar and M. Schwartz (eds), *Speed – Visions of an Accelerated Age*. London: The Photographers Gallery and Whitechapel Art Gallery, 16–24.

Miller, D. 2001: Driven societies. In D. Miller (ed.), *Car Cultures*. Oxford: Berg, 1–33.

Miller, P. and Rose, N. 1990: Governing economic life. *Economy and Society*, 19, 1–31.

Ministry of Information 1942: *Transport Goes to War: The Official Story of British Transport, 1939–42*. London: HMSO.

Mitchell, D. 1996: *The Lie of the Land: Migrant Workers and the California Landscape*. London: University of Minnesota Press.

Mitchell, D. 2001: The devil's arm: points of passage, networks of violence, and the California agricultural landscape. *New Formations*, 43, 44–60.

Mitchell, W. J. T. 1994: Introduction. In W. J. T. Mitchell (ed.), *Landscape and Power*. London: University of Chicago Press, 1–4.

[MKDC] Milton Keynes Development Corporation 1970: *The Plan for Milton Keynes (Volumes One and Two)*. Wavendon: Milton Keynes Development Corporation.

Moholy-Nagy, L. 1947 [1965]: *Vision in Motion*. Chicago: Paul Theobald and Company.

Montagu of Beaulieu, J. [The Editor] 1902: Editorial jottings: London to Birmingham by motor road. *The Car (Illustrated)*, 2(20), 8 October, 211.

Montagu of Beaulieu, J. [The Editor] 1903: Motor roads will be opposed. *The Motor Car Magazine*, 1(2), 133.

Montagu of Beaulieu, J. [The Editor] 1905a: The coming of the carway. *The Motor Car Magazine*, 3(18), November, 401–3.

Montagu of Beaulieu, Lord 1905b: Editorial jottings: the first Motor Road Bill. *The Car (Illustrated)*, 185, 6 December, 91.

Montagu of Beaulieu, Lord 1906a: Editorial jottings. *The Car (Illustrated)*, 190, 10 January, 261.

Montagu of Beaulieu, Lord 1906b: Editorial jottings. *The Car (Illustrated)*, 193, 31 January, 357.

Montagu of Beaulieu, Lord 1923a: Foreword. In Northern and Western Motorway, *Northern and Western Motorway: Direct Lines of Communication Between the Largest Cities and Business Centres of the Country*. London: Northern and Western Motorway, 3–6.

Montagu of Beaulieu, Lord 1923b: To the editor of *The Manchester Guardian* [letter, 6 November 1923]. In Northern and Western Motorway, *Northern and Western Motorway: Providing Direct Lines of Fast and Cheap Communication by Road Between the Largest Cities and Business Centres of the Country* [revised and extended edition]. London: Northern and Western Motorway, 42–3.

Montagu of Beaulieu, Lord 1923c: 'The motorway': a new trunk road [letter, *Daily Mail*, 27 October 1923]. In Northern and Western Motorway, *Northern and Western Motorway: Providing Direct Lines of Fast and Cheap Communication by Road Between the Largest Cities and Business Centres of the Country* [revised and extended edition]. London: Northern and Western Motorway, 44–7.

Montagu of Beaulieu, Lord 1923d: The railways and the motorway: a reply [letter, *Daily Mail*, 30 October 1923]. In Northern and Western Motorway, *Northern and Western Motorway: Providing Direct Lines of Fast and Cheap Communication by Road Between the Largest Cities and Business Centres of the Country* [revised and extended edition]. London: Northern and Western Motorway, 49–51.

Moorhouse, H. F. 1991: *Driving Ambitions: An Analysis of the American Hot Rod Enthusiasm*. Manchester: Manchester University Press.

Moran, J. 2005: *Reading the Everyday*. London: Routledge.

Moran, J. 2006: Crossing the road in Britain, 1931–1976. *The Historical Journal*, 49, 477–96.

Morgan, W. J. 1958: A nightmare now the H. P.'s off. *The Observer*, 16 November.

Morris, M. 1988: At Henry Parkes Motel. *Cultural Studies*, 2, 1–47.

Morse, M. 1998: *Virtualities*. Bloomington, IN: Indiana University Press.

The Motor 1959a: Motorway: *The Motor* guide to Britain's first long distance express highway [insert]. 116, 4 November, inserted between pages 508 and 509.

The Motor 1959b: Buffer for the motorway. 21 October, 339.

The Motor [J. L.] 1959c: Topical technics: an engineering notebook by J. L. 116, 9 September, 135.

The Motor 1959d: Reputations at stake. 116, 11 November, 517.

The Motor 1959e: Motorway in use. 116, 11 November, 518.

The Motor [Grand Vitesse] 1959f: This sporting side. 116, 9 September, 120–1.

The Motor 1960: Six months of M1. 117, 13 July, 898–900.

Motorway: An International Magazine for the Modern Motorist 1960: Summer issue, 1(1), 1–40.

Motorway 1961: Spring issue, 1(4), 1–34.

[MT] Ministry of Transport 1959: *London–Yorkshire Motorway. First Section: London–Birmingham*. London: HMSO.

[MTCA and COI] Ministry of Transport and Civil Aviation and the Central Office of Information 1958: *The Motorway Code*. London: HMSO.

[MTCA and COI] Ministry of Transport and Civil Aviation and the Central Office of Information 1959: *The Highway Code Including Motorway Rules*. London: HMSO.

Mulhern, F. 1996: A welfare culture? Hoggart and Williams in the fifties. *Radical Philosophy*, 77, 26–37.

Nairn, I. 1955: Outrage. *The Architectural Review*, 117, 363–460.

Nairn, I. 1956: Counter-attack. *The Architectural Review*, 120, 353–440.

Nairn, I. 1963: Look out. *The Architectural Review*, 133, 425–6.

Nairn, I. 1975: Outrage twenty years after. *The Architectural Review*, 158, 328–37.

Narotzky, V. 2002: Our cars in Havana. In P. Wollen and J. Kerr (eds), *Autopia: Cars and Culture*. London: Reaktion, 168–76.

Nead, L. 2000 *Victorian Babylon: People, Streets, and Images in Nineteenth-Century London*. London: Yale University Press.

Nead, L. 2004: Vehicles of the image c.1900. *Art History*, 27(5), 745–69.

New Society 1970: Artful accidents. 15, 9 April, 588.

Newby, R. F. and Johnson, H. D. 1963: London–Birmingham motorway accidents. *Traffic Engineering and Control*, 4, 550–5.

Newby, P. R. T. 2004: The contribution of Ordnance Survey to Britain's motorways, In P. Baldwin and R. Baldwin (eds), *The Motorway Achievement, Volume 1. The British Motorway System: Visualisation, Policy and Administration*. London: Thomas Telford, 305–66.

News Chronicle 1960: Is this the phantom van on the M1? 21 June, 13.

Newton, F. 1963: Two cheers for folk-song. *New Statesman*, 66(1689), 26 July, 119.

Nichols, P. 1975: *The Freeway: A Play in Two Acts*. London: Faber and Faber.

Nichols, P. 2000: *Diaries 1969–1977*. London: Nick Hern Books.

Nicholson, J. 2000: *A1*. London: Harper Collins Illustrated.

Nightingale, B. 1967: Rallying points. *New Society*, 9, 789.

Noble, D. 1969: *Milestones in a Motoring Life*. London: The Queen Anne Press.

Nockolds, H. 1950: *Roads: The New Way*. London: British Road Federation.

Normark, D. 2006: Tending to mobility: intensities of staying at the petrol station. *Environment and Planning A*, 38, 241–52.

Northern and Western Motorway 1923a: *Northern and Western Motorway: Direct Lines of Communication Between the Largest Cities and Business Centres of the Country*. London: Northern and Western Motorway.

Northern and Western Motorway 1923b: *Northern and Western Motorway: Providing Direct Lines of Fast and Cheap Communication by Road Between the Largest Cities and Business Centres of the Country* [revised and extended edition]. London: Northern and Western Motorway.

O'Connell, S. 1998: *The Car and British Society: Class, Gender and Motoring, 1896–1939*. Manchester: Manchester University Press.

O'Gorman, M. 1942: *Road Transport and the National Plan*. London: British Road Federation.

Opie, J. 2004: M1. In M. Kliege and J. Opie, *Julian Opie*. Nuremberg: Verlag für Moderne Kunst Nürnberg, 84.

Osborne, P. D. 2000: *Travelling Light: Photography, Travel and Visual Culture*. Manchester: Manchester University Press.

Osborne, R. H. 1960: The London–Yorkshire motorway: its route through the East Midlands. *The East Midland Geographer*, 13, 34–8.

Owen, O. 2000: Driving without a safety net. *The Observer*, 5 March, 20.

Parker, C. 1959a: Song of a Road. *Radio Times*, 145(1877), 1–7 November, 6.

Parker, C. 1959b: Song of a Road (letter). *The Listener*, 62(1600), 26 November, 938.

Parker, C. 1961: The relationship between vernacular speech and workers' song, and its implications for creative work in the new media of communication, with especial reference to the actuality Radio Ballad. Unpublished paper prepared for the International Symposium for Research into Workers' Songs, held at Prague 16–18.3.61 under the auspices of the Council of Czechoslovak Ethnologists and Folklorists [copy held in the Birmingham City Archives MS4000/1/3/1/4/2].

Parker, C. 1964: Singing of a motorway. *Granta*, 69(1236), 14 May, 2–4.

Parker, C. 1965: MacColl, against the tyranny of the nice guy. *Folk Music*, 1(2), 2–7.

Parker, C. 1973: On what authority? *New Statesman*, 26 January, 134.

Parker, C. 1975: The dramatic actuality of working-class speech. In W. Van der Will (ed.), *Workers and Writers: Proceedings of the Conference on Present-Day Working-Class Literature in Britain and West Germany, Held in Birmingham, October 1975*. Birmingham: Centre for Contemporary Cultural Studies, 98–105.

Parliamentary Debates, House of Commons 1900: Housing of the Working Classes Act (1890) Amendment Bill (Second reading). Fourth series, 83, cols 427–523.

Parliamentary Debates, House of Commons 1909: Development and Road Improvement Funds Bill (Second reading). 10, 6 September, cols 906–1042.

Parliamentary Debates, House of Commons 1923: Motorways Bill. 168, 16 November, col. 378.

Parliamentary Debates, House of Commons 1924a: London and Manchester motorway. 170, 11 March, col. 2138.

Parliamentary Debates, House of Commons 1924b: Motorways. 172, 9 April, cols 452–5.

Parliamentary Debates, House of Lords 1936: Trunk Roads Bill (second reading). 8 December, cols 656–86.

Parliamentary Debates, House of Commons 1943: Motorways (oral answers). 393, 10 November, cols 1129–30.

Parliamentary Debates, House of Commons 1944: Roads: Motorways (construction). 404, 15 November, cols 1956–7.

Parliamentary Debates, House of Commons 1946: Highway Development (Government programme). 422, 6 May, cols 590–5.

Parliamentary Debates, House of Commons 1948: Special Roads Bill – Second reading. 457, 11 November, cols 1737–848.

Parliamentary Debates, House of Commons 1955: Expanded road programme (Government's proposals). 536, 2 February, cols 1096–109.

Parliamentary Debates, House of Commons 1958: Motorways (traffic regulations). 592, 23 July, oral answers, cols 402–5.

Parliamentary Debates, House of Lords 1959a: Planning of motorways. 216, 7 May, cols 207–9.

Parliamentary Debates, House of Commons 1959b: Highway code. 609, 17 July, cols 733–818.

Parliamentary Debates, House of Commons 1959c: London–Birmingham Motorway. 603, 15 April, written answers, col. 85.

Parliamentary Debates, House of Commons 1960a: M.1 road (petrol stations). 632, 21 December, oral answers, cols 1292–3.

Parliamentary Debates, House of Commons 1960b: M.1 (bridges). 626, 6 July, written answers, col. 39.

Parliamentary Debates, House of Commons 1960c: London–Yorkshire Motorway. 627, 20 July, col. 489.

Parliamentary Debates, House of Commons 1961: Transport. M.1 (casualties). 635, 20 February, written answers, cols 23–4.

Parliamentary Debates, House of Commons 1965: 70 mph speed limit (vehicle testing). 721, 1 December, written answers, col. 192.

Parr, M. 1999: *Boring Postcards*. London: Phaidon.

Paul, D. 1959: Spoken word: smithereens. *The Listener*, 62(1598), 12 November, 847.

Paul, K. 1997: *Whitewashing Britain: Race and Citizenship in the Postwar Era*. Ithaca, NY: Cornell University Press.

Peach, H. H. and Carrington, N. (eds) 1930: *The Face of the Land*. London: George Allen and Unwin.

Pearce, L. 2000: Driving north/driving south: reflections upon the spatial/temporal coordinates of 'home'. In L. Pearce (ed.), *Devolving Identities: Feminist Readings in Home and Belonging*. London: Ashgate, 162–78.

Perrow, C. 1984: *Normal Accidents*. New York: Basic Books.

Peters, G. H. 1973: *Cost-Benefit Analysis and Public Expenditure* (Third Edition). London: The Institute of Economic Affairs.

Pettitt, L. 2000a: Philip Donnellan, Ireland and dissident documentary. *Historical Journal of Film, Radio and Television*, 20(3), 351–65.

Pettitt, L. 2000b: *Screening Ireland*. Manchester: Manchester University Press.

Pevsner, N. 1956: *The Englishness of English Art*. London: The Architectural Press.

Pevsner, N. 1960: *Buckinghamshire*. Harmondsworth: Penguin.

Pevsner, N. 1961: *Northamptonshire*. Harmondsworth: Penguin.

Pevsner, N. 1968: *Studies in Art, Architecture and Design, Volume 1: From Mannerism to Romanticism*. London: Thames and Hudson.

Pevsner, N. 1974: *Staffordshire*. Harmondsworth: Penguin.

Phillips, R. 1997: *Mapping Men and Empire: A Geography of Adventure*. London: Routledge.

Philo, C. and Parr, H. 2000: Institutional geographies: introductory marks. *Geoforum*, 31, 513–21.

Picken, S. 1999: Highways, by-ways and lay-bys: the Great British road movie. In J. Sargeant and S. Watson (eds), *Lost Highways: An Illustrated History of Road Movies*. London: Creation Books, 222–30.

Platt, E. 2000: *Leadville: A Biography of the A40*. London: Picador.

Plowden, W. 1970: MPs and the 'roads lobby'. In A. Barker and M. Rush, *The Member of Parliament and His Information*. London: George Allen and Unwin, 69–96.

Plowden, W. 1971 [1973]: *The Motor Car and Politics in Britain*. Harmondsworth: Pelican Books.

Pospisil, P. 1982: A motorway to nowhere. *The Geographical Magazine*, 54(7), 396–401.

Postgate, J. and Postgate, M. 1994: *A Stomach for Dissent: The Life of Raymond Postgate*. Keele: Keele University Press.

Postgate, R. 1961: *The Good Food Guide 1961–1962*. London: Cassell.

Postgate, R. (ed.) 1963: *The Good Food Guide 1963–1964*. London: Consumers' Association and Cassell.

Postgate, R. (ed.) 1965: *The Good Food Guide 1965–1966*. London: Consumers' Association and Cassell.

Postgate, R. (ed.) 1969: *The Good Food Guide 1969–1970*. London: Consumers' Association and Hodder & Stoughton.

Powers, A. 2001: The expression of levity. *Twentieth-Century Architecture*, 5, 47–56.

Powers, A. 2002: Landscape in Britain. In M. Treib (ed.), *The Architecture of Landscape*. Philadelphia: University of Pennsylvania Press, 56–81.

Powers, A. 2004: The heroic period of conservation. *Twentieth-Century Architecture*, 7, 7–18.

Priestley, J. B. 1934: *English Journey*. London: William Heinemann Ltd.

Priestley, J. B. 1935: The beauty of Britain. In *The Beauty of Britain: A Pictorial Survey*. London: B. T. Batsford Ltd, 1–10.

Prigogine, I. 1976: Order through fluctuation: self-organization and social system. In E. Jantsch and C. H. Waddington (eds), *Evolution and Consciousness: Human Systems in Transition*. Reading, MA: Addison-Wesley, 93–133.

Prigogine, I. and Herman, R. 1971: *Kinetic Theory of Vehicular Traffic*. New York: American Elsevier Publishing Company.

Proceedings of the Institution of Civil Engineers 1961: Discussion on the London–Birmingham motorway. 19, May–August, 61–119.

Pryke, H. S. 1958: Scars across the land [letter]. *The Times*, 6 June, 13.

[RAC] Royal Automobile Club 1960: *Royal Automobile Club Guide and Handbook 1960*. London: The Royal Automobile Club.

Radio Times 1959a: 'Song of a Road' advertisement. 145, 30 October, 42.

Radio Times 1959b: 'Song of a Road' advertisement. 145, 25 December, 34.

Raitz, K. (ed.) 1996: *The National Road*. London: Johns Hopkins University Press.

Ramsden, J. 1996: *The Winds of Change: Macmillan to Heath, 1957–1963*. London: Hodder and Stoughton.

[RBA] The Roads Beautifying Association 1930a: *Annual Report 1929–30*. London: The RBA.

RBA [The Roads Beautifying Association] 1930b: *Roadside Planting*. London: Country Life Ltd.

[RBA] The Roads Beautifying Association 1932: *Annual Report 1931–32*. London: The RBA.

[RBA] The Roads Beautifying Association 1935: *Seventh Report of The Roads Beautifying Association October, 1933–March, 1935*. London: The RBA.

[RBA] The Roads Beautifying Association 1936: *Eighth Report of The Roads Beautifying Association 1935–1936*. London: The RBA.

[RBA] The Roads Beautifying Association 1937: *Ninth Report of The Roads Beautifying Association 1936–1937*. London: The RBA.

[RBA] The Roads Beautifying Association 1938: *Tenth Report of The Roads Beautifying Association 1937–1938*. London: The RBA.

[RBA] The Roads Beautifying Association 1939: *Eleventh Report of The Roads Beautifying Association 1938–1939*. London: The RBA.

Relph, E. 1976: *Place and Placelessness*. London: Pion.

Revill, G. 2000: English pastoral: music, landscape, history and politics. In I. Cook, D. Crouch, S. Naylor and J. R. Ryan (eds), *Cultural Turns/ Geographical Turns: Perspectives on Cultural Geography*. Harlow: Prentice Hall, 140–58.

[RFAC] Royal Fine Art Commission 1959: *Sixteenth Report of the Royal Fine Art Commission, January 1958–August 1959 (Cmnd. 909)*. London: HMSO.

[RIBA] Royal Institute of British Architects 1939: *Road Architecture: The Need for a Plan. Handbook to the Exhibition Arranged by the Royal Institute of British Architects, 66 Portland Place, London W1*. London: RIBA.

Richards, S. 1994: Motorway express: Midland Red's initiative. *The Vintage Commercial Vehicle Magazine*, 10(61), 216–19.

Ritchie, B. 1997: *The Good Builder: The John Laing Story*. London: James and James.

Road Alert! 1997: *Road Raging: Top Tips for Wrecking Road Building*. Newbury: Road Alert!

Road International 1958: Britain's first motorway: a culmination of the British Road Federation's 25-year effort. 29, 24–31.

Roads and Road Construction 1935: Motorways for Britain: a radical plan to reduce road accidents. 1 October, 316.

Roads and Road Construction 1938: Roads in war and peace. 16, 355–7.

Roads and Road Construction 1962: The M1 motorway. 40, 288.

Robertson, S. 2007: Visions of urban mobility: the Westway, London, England. *Cultural Geographies*, 14, 74–91.

Robinson, R. 1959: Singing road songs. *The Sunday Times*, 8 November [copy held in Birmingham City Archives MS4000/2/74/2/2].

Rollins, W. H. 1995: Whose landscape? Technology, Fascism, and environmentalism on the National Socialist *Autobahn*. *Annals of the Association of American Geographers*, 85, 494–520.

Rolt, L. T. C. 1959: *The London–Birmingham Motorway*. London: John Laing and Son Limited.

Rolt, L. T. C. 1960 [1984]: *George and Robert Stephenson*. Harmondsworth: Penguin.

Rolt, L. T. C. 1962: A Laing company history, unpublished manuscript, CTIS library, Laing Technology Group, Hemel Hempstead.

Rolt, L. T. C. 1971: *Landscape with Machines: An Autobiography*. London: Longman.

Rolt, L. T. C. 1977: *Landscape with Canals: An Autobiography*. Gloucester: Alan Sutton.

Rolt, L. T. C. 1992: *Landscape with Figures: The Final Part of His Autobiography*. Stroud: Alan Sutton.

Ronay, E. 1981: *Just a Bite*. Harmondsworth: Penguin.

Roscoe, B. 1996: 'Bradford-on-Avon but Shell on the road': the heyday of motor touring through Britain's countryside. In C. Watkins (ed.), *Rights of Way*. London: Pinter, 89–99.

Rose, M. 2002: Landscapes and labyrinths. *Geoforum*, 33, 455–67.

Rose, M. and Wylie, J. 2006: Guest editorial: animating landscape. *Environment and Planning D: Society and Space*, 24, 475–9.

Rose, M. V. 1960: After 11 months of M1: good motorway driving rules are still flouted. *Birmingham Post*, 19 October.

Rose, N. 1996: *Inventing Our Selves: Psychology, Power, and Personhood*. Cambridge: Cambridge University Press.

Rose, N. 1999: *Powers of Freedom*. Cambridge: Cambridge University Press.

Rose, N. and Miller, P. 1992: Political power beyond the state: problematics of government. *British Journal of Sociology*, 43, 173–205.

Ross, K. 1995: *Fast Cars, Clean Bodies: Decolonization and the Reordering of French Culture*. London: The MIT Press.

Rotha, P. 1939: Films of purpose. *The Architectural Review*, 85, 133.

Routledge, P. 1997: The imagineering of resistance: Pollok Free State and the practice of postmodern politics. *Transactions of the Institute of British Geographers*, 22, 359–76.

Royal Commission on the Distribution of the Industrial Population 1940: *Report (Cmd. 6153)*. London: HMSO.

[RRL] Road Research Laboratory, Department of Scientific and Industrial Research 1965: *Research on Road Traffic*. London: HMSO.

[RRL and UB] Road Research Laboratory and University of Birmingham 1960: *The London–Birmingham Motorway: Traffic and Economics (Road Research Laboratory Technical Paper No.46)*. London: HMSO.

[RTS] Reclaim the Streets 1997: Reclaim the Streets. *Do or Die – Voices from Earth First!*, 6, 1–6.

Rugby Advertiser 1958a: New motor road: complaints at Dunchurch. 28 June.

Rugby Advertiser 1958b: Pre-Ice age fossils. 8 August.

Sachs, W. 1992: *For Love of the Automobile: Looking Back Into the History of Our Desires*. Oxford: University of California Press.

Saint, A. 1987: *Towards a Social Architecture: The Role of School-Building in Post-War England*. London: Yale University Press.

Saint, A. 1992: *A Change of Heart: English Architecture Since the War. A Policy for Protection*. London: Royal Commission on the Historical Monuments of England and English Heritage.

Saint, A. 2004: Williams, Sir (Evan) Owen (1890–1969). *Oxford Dictionary of National Biography*. Oxford: Oxford University Press (accessed online, 8 March 2007), http://www.oxforddnb.com/ view/article/51931.

St Johnston, E. 1959: Problems for five police forces. *The Guardian*, 2 November, 13.

Sampson, A. 1967: *Macmillan: A Study in Ambiguity*. London: Allen Lane/The Penguin Press.

Samuel, R. 1989: Ewan MacColl (obituary). *The Independent*, 30 October, 24.

Samuel, R. 1994: *Theatres of Memory, Volume 1: Past and Present in Contemporary Culture*. London: Verso.

Sanger, C. 2001: Girls and the getaway: cars, culture, and the predicament of gendered space. In N. Blomley, D. Delaney and R. T. Ford (eds), *The Legal Geographies Reader*. Oxford: Blackwell, 31–41.

Sardar, Z. 2002: The Ambassador from India. In P. Wollen and J. Kerr (eds), *Autopia: Cars and Culture*. London: Reaktion, 208–18.

Scharff, V. 1991: *Taking the Wheel: Women and the Coming of the Motor Age*. New York: The Free Press.

Schivelbusch, W. 1986: *The Railway Journey: The Industrialisation of Time and Space in the 19th Century*. Leamington Spa: Berg.

Schivelbusch, W. 1988: *Disenchanted Night*. London: University of California Press.

Schnapp, J. T. 1999: Crash (speed as engine of inviduation). *Modernism/Modernity*, 6, 1–49.

Schonfield, A. 1958: A time to build roads. *The Observer*, 7 September.

Schwarzer, M. 2004: *Zoomscape: Architecture in Motion and Media*. New York: Princeton Architectural Press.

Scott, D. 1999: New towns. In G. Collens and W. Powell (eds), *LDT Monographs No. 1: Sylvia Crowe*. Reigate: Landscape Design Trust, 47–67.

Scott-Giles, C. W. (ed.) 1946: *The Road Goes On*. London: The Epworth Press.

Seeger, P. (compiler) 2001: *The Essential Ewan MacColl Songbook: Sixty Years of Songmaking*. New York: Oak Publications.

Select Committee of the House of Lords on the Prevention of Road Accidents 1938: *Report (H.L. 192)*. London: HMSO.

Select Committee of the House of Lords on the Prevention of Road Accidents 1939: *Report (H.L. 2, 52)*. London: HMSO.

Self, W. 1993: Mad about motorways. *The Times*, Weekend section, 25 September, 1.

Self, W. 1994: Scale. In W. Self, *Grey Area and Other Stories*. London: Bloomsbury, 89–123.

Sennett, R. 1994: *Flesh and Stone*. London: Faber and Faber.

Seymour, J. 1964: The night trunkers. *The Geographical Magazine*, 37, 137–46.

Shand, J. D. 1984: The Reichsautobahn: symbol for the Third Reich. *Journal of Contemporary History*, 19, 189–200.

Sharpe, T. 1975 [1977]: *Blott on the Landscape*. London: Pan Books.

Sheail, J. 1981: *Rural Conservation in Inter-War Britain*. Oxford: Clarendon Press.

Shell Touring Service 1961: *Road Map. Sheet No. 1: South West England and South East England*. London: Shell-Mex and BP Limited and George Philip and Son Ltd.

Sheller, M. 2004: Automotive emotions: feeling the car. *Theory, Culture, and Society*, 21(4–5), 221–42.

Sheller, M. and Urry, J. 2000: The city and the car. *International Journal of Urban and Regional Research*, 24, 727–57.

Sheller, M. and Urry, J. 2006a: Introduction. In M. Sheller and J. Urry (eds), *Mobile Technologies of the City*. London: Routledge, 1–17.

Sheller, M. and Urry, J. 2006b: The new mobilities paradigm. *Environment and Planning A*, 38, 207–26.

Sherman, P. 1991: Counties on the fast track. *Chartered Surveyor Weekly*, 23 May, 36–7.

Shields, R. 1991: *Places on the Margin: Alternative Geographies of Modernity*. London: Routledge.

Shipp, H. 1939: Road architecture. *Landscape and Garden*, 6(1), 20–2.

Sibley, D. 1994: The sin of transgression. *Area*, 26, 300–3.

Sibley, D. 1995: *Geographies of Exclusion*. London: Routledge.

Sillitoe, A. F. 1973: *Britain in Figures: A Handbook of Social Statistics* (Second Edition). Harmondsworth: Penguin.

Sinclair, I. 1999: *Crash*. London: British Film Institute.

Sinclair, I. 2002: *London Orbital: A Walk Around the M25*. London: Granta Books.

[Sir OWP] Sir Owen Williams and Partners 1957: *London–Yorkshire Motorway (South of Luton–Watford Gap–Dunchurch Special Road). Landscape Report and Model*. Unpublished report prepared by A. P. Long for the Ministry of Transport and Civil Aviation. A copy is held in the offices of Owen Williams in Birmingham.

[Sir OWP] Sir Owen Williams and Partners 1960: *London–Yorkshire Motorway: South of Luton–Watford Gap–Dunchurch–Crick Special Road. Interim Report at End of First Planting Season.* Unpublished report prepared by Sir William Ling Taylor for the Ministry of Transport. A copy is held in the offices of Owen Williams in Birmingham.

[Sir OWP] Sir Owen Williams and Partners 1961: *London–Yorkshire Motorway: South of Luton–Watford Gap–Dunchurch Special Road. Report on Widening and Reconstruction of Grass Hard Shoulders, 12th September 1960–21st May 1961.* Unpublished report prepared for the Ministry of Transport. A copy is held in the offices of Owen Williams in Birmingham.

[Sir OWP] Sir Owen Williams and Partners 1973: *London–Yorkshire Motorway (South of Luton to Doncaster By-pass M1, M45, M18).* London: Sir Owen Williams and Partners.

[SJC] Standing Joint Committee of the RAC, AA and RSAC 1944: *Post-War Roads and Traffic: A Memorandum.* London: Standing Joint Committee of the RAC, AA and RSAC.

Smith, A. 1958: Fossils 130 m. years old on motorway route. *Daily Telegraph,* 9 December.

Smithson, A. 1983: *AS in DS: An Eye on the Road.* Delft: Delft University Press.

Special Roads Act, 1949. In *The Public General Acts and Measures 12&13 Geo.6* (Chapter 32). London: HMSO.

Spence, B. 1959: Inaugural address of the president. *RIBA Journal,* 67(2), 36–8.

Spencer, H. 1961: Comments. *Design,* 152, 61.

Spens, M. 1992: *Gardens of the Mind: The Genius of Geoffrey Jellicoe.* Woodbridge, Suffolk: Antique Collectors' Club.

Spens, M. 1994: *The Complete Landscape Designs and Gardens of Geoffrey Jellicoe.* London: Thames and Hudson.

The Sphere 1959: Danger on the motorway. 239, 14 November, 265.

Spitta, M. 1952: A quarter of a century of highway planting. *Journal of the Royal Horticultural Society,* 77, 4–12.

Spurrier, R. 1959: Caution – road works. *The Architectural Review,* 125, 242–6.

Spurrier, R. 1960a: Road-style on the motorway. *The Architectural Review,* 128, 406–11.

Spurrier, R. 1960b: The service-area problem. *The Architectural Review,* 128, 411–16.

Spurrier, R. 1961: Better bypasses. *The Architectural Review,* 130, 229–35.

Stamp, G. 1986: Sir Owen Williams and his time. In G. Stamp (ed.), *Sir Owen Williams 1890–1969.* London: The Architectural Association, 7–11.

Starkie, D. 1982: *The Motorway Age.* Oxford: Pergamon.

Steering Group and Working Party appointed by the Ministry of Transport 1963: *Traffic in Towns: A Study of the Long Term Problems of Traffic in Urban Areas.* London: HMSO.

Sterne, E. [County Planning Officer] 1952: *Bedfordshire County Council, County Development Plan 1952: Written Analysis Prepared in Accordance with the Provisions of the Town and Country Planning Act, 1947.* Bedford: Bedfordshire County Council.

Strathern, M. 1996: Cutting the network. *The Journal of the Royal Anthropological Institute,* 2 (N.S.), 517–35.

Strauss, S. and Orlove, B. (eds) 2003: *Weather, Climate, Culture.* Oxford: Berg.

The Sunday Telegraph 1961: AA to fight on for M.1 danger lights. 12 November.

The Surveyor and Municipal and County Engineer 1959: London–Birmingham motorway: landscaping recommendations. 118, 20 June, 623.

Tate Gallery 1992: *Richard Hamilton.* London: Tate Gallery.

Taylor, E. G. R. 1949: The geographical basis of a county plan. *Journal of the Town Planning Institute,* 35, 49–52.

Taylor, E. G. R., Daysh, G. H. J., Fleure, H. J. and Smith, W. 1938: Discussion on the geographical distribution of industry. *The Geographical Journal,* 92(1), 22–39.

Taylor, N. 2003: The aesthetic experience of traffic in the modern city. *Urban Studies,* 40, 1609–25.

Team Spirit: The Monthly News Sheet Issued by John Laing and Son Limited 1951: Terence Cuneo the artist at work, 54, 7–8.

Team Spirit 1958a: Artist in a helicopter. 146, 7.

Team Spirit 1958b: Keeping in touch: communications system on the motorway. 146, 7.

Team Spirit 1958c: Workers in the team no.127 – John Alan Pymont. 142, 2.

Team Spirit 1958d: Workers in the team no.122 – John J. G. Michie. 137, 2.

Team Spirit 1958e: Homes on wheels. 145, 5.

Team Spirit 1958f: Minister visits the motorway. 141, 5.

Team Spirit 1958g: Editorial. 141, 2.

Team Spirit 1959a: Premiere of motorway film. 158, 10.

Team Spirit 1959b: Editorials. 150, 1.

Team Spirit 1959c: Editorials. 153/4, 1.

Team Spirit 1959d: Minister visits motorway. 153/4, 8.

Team Spirit 1959e: Workers in the team no.134 – Walter Hunt. 149, 2.

Team Spirit 1959f: Motorway luncheon. 158, 9.

Team Spirit 1960a: Motorway film shows. 161, 10.

Team Spirit 1960b: Company films. 164, 7.

Team Spirit 1960c: [Untitled note on *Song of a Road*]. 169, 12.

Tendler, S., McGrory, D. and Eason, K. 1997: Bomb chaos on motorways. *The Times,* 4 April, 1.

Tester, K. (ed.) 1994: *The Flâneur.* London: Routledge.

Thacker, A. 2000: E. M. Forster and the motor car. *Literature and History,* 9(2), 37–52.

Thomson, J. M. 1969: *Motorways in London: Report of a Working Party.* London: Duckworth.

Thornton Butterworth 1938: Publishers' note. In *Germany Speaks: By 21 Leading Members of Party and State.* London: Thornton Butterworth Ltd, 11–12.

Thrift, N. 1990: Transport and communication 1730–1914. In R. A. Dodgson and R. A. Butlin (eds), *An Historical Geography of England and Wales* (Second Edition). London: Academic Press, 453–86.

Thrift, N. 1995: A hyperactive world. In R. J. Johnston, P. J. Taylor and M. J. Watts (eds), *Geographies of Global Change*. Oxford: Blackwell, 18–35.

Thrift, N. 1996: *Spatial Formations*. London: Sage.

Thrift, N. 1999: Steps to an ecology of place. In D. Massey, J. Allen and P. Sarre (eds), *Human Geography Today*. Cambridge: Polity Press, 295–322.

Thrift, N. 2004: *Driving* in the city. *Theory, Culture, and Society*, 21(4–5), 41–59.

Thrift, N. and French, S. 2002: The automatic production of space. *Transactions of the Institute of British Geographers*, 27, 309–35.

Thwaite, B. H. 1902a: Why not a motor-car way through England? *The Nineteenth Century*, 52 (306), 305–8.

Thwaite, B. H. 1902b: Motor-car ways. *The Car (Illustrated)*, 20, 8 October, 218.

Thwaite, B. H. 1903: Concerning motor-ways: a chat with Mr. B. H. Thwaite. *The Car (Illustrated)*, 76, 4 November, 327–8.

The Times 1928: Beautifying the roads – ceremony on Kingston By-pass. 1 November, 11.

The Times 1929a: London to Brighton motor way: opposition of county councils. 9 February, 6.

The Times 1929b: London–Brighton motor road: syndicate's bill withdrawn. 12 February, 11.

The Times 1937a: German trunk roads – British delegation invited – 1000 mile tour. 15 September, 14.

The Times 1937b: Trunk roads in Germany – departure of English delegation. 25 September, 8.

The Times 1937c: German motor roads – British delegation in Berlin. 27 September, 14.

The Times 1938a: Road delegation report – making of motorways urgent. 27 January, 9.

The Times 1938b: 1000 miles of motorway – County Surveyors' Society scheme. 2 August, 7.

The Times 1938c: Demand for modern motor roads – a new organisation. 18 November, 18.

The Times 1938d: London-to-Birmingham Road scheme: Chalfont St Giles protest. 6 May, 10.

The Times 1938e: Germany as a road builder. 17 January, 14.

The Times 1938f: Reich troops cross the frontier. 12 March, 12.

The Times 1938g: Reich motor roads. 8 April, 15.

The Times 1939: Modern Roads Movement. 1 August 1939: 14.

The Times 1943: Plans for motor roads. 22 May, 2.

The Times 1954: Obituary: Mr W. Rees Jeffreys. 27 November, 6.

The Times 1955: £8 for a village Hampden? – Cottages hoping to move a motor road. 12 December.

The Times 1958a: A scar foretells an English Autobahn [picture]. 4 June, 20.

The Times 1958b: Road signs 20 ft. high – designs for new motorways – legible at 70 mph. 2 December, 10.

The Times 1959a: Relating the motorways to landscape: joint planning urged. 15 April, 6.

The Times 1959b: Popular art for a cultural minority. 6 November, 4.

The Times 1959c: Compulsory vehicle testing. 1 September, 12.

The Times 1959d: Minister 'appalled' by new motorway driving. 3 November, 8.

The Times 1959e: Overtaking technique key for drivers on M1. 10 November, 4.

The Times 1959f: Industrial films. 14 December, 14.

The Times 1959g: Help from the air for motorists [photograph]. 20 October, 22.

The Times 1959h: First motorway fine imposed. 30 November, 6.

The Times 1959i: 5,000 cars an hour on motorway. 9 November, 10.

The Times 1959j: Trips to see motorway. 6 November, 6.

The Times 1959k: Obituary: Mr J. M. Hawthorn. 23 January, 12.

The Times 1959l: Mr M. Hawthorn killed. 23 January, 10.

The Times 1959m: London–Birmingham coach time cut by over two hours. 3 November, 8.

The Times 1959n: 85 mph coaches for new motorway. 3 September, 10.

The Times 1959o: Motorway 'hard shoulder' slips. 4 November, 9.

The Times 1959p: Oakley give up land to avoid motorway. 6 August, 15.

The Times 1960a: Magistrates reject plea for drinks at M1 restaurant. 11 February, 6.

The Times 1960b: Holiday makes new records. 18 April, 4.

The Times 1961a: Experts against having crash barrier to divide M1 lanes. 27 October, 15.

The Times 1961b: Nearly £11 million orders for Jaguar in US. 5 April, 8.

The Times 1961c: £1 m. worth of goods stolen yearly by lorry thieves. 26 August, 4.

The Times 1962: Dr Wilfrid Fox (obituary). 24 May, 25.

The Times 1964a: House of Commons. Driving folly in motorway fog. 23 January, 16.

The Times 1964b: Warning sign trial on M5. 6 February, 7.

The Times 1964c: Motorway speed test talks. 16 June, 7.

The Times 1965a: Expert study crime rise in Britain. 25 November, 11.

The Times 1965b: House of Commons: Saving motorists from themselves. 11 November, 16.

The Times 1965c: 70 mph limit for four months. 25 November, 12.

The Times 1967: Jambaraya [*sic*] on the motorway. 23 May, 23.

The Times 1970a: Mr Harry Weedon [obituary]. 20 June, 12.

The Times 1970b: Twelfth death as car crosses M1 reservation. 21 August, 1.

The Times 1970c: Minister agrees to motorway barriers. 28 August, 1.

The Times 1972: Threat of strike over BBC dismissal. 17 November, 1.

Todt, F. 1938: The motor highways built by Herr Hitler. In *Germany Speaks: By 21 Leading Members of Party and State*. London: Thornton Butterworth Ltd. 251–76.

Traffic Signs Committee [Ministry of Transport] 1963: *Report of the Traffic Signs Committee 18th April 1963.* London: HMSO.

Transport Management 1927: Roads specially reserved for motor cars. October [copy held in Rees Jeffreys Archive 16/4].

Tritton, P. 1985: *John Montagu of Beaulieu 1866–1929: Motoring Pioneer and Prophet.* London: Golden Eagle/George Hart.

Troubridge, Lady and Marshall, A. 1930: *John Lord Montagu of Beaulieu: A Memoir.* London: Macmillan and Co. Ltd.

Trunk Roads Joint Committee of The Council for the Preservation of Rural England and The Roads Beautifying Association 1937: *Report of the Trunk Roads Joint Committee.* London: CPRE.

Turner, D. 1962: *Semi-Detached: A Comedy.* London: Heinemann.

Turner, G. 1964: *The Car Makers.* Harmondsworth: Penguin.

Turton, B. J. 1978: The road that started at Preston. *The Geographical Magazine,* 50(7), 452–6.

Tyme, J. 1978: *Motorways Versus Democracy.* London: Macmillan.

Urry, J. 2000: *Sociology Beyond Societies.* London: Routledge.

Urry, J. 2002: Mobility and proximity. *Sociology,* 36, 255–74.

Urry, J. 2003a: Social networks, travel and talk. *British Journal of Sociology,* 54, 155–75.

Urry, J. 2003b: *Global Complexity.* Cambridge: Polity.

Urry, J. 2004a: Connections. *Environment and Planning D: Society and Space,* 22, 27–37.

Urry, J. 2004b: The 'system' of automobility. *Theory, Culture, and Society,* 21(4–5), 25–39.

Venturi, R., Scott Brown, D. and Izenour, D. 1972: *Learning from Las Vegas.* London: The MIT Press.

Virilio, P. 1986: *Speed and Politics: An Essay on Dromology.* New York: Semiotext(e).

Virilio, P. 1991: *The Aesthetics of Disappearance.* New York: Semiotext(e).

Virilio, P. 2000: *Polar Inertia.* London: Sage.

Virilio, P. 2002: *Unknown Quantity.* London: Thames and Hudson.

Wagstaff, S. 1966: Talking to Tony Smith [interview]. *Artforum,* 5(4), 14–19.

Walker, L., Butland, D. and Connell, R. W. 2000: Boys on the road: masculinities, car culture, and road safety education. *The Journal of Men's Studies,* 8(2), 153–69.

Wall, D. 1999: *Earth First! and the Anti-Roads Movement.* London: Routledge.

Walton, J. K. 2000: *The British Seaside.* Manchester: Manchester University Press.

Wartime Journal of the Institute of Landscape Architects 1944: Exhibitions: British Road Federation. 5, 18–19.

Waters, C. 1997: 'Dark strangers' in our midst: discourses of race and nation in Britain, 1947–1963. *Journal of British Studies,* 36, 207–238.

Watson, I. 1983: *Song and Democratic Culture in Britain: An Approach to Popular Culture in Social Movements.* London: Croom Helm.

Way, J. M. 1970: Wildlife on the motorway. *New Scientist*, 47, 10 September, 536–7.

Wells, D. V. 1970: History of the Roads Beautifying Association. *The Arboricultural Association Journal*, 1(11), 295–306.

Wells, H. G. 1902: *Anticipations of the Reaction of Mechanical and Scientific Progress upon Human Life and Thought*. London: Chapman and Hall.

Wernick, A. 1991: *Promotional Culture*. London: Sage.

Wharton, M. and ffolkes, M. 1958: A land fit for traffic to live in. *Daily Telegraph*, 31 July [copy held in the Owen Williams archive].

Whitelegg, J. 1997: *Critical Mass: Transport, Environment and Society in the Twenty-First Century*. London: Pluto Press.

Whiteley, N. 2002: *Reyner Banham: Historian of the Immediate Future*. London: The MIT Press.

Williams, B. R. 1965: Economics in unwonted places. *The Economic Journal*, 75(297), 20–30.

Williams, Sir E. O. 1956: The motorway and its environment. *Journal of the Royal Institute of British Architects*, 63, 131–4.

Williams, Sir E. O. 1958a: 'Motorways and land use': Sir Owen Williams' paper at the RICS [printed extracts of speech]. *The Builder*, 7 February, 269.

Williams, Sir E. O. 1958b: Motorways and land use. *Civil Engineering and Public Works Review*, 53(621), 305 and 307.

Williams, V. 2002: *Martin Parr*. London: Phaidon.

Williams-Ellis, C. 1928 [1996]: *England and the Octopus*. London: CPRE.

Williams-Ellis, C. 1930: Introduction. In H. Peach and N. Carrington (eds), *The Face of the Land*. London: George Allen and Unwin, 11–24.

Williams-Ellis, C. 1978: *Around the World in Ninety Years*. Portmeirion: Golden Dragon Books.

Williamson, T. 2003: The Fluid State: Malaysia's National Expressway. *Space and Culture*, 6(2), 110–31.

Wilson, A. 1992: *The Culture of Nature*. Oxford: Blackwell.

Winter, E. 1961: Record catch. *English Dance and Song*, New Year special, 27–30.

Wolff, J. 1993: On the road again: metaphors of travel in cultural criticism. *Cultural Studies*, 7, 224–39.

Wollen, P. 2002: Automobiles and art. In P. Wollen and J. Kerr (eds), *Autopia: Cars and Culture*. London: Reaktion, 25–50.

Wollen, P. and Kerr, J. (eds) 2002: *Autopia: Cars and Culture*. London: Reaktion.

Wolmer, Lord 1937: Motor roads in Germany – Impressions of a tour – a profitable example. *The Times*, 7 October, 15–16.

The Wolverton Express 1958a: Farmer Williams wins with a road block. 12 May.

The Wolverton Express 1958b: No rest from the noise. 16 May.

The Wolverton Express 1958c: Will have to bring gravel sixty miles if fail appeals. 18 July.

The Wolverton Express 1958d: Effect of the motorway: Newport has a traffic problem. 10 October.

The Wolverton Express 1958e: Farmers appeal to motorway sightseers. 16 May.

The Wolverton Express 1958f: Years driving ban on motorway man. 29 September.

Woodward, R. 2004: *Military Geographies*. Oxford: Blackwell Publishing.

Working Party 1972: *How Do You Want to Live? A Report on the Human Habitat*. London: HMSO.

Wright, P. 1995: *The Village That Died for England: The Strange Story of Tyneham*. London: Jonathan Cape.

Wylie, J. 2002: An essay on ascending Glastonbury Tor. *Geoforum*, 33, 441–54.

Wylie, J. 2005: A single day's walking: narrating self and landscape on the South West Coast Path. *Transactions of the Institute of British Geographers*, 30, 234–47.

Wylie, J. 2006: Depths and folds: on landscape and the gazing subject. *Environment and Planning D: Society and Space, 24, 519–35.*

Yeomans, D. and Cottam, D. 2001: *The Engineer's Contribution to Contemporary Architecture – Owen Williams*. London: Thomas Telford.

Young, D. 2001: The life and death of cars: private vehicles on the Pitjantjatjara lands, South Australia. In D. Miller (ed.), *Car Cultures*, Oxford: Berg, 35–57.

Zeller, T. 1999: 'The landscape crown': landscape, perceptions, and modernizing effects of the German Autobahn system, 1934 to 1941. In D. E. Nye (ed.), *Technologies of Landscape*. Amherst: University of Massachusetts Press, 218–38.

DISCOGRAPHY

Black Box Recorder 2000: 'The English Motorway System'. On *The Facts of Life*, CD album, Nude Records, Nude 16CD.

Roy Harper 1977: 'Watford Gap'. On *Bullinamingvase*, LP album, EMI Records SHSP4060.

Ewan MacColl, Charles Parker and Peggy Seeger 1999: *Song of a Road*. CD album, Topic Records, TSCD802.

St Etienne 1994: *Like a Motorway*, 7 inch single. Heavenly Records, HVN 40.

Ted Taylor Four 1960: *M1*, 7 inch single, Oriole Records 1573.

FILMS

Butterfly Kiss 1995: Written by Frank Cottrell Boyce. Directed by Michael Winterbottom.

Charlie Bubbles 1967: Written by Shelagh Delaney. Directed by Albert Finney.

Crash 1996: Film written and directed by David Cronenberg, based on the novel by J. G. Ballard.

Hell Drivers 1957: Rank/Aqua. Directed by C. Raker Endfield.

The Hi-Jackers 1963: Written and directed by Jim O'Connolly.

The Irishmen: An Impression of Exile 1965: 16 mm film, produced and directed by Philip Donnellan for the BBC. Video copy held in Birmingham City Archives.

London Orbital: A Film by Iain Sinclair and Chris Petit 2002: Channel Four Television, UK. Available on DVD by Illuminations.

Major Road Ahead 1958: 16 mm film, produced by John Laing and Son Limited. Copy viewed in CTIS library, Laing Technology Group, Hemel Hempstead.

Motorway 1959: 16 mm film, produced by John Laing and Son Limited. Copy viewed in CTIS library, Laing Technology Group, Hemel Hempstead.

Radio On 1979: Directed by Chris Petit.

Week-End 1967: Written and directed by Jean-Luc Godard.

Index

Italicized numbers refer to figures.

A1 20, 61, 244n
A5 85, 184–5, 190, 193, 203
A30 61, 207
A34 20, 207
A40 20, 210
AA, *see* Automobile Association
Abercrombie, Patrick 29–30
ability 149–52, *150, 151*
Abrams, Mark 180–1
Absolute Beginners (book) 181
accidents, *see* traffic accidents and
    crashes
Adshead, S. D. 29–30
advertising 30, 37, 49, 122, 149
Advisory Committee on Landscape
    Treatment of Trunk Roads
    73, 80–3, 85, 89–97, 100, 172,
    178
  formation of 81
  membership 81–2
  M1 and 81–5, 89–97, 100
  planting policy 81, 89–90, 96
  service areas 91–3, 96–7, 178
  signs 99–100
Advisory Committee on Traffic Signs
    for Motorways 97–102
  David Kindersley and 100–1
  experiments by 98–101
  formation of 99

Jock Kinneir and 98–9, 100, 102
Margaret Calvert and 98–9, 102
membership 98
see also signs
aerial viewpoint 149, 164
  of motorway construction 111–16,
    114
aesthetics 12–15, 50, 79, 80, 83, 86,
    101, 104, 166, 214–15
  landscape 37, 54–5
  motorways and 14–15, 32, 37, 63,
    80, 85–7, 89, 101–2, 166,
    214–15, 244n
  technology 104
agriculture 25, 27, 29–30, 45, 61,
    70–1, 110, 202, 231n, 234n
airports 5, 10, 47, 210, 214
  Gatwick 98
  Heathrow 210
  Third London Airport 203
alcohol 47, 179–80, 242n
Americanization 149, 172, 180
  British culture 126–7, 172, 180
  folk culture 126–7
Anderson, Sir Colin 98, 100, 235n
Anglo-German Fellowship 33
anthropology 3, 11, 20
Appleton, Jay 67–8
Appleyard, Donald 2, 14–15

architecture   2, 15, 18, 36, 44, 51,
    56, 74, 83, 86–7, 91, 93, 205,
    226n, 233n
  Brutalism   15, 86, 93
  Festival of Britain and   80
  functionalism   50, 83, 86–7, 89,
    91–4, 97, 102, 213
  geography and   1, 74
  M1 structures   83, 85–7, 89, *85*
  modernism   2, 21, 49–50, 69, 83,
    85–7, 89, 90–4, 180–1, 184
  picturesque   79–80, 93–4
  service areas   90–7, *92*, *95*, 206–7,
    241n
  vernacular   2, 18, 91, 93, 215,
    226n
Architects' Journal, The   83, *88*, 93
Architectural Review, The   50, 74,
    78–80, *78*, *79*, 93–6, *95*, 101, 215,
    231n, 233n
  editors of   80
  picturesque planning   77–80, 93
  Raymond Spurrier and   77–9, 80
  'Roads' issue   50
art   1–3, 15–16, 75–7, *78–9*, 80, 192,
    205–6, 212–13, 235n
Ashley, Wilfrid   51
Aston Martin   172–3
Augé, Marc   10–11, 17, 177–8, 201,
    210–12, 214
Austria   33–4, 40
Autobahnen   18, 31–40, 42–3, 49, 81,
    83, 85
  1937 British tour of   32–3, 35–8,
    40, 42
  British Minister tours   39–40
  British reactions to   33–40, 42–3,
    172
  design and planning   18, 33, 36,
    49, 83, 85
  Hitler and   31–4
  landscaping and planting   18, 32–3,
    36–8, 49, 81, 83, *84*
  militarism   33–5, 37
  national plans   31
  Nazi propaganda and   32, 38
  signs on   98

automobile(s), *see* car(s)
Automobile Association (AA)   24, 38,
    45, 51, 145, 147, 155, 160–2,
    161, 167, 176, 194, 219, 230n
  German Roads Delegation (1937)
    and   32, 38
  Guide to the Motorway   145, *146*,
    147
  M1 patrols   152, 160–2, *161*, *163*,
    167
  Roads Beautifying Association
    and   51, 230n
  Trunks Roads Act and   38
automobility   6–8, 19, 189, 225n
Automotive Products Associated
    Limited   149–52, *151*

Baker, Margaret   177–8, 218
Balfour, Arthur   25
Ballard, J. G.   19, 192, 209–10, 216
  Concrete Island   19, 209–10, 244n
  Crash   19, 192, 209–10
  dystopian narratives   209–10, 216
banality   213
Banham, Reyner   1–3, 14–15, 48, 79,
    80, 86, 87, 93, 94, 204, 208,
    224n, 226n, 233n
  Brutalism   86, 93
  criticizing M1   86, 233n
  on modern architecture   79, 86
  on movement and vision   14
Barlow Commission   45
Barnes, Alfred   61–3, 66
Baudrillard, Jean   8, 12
BBC   110, 124, 159, 217
  Irishmen, The   139–40, 238n
  motorway radio programmes
    124–39, 159, 166–7
  motorway television
    programmes   129, 156
  Radio Ballads   110, 124–39,
    237n–8n
  Song of a Road   110, 124–39,
    237n–8n
  see also Donnellan, Philip; Parker,
    Charles
Bean, William J.   51

Bedford, Duke of   177
Bedfordshire   70, 73, 111, 158–9,
    162, 202
Beeching, Richard   239n
Beesley, Michael   188
Belloc, Hilaire   28–9
Betjeman, John   86, 205–6
Birmingham   2, 27, 39, 73, 91, 93,
    116, 127, 154, 159, 166, 168–9,
    173–4, 240n
Black Box Recorder   19
black-out   44
Blackwood Hodge   106, *107*
blame, *see* traffic accidents and
    crashes
Blizard, George Pearce   29, 31
Blue Boar   91, 179, 185–6, 206,
    235n
body   44, 52, 65–6, 76, 109, 113,
    139, 192, 209–10
  construction workers   106, 109,
    113, 118–19, 122–3, 140
  driving and   7–9, 11–12, 14, 76,
    149, 156, 192
  traffic and bodily metaphors   44,
    63–6, 112
  see also embodiment
books about motorways   18–19,
    168–71, 208–10, 241n
  children's books   168–71, *168*, *170*,
    241n
  dystopian novels   19, 208–10
  I-Spy on the Motorway   218
  *Man who Travelled on Motorways,
    The*   19
  see also Ballard, J. G.
boredom   10, 19, 89, 184, 212–15,
    218
  Boring Postcards   184, 214–15
  motorway driving and   10, 89,
    212–13, 218
botanical surveys of M1   216
Boumphrey, Geoffrey   35, 65
Bowes-Lyon, Sir David   81, 89, 93,
    100–1
Boyd-Carpenter, John   66
Bracewell, Michael   18–19, 213–14

breakdown of vehicles   143, 154,
    159–62, *163*, 178, 190, 217
Bressey, Charles   36, 43, 51, 229n
bridges on M1   86–90, *85*, 92,
    147, *148*, 170, 195, 197–9,
    218
  accidents and   195, 197–9, 233n
  central supports   84, *148*, 197,
    233n
  cost of   43, 84, 233n
  criticisms of   86, 100–1, 197, 215
  design of   83, 85–6
  farm access   87, 202
  footbridges   91–2, *92*
  heritage and   215
  over-bridges   83–86, 90, *85*, 114,
    147, *148*, 170, 195, 197–9
  railway bridges   87, 233n
  under-bridges   87, *183*
Brighton   24, 26, 29–30
British Broadcasting Corporation, *see*
    BBC
British Industrial Plastics (BIP)
    173–4, *175*
British Railways   66–7, 99, 173
British Road Federation (BRF)
    38–43, 45, 50, 56–9, 63, *64*, 66,
    220, 229n, 242n
  Case for Motorways, The   63, *64*
  Crisis and the Roads   65
  Economics of Motorways   63
  exhibitions by   50, 56–9
  formation of   39
  German Roads Delegation and
    32
  landscaping   56–9
  lobbying parliament   59
  motorways and   39–43, 45, 56–9
  Motorways for Britain   56–9
  National Motor Roads   39
  Road Way to Recovery   63
  see also Brunner, Christopher;
    Jellicoe, Geoffrey
British Union of Fascists (BUF)   33
Britishness   106–9, 123, 131, 139,
    171–2, 200
Brodsly, David   18, 20

Brooks, Tony   171–3, 181
Brunner, Christopher   63–5
Brutalist architecture   15, 86, 93
Buchanan, Colin   239n
Buckinghamshire   56, 86, 93, 111,
     143, 158, 202–4, 228n
Burgin, Leslie   39–40
Burra, Edward   205–6
bypasses   48, 211
   Doncaster   66, 69
   Kingston   30, 51–3
   Mickleham   77
   Newbury   207
   Oxford   77
   Preston   66, 97, 99–100, 141
   St Albans   66, 69, 72, 73, 91, 104,
     129, 141, 153, 172, 182

cafés   205–6
   lorry drivers and   131, 184
   service areas   167, 179–82, 184–6,
     206
   transport cafés   91, 131, 184–6,
     184, 242n
   Soho   181
   teenagers and   181–2
campaigns   30, 70–1
   against M1 routes   71, 204
   anti-car   16–17
   anti-road construction   70, 204,
     207–8
   urban conservation   204–6
   see also protests
car(s)
   accessories for   147–52, 150, 151
   consumption and   6–7, 11
   geography and   7
   maintenance   147–52
   manufacturing   6–7, 149, 162
   ownership levels   25, 27, 147
   sociology and   7
caravan sites   121–2
Carrington, Noel   49, 98–100
Carter, Bruce   169–71, 170, 241n
Casson, Sir Hugh   98, 100
Castle, Barbara   192

Caterpillar construction
   machines   106–9, 108
central reservation (median strip)   10,
     195–7, 195
   accidents and   195–7, 195
   bridges columns on   197–9
   crash barriers on   195–6
   planting vegetation on   89
   vehicles crossing   195–7, 195
Charnwood Forest   71, 204, 215
children   133–4, 174, 218
   motorway books   19, 168–71, 168,
     170, 174, 241n
   motorway toys   19, 174
   of construction workers   133–4
   radio programmes for   174
Chubb, Lawrence   29
cinema,
   driving and cinematic gaze   13, 96
   see also films
circulation, see body; traffic
   congestion
cities   12, 14, 25, 205
   see also urban
citizenship   6
Civic Trust   205
Clements, R. G. H.   36–8, 42
Clough, Sidney, see Sidney Clough
   and Sons Limited
coaches   47, 180
   Midland Red motorway
     express   173–4, 175, 241n
coffee bars   172, 181
collisions, see traffic accidents and
   crashes
Colvin, Brenda   48, 55–7, 73–4, 79,
     80, 82–4, 86–7, 89, 100, 212
   and Institute of Landscape
     Architects   55, 80–2
   Land and Landscape   56, 73–4
   remarks on M1   83, 86, 87, 89, 100
   Roads Beautifying Association
     and   55–6
   Trees for Town and Country   80
Commons and Footpaths Preservation
   Society   29

Communism   110, 125, 127, 137
complexity   6, 224n–5n
conduct
   codes of conduct   144–5, 153, 158,
      176
   construction workers   115
   drivers   7, 9, 16, 47, 98, 111,
      143–5, 153–9, 167, 197, 225n
   service area users   178
congestion, see traffic congestion
conservation   204–6, 216
Conservation Society   207
Conservative Party   25, 61, 66, 70,
      152–3, 179, 196, 198
construction sites
   landscape   90, 117
   M1 motorway   90, 104–24, 128–40
   spectacle of   112
consulting engineers for M1, see Sir
      Owen Williams and Partners
consumption   6–7, 11, 21, 23, 46,
      162
   driving and   6–7, 10–11, 152
   mass   23, 206–7
   motorways as space of   10–11, 38,
      94, 97, 152, 162–86, 190, 206,
      210–11
   service areas and   94, 96, 178–86,
      206, 241n
contracting engineers for M1, see John
      Laing and Son Limited; Tarmac
      Civil Engineering Company
Cook, Frederick   40, 45–6
Cooke, Sir Stenson   33
corporate spaces   105, 113, 115,
      119–22
cost-benefit analysis   186–9
Council for the Preservation of Rural
      England (CPRE)   29, 47–9,
      53–5, 61, 71, 80–1, 93, 221, 232n
Council for the Preservation of Rural
      Wales (CPRW)   80
Country Landowners'
      Association   70–1, 231n
County Surveyors' Society   40–3, 41,
      65

Coventry   28
Coventry Cathedral   21, 85, 240n
Crary, Jonathan   8, 10, 156
Crash (book by J. G. Ballard)   192
Crash (film by David
      Cronenberg)   192
crashes, see traffic accidents and
      crashes
Cresswell, Tim   3–6, 18, 133, 224n
Crick   73, 203
crime
   books about M1 and   168–71
   effects of M1 on   159–60
   gangs   159–60
   lorry theft   160
   motorway crime   159–60, 168–71
Crome, John   76–7, 77
Cross, Andrew   211–13, 244n
Crowe, Sylvia   48, 73–7, 79–80,
      82–85, 87, 88, 100–1, 233n
   and Institute of Landscape
      Architects   81–2, 101
   comments on M1   83, 84, 85–9,
      88
   comments on motorway
      signs   101–2
   Landscape of Roads, The   74–5, 77,
      83, 84
   on modern landscapes   74–6, 79,
      83
Crutchley, Brooke   99–100, 101
Cullen, Gordon   79–80, 101
Cuneo, Terence   113–15, 121
   M1 painting   113–15, 114, 121
   railways   113–15

dance   1–3, 109
Daniels, Stephen   13–14, 76, 106,
      113
dazzle   73, 173, 191
death   116, 168, 171, 177, 180, 186,
      192–9, 217
defence   228n
   motorways and   35, 39–40, 50, 66
democracy   34, 49, 207
Department of Environment   206

depressed regions 28, 52–3
'depression, the' 27–8, 30
design 74, 80
  landscape architecture and
    design 2–3, 9, 36–8, 55–9, 73,
    83, 91
  motorway signs 97–9, 101–2
  motorways and 79
  roads and 46–51
  see also architecture; functionalism;
    landscape architecture and design;
    modernism
Design and Industries Association
    (DIA) 47–9
detachment
  driving and 11, 212
Dial M for Murder (play and
    film) 193
Diana, Princess of Wales 217
DIRFT (Daventry International Rail
    Freight Terminal) 203, 211–12
distance, perceptions of 156–7
distraction, driving and 8, 10, 90
distribution centres 203, 243n
documentaries, see films; television
domesticity 121
Donnellan, Philip
  Irishmen, The 139–40, 238n
driving
  attention and 8–9, 73, 145, 156,
    204
  black-out and 44
  cognition and 8
  distraction and 8, 10, 90
  embodiment 1–3, 11–12, 14, 16,
    76, 80, 142
  freedom and 8, 19, 144, 162, 166,
    168, 191, 208
  landscape and 3, 10, 12, 16, 73,
    90, 93–4, 96–8
  motorway driving 16, 143–62, 204
  practices of 3, 6–8, 9–12, 15–16,
    143–62, 226n
  vision and 3, 8, 12–16, 75, 76, 80,
    89, 145–9, 150, 151, 155–6, 212,
    218

dwelling 3, 5, 7
dystopia(s)
  motorway verges 208–10
  motorways and 19, 208–10

East Anglia 39
East Midlands 68–9
ecology 37–8, 81, 207, 215–16, 232n
  motorway verges 215–16
  roadsides 55, 91, 97
economics
  accidents and 187–8
  calculation 187–9
  construction 27–9, 43, 71
  motorways and 63, 186–9, 203
  see also cost-benefit analysis
elections 29–30, 152
embankments
  motorways 37, 83, 84, 87, 90,
    93–4, 97–8, 101–2, 209
embodiment 1–3, 11–12, 14, 16, 76,
    80, 142
  skills and practices 2–3, 7–8, 12,
    14, 142, 144–5, 213
  technological enhancements
    11–12
  see also body
emotions 11, 19, 218
  anger 11, 164, 207
  driving and 11, 145
  excitement 11, 19, 24, 96, 164,
    169, 200, 213, 218
  fear 11, 24, 154, 164, 169, 200
  motorways and   145, 154, 164,
    169, 213, 218
empire 33–4, 46, 113, 123, 131,
    228n
  construction companies and 113
  roads and 34, 228n
engineering, see construction sites;
    Institution of Civil Engineers
    (ICE); Institution of Highway
    Engineers (IHE); John Laing and
    Son Limited; Road Research
    Laboratory (RRL); Sir Owen
    Williams and Partners

engineers   9, 16, 44, 50, 69–70, 77,
    79, 82, 120, 128, 168
England   48–50, 53, 66–7, 74, 123,
    130, 134, 171–2, 205, 209,
    211–12, 244n
English Heritage   215, 226n, 245n
Englishness   49–50, 81, 83,
    93–4, 97, 99, 102, 117, 126,
    171–2, 200, 205, 209, 211–12
environment   6–8, 16–17, 207
    Department of the
        Environment   206
    driving and   6–8, 17, 204–6
    Friends of the Earth   206
    Greenpeace   206
    motorway construction and   204–8
ethnicity   7, 12, 81, 113–14, 122–4,
    139, 145
    Britishness and Englishness   122–4
    construction workers   109, 122–4,
        139
Europe   172
    road numbering across   93
    studies of signs in   98
evacuation   35, 43, 46, 50
everyday, the
    driving and   10, 171
    motorways and   10, 101, 171, 200,
        214, 216, 243n
experiments   21, 40, 42, 98–101, 141,
    163
    accidents   197–9
    motorway construction   141, 186
    motorway design   186, 200
    signs   98–101
externalities   7, 188–9

Fairbrother, Nan
    M1 and   215–16
farmers   24, 71, 128, 138, 202, 217
    complaints about sightseers   112
    objecting to M1 route   70, 128,
        138
    see also landowners
fascism   31, 33–4
fear, see emotions

Festival of Britain   21, 73, 80
films   18–19, 32, 44, 50–1, 128,
    139–40, 158, 192, 217, 243n
    3 Hours from Here   211–13
    Butterfly Kiss   19
    Charlie Bubbles   19
    Crash   192
    featuring the M1   11, 19–21, 111,
        119–20, 123, 125, 128, 139–40,
        217
    Hell Drivers   111, 236n
    Hi-Jackers, The   160
    Irishmen, The   139–40, 238n
    London Orbital   226n–7n
    Major Road Ahead   111
    Motorway   119–20, 123
    of motorway construction   11,
        19–21, 123, 125, 128, 139–40,
        217
    public information films   158
    Radio On   19
    road movies   18–19
    Roads Across Britain   50–1
    Week-End   192
fog
    accidents in   192, 194
    motorways and   192, 194
    warning signs   194
folk culture   124–5
    and industrial heritage   124–6
    and working classes   125–6
folk music and song   124–40
    British revival   125–6
    Irish   129–31, 134–5, 139–40
    romanticism   133–4
    Scottish   125–6, 134–5, 139
    see also MacColl, Ewan; Parker,
        Charles; Seeger, Peggy; Song of a
        Road (BBC Radio Ballad)
food   46, 121
    Good Food Guide, The   176–7,
        182
    service areas and   178–82, 184–6,
        206–7
Ford   6, 149, 154, 159–60, 185,
    198

Forte
  Newport Pagnell service area   91,
    93, 94, 96, 179–86, *183*,
    235n
  Soho coffee bars   181
Foucault, Michel   115, 142, 144,
    158
Fox, Wilfrid   51, 53–5, 80–1, 230n,
    232n
fox hunting, motorways and   202
Fraser, Tom   192
freedom, driving and   8, 19, 144,
    162, 166, 168, 191, 208
Freeway, The (play)   19, 208–10
freeways   1–3, 8, 20, 208–10
  Los Angeles   1–2, 12, 20
Friedberg, Anne   13
Friends of the Earth   206
functionalism   73, 89, 91–4, 102,
    213
  architecture   50, 83, 86–7, 89,
    91–4, 97, 102, 213
  design   50, 79–102
  'road-style'   93–4, *95*
  roadside planting   53, 73, 91, 97
  service areas   91–4, 96, 97

games   178, 218
garden design   74
  Institute of Landscape Architects
    and   56, 74
  see also landscape architecture and
    design; roadside planting; trees
    and shrubs
gender   7, 12, 133, 138–40, 225n
  children's books   168–71
  construction site   90, 113–14,
    120–2, 133, 139–40, 238n
  driving and   7, 12, 120, 145, 149,
    150, 167–74, 225n
  see also masculinity
Geographical Magazine, The   33–4, *72*,
    83, 203
geography   3–5, 7, 9–11, 18, 20, 42,
    45, 67–9, 99, 116, 156–7
  driving and   11, 16, 100

motorways and   10, 16, 39, 67–9,
    156–7
geology   83, 132–3
German Roads Delegation
  1937 tour   32–3, 35–8, 40, 42
  reports   35–8, 42
  see also Autobahnen
Germany   18, 31–40, 42–4, 49–50,
    76, 81, 83, *84*, 98, 155, 158, 172,
    191, 227n–8n
  see also Autobahnen; German Roads
    Delegation; Nazi regime
Giles (cartoonist)   *184*, 184–5
Gill, Eric   100
Glanville, Sir William   98, 186–7
Good Food Guide, The   176–7, 182
government and
    governmentality   225n
  conduct   9, 142, 153, 158–9, 197
  construction workers   115
  drivers and   9, 142, 144, 153,
    158–9, 197, 225n
  Foucault and   115, 142, 144, 158
  self-government   142–7, 153
  technologies of   97, 98, 102,
    115–16, 121, 142, 144, 158
Greenpeace   206
Gresham-Cooke, Roger   40–2, 229n
guide books   47, 174–8, 218
  Discovering M1   177–8, 218
  Good Food Guide, The   176–7, 182
  Guide to the Motorway (AA)   145,
    146, 147
gypsies   6, 122, 127

Halprin, Anna   3
Halprin, Lawrence   1–3, 48
Hamilton, Richard   14–15, 225n
hard shoulder   158–9, 190, 194, 213
Harper, Roy   206–7
Harris, Alan   83, 86, 87
Harry W. Weedon and Partners   93
Harvey, David   157
Hatfield and the North   244n
Hawthorn, Mike   171
health   45–6, 63

and driving   156
and nation's roads   63–6
metaphors   63–6
helicopter   111–16, *114*, 120, 234n
Hell Drivers (film)   111, 236n
heritage
conservation   204–6
European Architectural Heritage
    Year   205
motorway bridges   215, 245n
see also Betjeman, John;
    conservation; preservation
Highway Code, The   9, 144–7, 158,
    174
see also conduct; *Motorway Code*
Hi-Jackers, The (film)   160
history   17, 20, 124
Hitler, Adolf   31–4
Hoggart, Richard   126–7, 180
Hollowell, Peter   7, 185–6
home   121–2, 128, 133–4
horticulture   30, 51–6, 81
exotic species   53, 55
native species   81
Roads Beautifying
    Association   51–5
see also landscape; roadside planting;
    trees and shrubs
Hoskins, W. G.
M1 and   215
Making of the English Landscape,
    The   215
hostels   122, 129, 131–2, 137
conditions of hostels for
    labourers   131–2, 137
see also Youth Hostels Association
hybridity
vehicle driver   8, 142–3, 193

immigration   21, 123, 131
India   11, 20, 120, 123, 130, 135
India Tyres   149, *150*, 167
industrial estates   52, 203, 210–12
Institute of Landscape Architects
    (ILA)   55–7, 73, 80–2, 101,
    230n–2n

formation of   230n
journals and magazines   55, 57
reconstruction   73
Roads in the Landscape   80,
    232n
roadside planting   55–9, 91, 97
see also Colvin, Brenda; Crowe,
    Sylvia; Jellicoe, Geoffrey;
    landscape architecture and design
Institution of Civil Engineers
    (ICE)   26, 44, 136, 221
Institution of Highway Engineers
    (IHE)   45, 63, 229n
IRA and M1   216
Ireland
construction workers   123, 129–31,
    133–5, 137, 139–40
Irishmen, The (film)   139–40,
    238n
migration   139, 238n
music and song   129–31, 133–4,
    139–40
I-Spy on the Motorway   218
Italy   31, 49, 130, 170, 172
Italianization of style   172

Jacobs, Jane   225n
Jacobs, Jane M.   224n
Jackson, John B.   2, 48, 226n
Jaguar   154, 171, 173, 243n
Jamaica   120, 122, 130–1, 182
jazz   125–6, 135, 139
Jeffreys, William Rees   220, 227n
Jellicoe, Geoffrey   56–9, 73, 76–7,
    79–82, 87
and British Road Federation   56–9,
    58, 63, *64*
and Institute of Landscape
    Architects   56
and Royal Fine Art
    Commission   81
Case for Motorways, The   59, 63,
    64
M1 and   81
Motorways for Britain   57–9
Joad, Cyril E. M.   47, 112

John Laing and Son Limited   71,
    103–6, 125, 128–9, 136, 158,
    220, 227n
  contractors for M1
    construction   103–6, 227n
  corporate spaces   105, 113, 115,
    119–22
  films by   111, 119–20, 128, 139
  *London to Birmingham Motorway,
    The*   71, 105–6, 109, 111–13,
    115, 117–18, 120–3, 139
  Major Road Ahead (film)   111, 139
  Motorway (film)   119–20, 139
  public relations   105, 110–12, 120,
    125, 128–9, 136–7, 139
  Team Spirit staff newsletter   85,
    111, 114–15, 120–3, 139

Kaplan, Caren   5
Katz, Jack   7–8, 11–12, 145, 155,
    164, 193
Kegworth, M1 air crash   217
Kent   39, 218
kinaesthetics   1–3, 12, 16
Kindersley, David   100–1
Kinneir, Jock   98–9, 100, 102
kitsch   19, 214–15

labour   21, 106, 155, 120, 122–4,
    129, 133, 138–9
Labour Party   50, 61, 63, 192, 204
Laing construction, *see* John Laing and
    Son Limited
Lancashire   40, 45, 52, 57, *58*, 66,
    141, 159
landowners   70–1
  object to M1 route   70–1
  see also Country Landowners'
    Association; farmers; Bedford,
    Duke of
landscape   1, 18, 20, 51, 55–9, 67,
    73–6
  Autobahnen and   18, 32–3, 36–8,
    49, 81, 83, *84*, 85
  Chinese conceptions   76
  construction   3, 103–9, 215–16

driving and   3, 10, 12, 16, 100
  English   59, 74–6, 79–81, 83–4, *84*,
    93–4, 97, 99, 102, 116, 123, 163,
    171–2, 200, 205, 211–13, 215–16
  mobility and   1–3, 6, 13–16, 55–6,
    60, 83–5, 87, 102, 177
  motorways and   10, 12, 14–16,
    32–3, 36–8, 46, 50, 55–9, 61–3,
    64, 73, 76, 79–83, 87, 90, 93–4,
    97–8, 101–2, *75*, 157, 163, 166,
    171–2, 177–8, 184, 189, 207,
    215–16
  painting   13, *75*, *77*, 113–15, *114*
  planting of   9, 37–8, 51, 53–5, 77,
    80–1, 89–90, 96, 215–16
  quarries and   103–11
  scars across   61, 112, 115
  sculpture and   1, 75
  service areas   90–1, 93–4, 96–7
  speed and   13, 47, 55, 60, 74–5,
    75, 89–90, 100, 163
Landscape Advisory Committee, *see*
    Advisory Committee on
    Landscape Treatment of Trunk
    Roads
landscape architecture and design
    2–3, 9, 36–8, 55–9, 73, 83, 91,
    197, 212, 215
Langley-Taylor, George   80, 93, 101
Larkin, Philip   206, 244n
Latour, Bruno   9, 11, 119, 142, 149,
    152
Laurier, Eric   157, 203, 213, 217
Law, John   119, 193
Lawrence, Sir William   30, 51
Leeds   167, 203, 216
Leicester   48, 120, 217
Leicester Forest East service area   182
Leicestershire Trust, The   71
leisure   71, 139
  driving and   65, 163–4, 167, 189, 206
Leonard Manasseh and
    Partners   94
Lloyd, A. L.   125–6, 135
Lloyd, Thomas I.   66, 231n
Lloyd-George, David   26

Lomax, Alan 125–6
London 36, 39–40, 45, 56, 116, 141,
    145, 159, 166, 168, 180
  Irish communities in 139
  metropolitan culture 172
  motorways in 204, 209–10
  North 98, 139–40, 172, 185, 207
  roads in 172, 207
  Soho 181
  Uxbridge 27
  West 209–10
  Westway 18, 244n
  see also Victoria Line (London
    Underground)
London–Birmingham Motorway, see
    M1 motorway
London–Birmingham Motorway, The, by
    L. T. C. Rolt 105–6, 109,
    111–13, 115, 117–18, 120–3, 128
London Orbital (book) 19–20
London Orbital (film) 226n–7n
London Orbital motorway, see M25
    motorway
London Transport 44, 164
LondonYorkshire Motorway, see M1
    motorway
Long, Archibald P. 82, 89, 96, 233n
lorry/lorries 25, 27, 106, 110–11,
    131, 158, 172–3, 184–6, 190,
    194, 197–8, 205–6, 236n, 242n
  driving 100, 111, 131, 145, 172,
    184–6, 194, 197–8, 211–12, 225n
  impact on landscape 110–11,
    205–6
  phantom van on M1 197–8
  service areas and 94, 184–6
  transport cafés and 131, 184–6
Los Angeles 1–2, 12, 20
Lupton, Deborah 8, 149, 152, 164,
    193
Luton 73, 91, 111, 153, 182
Lynch, Kevin 2, 14–15

M1 motorway
  air crash, Kegworth 217
  botanical survey 216

  breakdowns on 143, 154, 159–62,
    163, 178, 190, 217
  bus trips 164
  Charnwood Forest and 71, 204,
    215
  coach services 173–4
  construction 103–40
  consumption of 3, 10–11, 94, 96,
    152, 162–86, 206
  corridor 67, 160, 176–7, 202–4,
    216
  Crick to Doncaster sections 68–9
  driving along 143–99
  early proposals for London–
    Birmingham Motorway 25–7,
    40, 42, 228n
  economic studies of 63, 186–9,
    203
  experiments on 186–99
  music about 164–6
  opening 106, 141, 152–4, 239n
  policing 158–60, 162, 170, 192,
    194, 196
  route of 67, 70–1, 72, 81
  St Albans Bypass section 66, 72,
    73, 91, 104, 129, 141, 153, 172,
    182
  scientific studies of 186–99
  see also landscape; motorway service
    areas
M2 motorway 218
M3 motorway 207
M4 motorway 61, 204, 210
M5 motorway 61
M6 motorway 61, 203–4, 217–18
  Preston Bypass section 21, 97,
    99–100, 141
  Spaghetti Junction (Gravelly
    Hill) 2
M10 motorway 172, 227n
M11 link road 207
M25 motorway 19–20, 61, 226n–7n
  see also London Orbital (book and
    film)
M27 motorway 207
M40 motorway 212–13, 244n–5n

M62 motorway   61
M77 motorway (Pollok)   207
Mabey, Richard   216
McClintock, Anne   123
MacColl, Ewan (b. Jimmie Miller)   110,
   124–40, 223, 237n–8n
   BBC and   125
   Communism   125
   early life   125–6
   folk song revival   125–7
   Irish music and song   129–31,
     133–4, 139–40
   Irishmen, The   139–40, 238n
   Parker, Charles and   110, 124,
     237n
   Radio Ballads   124–5, 237n–8n
   Seeger, Peggy and   132, 237n
   Song of a Road   110, 124–5,
     237n–8n
   Theatre Workshop and   125
MacEwen, Malcolm   202
MacInnes, Colin   181
Macmillan, Harold   21, 141, 152–3
maintenance
   motorways   190–1
   vehicles   147–52
Manzoni, Herbert   65–6
maps   16–17, 116–20, 132, 145,
   174–8, 217
   authority and   71, 119–20
   driving and   177–8
   Ordnance Survey   132, 174
   of M1   145, 147, 166–7, 174
   of proposed motorway
     networks   *41, 62*
Marples, Ernest   102, 152–4, 171,
   179, 190–3, 239n
Martin, Robert   168–9, *168*, 241n
masculinity
   children's books and   168–71
   driving and   149, *150*, 167–74,
     225n
   labouring and   106, 109, 113–14,
     128, 133, 138–40, 238n
Massey, Doreen   6, 20, 224n
materiality   1, 5–7, 71, 104, 111, 116,
   118–19, 123, 142, 225n

Matless, David   21, 29–31, 38, 47–9,
   73, 81, 177, 206, 244n
Michael, Mike   6, 11, 142, 164, 193
Midland Red Motorway Express, *see*
   coaches
Midlands of England   45, 48, 83,
   145, 204
military   44
   engineering and   116
   metaphors during
     construction   116–19, 236n
   motorways and   26, 33–5, 37, 39,
     50
   see also Autobahnen
Miller, Daniel   7–8, 11, 189
Miller, Jimmie, *see* MacColl, Ewan
Milton Keynes   203, 217
Ministers of Transport, *see* Ashley,
   Wilfrid; Barnes, Alfred;
   Boyd-Carpenter, John; Burgin,
   Leslie; Castle, Barbara; Fraser,
   Tom; Marples, Ernest; Morrison,
   Herbert; Peyton, John; Watkinson,
   Harold
Ministry of Agriculture, Fisheries and
   Food   110, 231n
Ministry of Information   44, 117
Ministry of Transport   27, 35–6,
   39–40, 42, 44, 46, 49–52, *62*, 66,
   69–70, 82, 90, 101, 104, 120,
   128, 137, 139, 159, 176, 178,
   185, 229n, 231n
mirrors
   driving and use of   144, 155–6,
     173, 212
Mitchell, Don   3–4, 103, 140
Mitchell, W. J. T.   3
mobilities   4–6, 16, 76, 111, 121–2,
   124, 164, 224n, 226n
Modern Roads Movement   40, 229n
modernism   29, 153, 171–2, 186,
   200, 213–15
   architecture   2, 21, 49–50, 69,
     83–7, 91–7, 180–1, 184
   design   21, 48, 50, 97–102, 173–4,
     180–1, 235n
   European   69, 99–100, 171–2

'fitness for purpose' 50
landscape 2, 53–4, 74–102, 162–4,
    171–2, 213–15
motorway 73–6
signs 97–102
see also Williams, Sir (Evan) Owen
modernity 19–20, 32, 48, 53–4, 87,
    105, 141, 147, 153, 162–4, 166,
    172, 184, 186, 212, 215
and supermodernity 10, 211
Montagu of Beaulieu, John (later
    Lord) 25–9, 51, 222, 227n
Morrison, Herbert 50
Morse, Margaret 6–8, 10, 13
motor cars, see car(s)
motorbikes 34, 144, 181, 225n
motorway(s)
    children's books 19, 168–71, 168,
        170, 174, 241n
    conduct on 9, 143–5, 153, 158,
        176
    defence and 35, 39–40, 50
    early proposals 25–31, 40, 41, 42,
        228n
    ecology of 215–16
    films of 11, 19–21, 123, 125, 128,
        139–40, 217, 226n–7n
    Germany 18, 31–40, 42–3, 49, 81,
        84, 83–5
    Italy 31
    Lancashire 40, 42, 57, 66, 97–100,
        141
    landscape architecture and 2–3, 9,
        36–8, 55–9, 73, 83, 91, 197, 212,
        215
    London 204, 209–10
    London to Brighton 26, 29–30,
        46–7
    military and 26, 33–5, 37, 39, 50
    parliamentary bills and acts 26,
        29–30, 61–3, 66, 231n
    postcards of 182–3, 183, 214–15
    private motorways 27–31
    protests 16–18, 192, 205, 207–8,
        216, 225n
    tolls 25, 193
    USA 1–3

    see also Autobahnen; bypasses; M1
        motorway; motorway service areas
    Motorway (television docu-soap) 217
    Motorway (Laing film) 119–20, 123
    Motorway Code 143–5, 155, 158–9,
        176
    see also conduct: codes of conduct;
        Highway Code, The
    motorway driving, see driving
    'motorway madness' 144, 162
    motorway service areas 18, 83,
        178–86, 241n–3n
    Advisory Committee on Landscape
        Treatment of Trunk Roads
        91–7, 178
    alcohol licences 179–80
    architecture 91–7, 206–7, 241n
    cafés 167, 179–86, 206
    celebrity 181–2
    design 90–4, 92, 96–7, 234n
    landscaping of 90, 94–7
    layout 91, 93–4
    Leicester Forest East 182
    location 90, 234n
    lorry drivers and 96, 184–6
    national imagination 204
    Newport Pagnell 90–1, 93–6, 95,
        176, 178–82, 183, 185, 192, 204,
        214–15
    petrol brands 178–9
    planting vegetation in 96–7
    Royal Fine Art Commission 91,
        178
    Services, The by Peter Kay 217
    teenagers and 180–1
    Toddington 91, 204
    Trowell 204, 242n
    restaurants 179–82, 183
    Rothersthorpe 91, 204
    US influences 90–1
    Watford Gap 91–4, 92, 96, 167,
        178–80, 185–6
    see also Blue Boar; cafés: transport
        cafés; Forte
Motorway Services Limited (Forte
    and Blue Star garages) 91, 179,
    235n

movement   1–5, 12, 15, 74, 83, 85,
    89, 94, 103, 116, 119, 122, 214
mundane   6, 8, 10, 20, 97, 142, 200,
    211–12, 214–15, 217
music
    about motorways   124–40, 164–6,
        165, 206–7
    driving and   11, 149, 213
    see also folk music and song; noise

Nairn, Ian   74, 86–7, 101–2, 208,
    215, 231n
    discussing motorways   86–7, 208
    motorway signs   99–102
    Outrage   74, 86, 215, 231n
National Farmers' Union   71, 112,
    231n
National Parks Commission   71, 98
National 'Safety-First'
    Association   36, 42
National Trust   54, 61
nationality of labourers   120, 122–3,
    129–31, 133–5, 137, 139–40
Nature Conservancy   71
Nazi regime
    Germany   18, 31–8, 81, 227n–8n
    motoring and   228n
    motorways and   18, 31–8, 81
new towns   21, 203
Newport Pagnell   111–12, 121, 160,
    203, 217
    construction headquarters   111,
        117–18
    traffic in   110
Newport Pagnell Service Area   91,
    93–6, *95*, 176, 178–82, *183*, 185,
    192, 214–15
    see also Forte; Motorway Services
        Limited (Forte and Blue Star
        garages); Sidney Clough and Sons
        Limited
newspapers
    advice on motorway driving   145–7,
        149–58
    coverage of M1
        construction   109–12

coverage of M1 opening   152–5
    reports of motorway driving   154–8
Nichols, Peter   19, 208–10, 216
noise   110, 149, 212
non-places   10–11, 17, 19, 177, 201,
    210–14, 225n
non-representational theories and
    practices   1, 15, 74, 102, 224n
north–south journeys   243n–4n
North–South divide   68, 204,
    243n–4n
Northampton   112, 145, 202
Northamptonshire   67, 70–1, 73,
    86–7, 112, 117, 158, 202, 204,
    217
Northern and Western
    Motorway   27–9, *28*, 40, 47
nostalgia   19, 181, 214–15
notation of movement   3, 14–15
Nottingham   48, 69, 86, 123, 229n,
    242n

O'Connell, Sean   7, 11–12, 47, 164,
    167, 177, 225n
oil crisis   205
ontology   4, 6, 8, 10
    and driving   4, 8, 10, 145
Opie, Julian   211–13, 244n
    and M1   213
    and M40   212–13, 244n
opposition, *see* campaigns; protests
ornamental planting, *see* trees and
    shrubs
Osborne, Richard H.   68–9
overtaking   144, 155

'panoramic perception'   13–14
Parker, Charles   110, 124–5, 127–40,
    219, 237n–8n
    and *The Irishmen*   139–40, 238n
    and *Song of a Road*   110, 124–5,
        219, 237n–8n
    biography of   127
    politics   127
    relations with BBC   127–8
parkways   18, 31, 49, 51, 57

parliament   24–6, 29–30, 32, 35, 46,
    61–3, 66, 82, 110–11, 143–4,
    179, 192–3, 198, 230n, 243n
Parr, Martin   182–4, 214–15
passengers   7, 11–16, 213, 218
Pedestrians' Association   36
performance   32, 71, 104, 109, 118,
    149, 152, 177, 186, 218
    driving as a   50, 149, 152
Perrow, Charles   152, 193, 210
Petit, Chris   19, 226n
petrol and petrol companies   90,
    178–9, 205
    advertising   49, 179
    BP   49, 158, 178–9
    Esso   179
    Fina   178–9
    Mobil   179
    oil crisis   205
    rationing   44, 205
    Regent   178–9
    service areas   91, 96, 178–9
    Shell   49, 63, 158, 178–9
petrol stations   18, 48–9, 211,
    241n
Pevsner, Nikolaus   79, 86–8
    Architectural Review, The   79
    Buckinghamshire guide   86
    M1 structures   86–8
    Northamptonshire guide   86
    picturesque   76, 79–80
Peyton, John   196
phantoms of the M1
    motorway   197–9
Philo, Chris   119, 203
photography   3, 20, 77, 117, 145,
    167, 173, 213
    aerial photography   112–13
    motorway construction   105, 112,
    117
picturesque   76, 79–80, 93–4,
    233n–4n
    and movement   76, 80
    landscape   76, 79–80, 233n–4n
    motorways   80, 94, 97
    townscape   79–80, 96

place   1–2, 5–6, 9–11, 17, 153, 177,
    211, 213–14, 218
    sense of   2, 217
    see also non-places; placelessness
placelessness   9–11, 17, 19, 201,
    211–12, 225n
planner-preservationism   29–30, 47–9,
    73
planning   16, 29–30, 32, 35, 39,
    43–4, 51, 56–7, 70–1, 79, 93,
    117, 119–20, 138, 205
plants, *see* roadside planting; trees and
    shrubs
Poland   38, 81, 123, 130, 134
police   24, 152, 158–60, 162, 167,
    170, 192, 194, 208, 217
    accidents and   194, 196
    advice on motorway driving   159,
    192
    and speeding   24, 192
    cars   159
    motorbikes   159
    motorway crime   159–60, 170
politics
    Communism   125, 127, 137
    democracy and motorways   34, 49,
    207
    dictatorship and motorways   33–4,
    49
    folk song   125–7
    Nazism   18, 31–8, 81, 227n–8n
    see also Conservative Party; Labour
    Party; parliament
Porter, Michael   82
postcards   214–15
    kitsch   214
    Martin Parr   184, 214–15
    of motorway service areas   182,
    183, 214–15
    of motorways   182–3, *183*, 214–15
    nostalgia   214–15
Postgate, Raymond   176–7, 182
Powers, Alan   79, 204
practices, *see* driving: practices of; non-
    representational theories and
    practices

preservation
    architectural   215, 245n
    and planning   29–30, 47–9, 73
    see also conservation; Council for
        the Preservation of Rural
        England (CPRE); landscape;
        planner-preservationism
Preston Bypass Motorway, see M6
    motorway
Priestley, J. B.   47–8, 211
Prigogine, Ilya   224n–5n
privatization of space   6, 8, 16–17
protests   16–18
    against 70 mph speed limit   192
    against motorways and roads   18,
        205, 207–8, 225n
    on M1 by miners   216
    see also campaigns
psychology   11, 79, 96, 100
public inquiries   70, 110, 207
public relations media   71, 104–6,
        109–13, 115, 117–25, 128–9,
        136–7
public space   17, 73

quarries   110–11

RAC, see Royal Automobile Club
race, see ethnicity; racism
racing driving   152, 169–74
    Britishness   171
    masculinity   169–71, 173
    Mike Hawthorn   171
    Stirling Moss   171
    Tony Brooks   171–3
racism   123, 131, 140
    'colour bar, the'   123, 131
    construction workers'
        experiences   131, 140
    Irish workers   123, 140
    Jamaican worker   140
    race riots   123
radio   44, 149, 164
    BBC Radio Ballads   124–40,
        237n–8n
    programmes about M1   124–40,
        160, 166–7

Radio Times   135, 136
Song of a Road   124–40, 136,
        237n–8n
Railway Conversion League   66–7
railways   6, 13–14, 27–8, 39, 43–4,
        66–7, 68, 93, 105–6, 113–15, 122,
        157, 173, 176–7, 203, 225n,
        231n, 239n
    closures   66–7, 239n
    competition from road
        transport   39, 173
    construction of   122
    conversion to roads   66–7
    electrification   173, 231n
    landscape and   6, 12–14, 93,
        177
    London to Birmingham
        Railway   68, 105–6, 119, 121,
        173, 203, 233n
    M1 railway bridges   233n
    modernization   173, 231n
    Terence Cuneo paintings   113–15,
        114
rearmament   32, 34–5, 66
    and motorways   34–5, 66
Reclaim the Streets   16–17, 225n
reconstruction   21, 44–6, 50, 56–9,
        61, 65–6, 73, 80, 106–9, 108,
        200
    after World War II   21, 50, 56–9,
        106–9
    landscape architects and   56–9,
        73
    motorways and   44–6, 50, 57–9,
        61, 66, 106–9, 108
    planning   44–6
regions   4, 20, 30, 37, 45, 67–9,
        202–4, 211
    East Midlands   68–9
    landscape and   37, 74
    motorways and development   20,
        45, 67–9, 202–4
    South-East England   45,
        202–4
Relph, Edward   7, 9–10, 17, 201,
        210
Repton, Humphry   76

resistance, *see* campaigns; protests
restaurants, *see* motorways service
    areas
ribbon development   27, 29–30, 36,
    46, 48–50
Richards, J. M.   79
rivers   117
Road Haulage Association   145
road rage   11
Road Research Laboratory (RRL)   98,
    100–1, 186–95, 197–8
Roads Beautifying Association
    (RBA)   51–9, 77, 80–1, 220,
    230n–2n
    formation   51
    membership   51–2
    see also Fox, Wilfrid; Rothschild,
    Lionel de
Roads Campaign Council   66
roadside planting   9, 37–8, 51, 53–5,
    73–81, *75, 79,* 89–90, 96–7,
    215–16
Rolt, Lionel Thomas Caswell
    (L. T. C.)   71, 105–6, 109,
    111–13, 115, 117–18, 120–3, 139
Ronay, Egon   206
Rose, Nikolas   116, 142, 187
Rotha, Paul   50–1, 229n
Rothersthorpe service area   91, 204
Rothschild, Lionel de   51, 54
roundabouts   49–50, 212
    accidents and   195–7, *196*
    motorways and   195–7, *196*
Royal Automobile Club (RAC)
    25–6, 36, 42, 45, 65, 145, 176,
    222
    and German Roads Delegation   32,
    36
    Know Your Motorway booklet
    145
    membership handbook   176
    motorway proposals   42
Royal Fine Art Commission   71,
    81–3, 85, 89, 91, 98, 178
Royal Forestry Society of England and
    Wales   80, 89
Royal Horticultural Society   51, 81

Royal Institute of British Architects
    (RIBA)   44, 50, 65, 70, 82, 85,
    91, 229n
Royal Society for the Prevention of
    Accidents (ROSPA)   98

Sachs, Wolfgang   7–8, 32, 156
safety   36, 42, 46, 192
    see also traffic accidents and crashes
Saint, Andrew   69, 215
Salisbury, Edward J.   54
Samuel, Godfrey   82
Samuel, Raphael   125–6, 152
satire   144–5, 193, 207, 209
Schivelbusch, Wolfgang   6, 10, 13,
    18, 157
science
    and motorways   42, 100–1, 153,
    186–200
    experiments   42, 98, 100–1,
    186–99
sculpture   1, 3, 47–8, 79
Seeger, Peggy   110, 124–5, 129,
    132–5, 137–40, 223, 237n–8n
Select Committee of the House of
    Lords on the Prevention of Road
    Accidents   42–3
Self, Will   18–19, 245n
sense of place, *see* place
senses   12, 75–6, 147, 157, 213
    and driving   7, 12–16, 147, 149,
    155–7, 213, 218
    kineaesthetics   1–3, 12, 155, 213
    sound and hearing   12, 147–9
    visuality   3, 12–16, 25, 47, 57, 73,
    75–6, *75,* 83, 85, 89, 147, 155–7,
    159, 208, 225n–6n
service areas, *see* motorway service
    areas
Seymour, John   185–6
Sharpe, Tom   207
Shell petroleum company   49, 63,
    158, 174–6, 178–9
Sheller, Mimi   5–8, 142, 145, 147,
    156, 164
shopping centres   210–12, 214
Sibley, David   5–6, 122

Sidney Clough and Sons Limited   93
signs   9, 77, *78*, 97–102, 145, 177,
    218, 243n
  colour   97–8, 100
  lettering   97, 99–101
  lower case v. upper case   99–101
  motorway   97–102, 145, 212, 218,
    243n
  road   44, 77, *78*, 101–2, 212
  serifs   99–101, 236n
  size of   97, 100
  see also Advisory Committee on
    Traffic Signs for Motorways;
    Worboys Committee
Sinclair, Iain   19–20, 192, 210,
    226n–7n
  see also London Orbital (book and
    film)
Sir Owen Williams and Partners
    69–71, 81–7, *88*, 89–96, 110, 115,
    118, 129, 185, 197, 215, 220,
    227n
Smith, Tony   15
Smithson, Alison   15–16, 93
Soar Valley, Leicestershire   71
Society for Checking the Abuses of
    Public Advertising (SCAPA)   47
Society of Motor Manufacturing and
    Trading (SMMT)   42, 63, 191,
    229n
sociology   7, 11, 17, 20, 142, 157
  car   7
  crime   159–60
  lorry driving   7, 185–6
  mobilities   4–7
Song of a Road (BBC Radio
    Ballad)   110, 124–39, *136*,
    237n–8n
  authorship   237n
  conceptualization of   124–5, 128–9
  journalistic reception of   134–7
  Laing's response to   128–9, 136–7
  see also MacColl, Ewan; Parker,
    Charles; Seeger, Peggy
South Africa   113–14, 120, 123
South-East England   21, 141

spatialities of driving   145, 155–8
Special Roads Bill and Act (1948 and
    1949)   61–3, 66, 231n
spectacle of motorway   149, 164,
    182–4, 200, 240n
speed   13, 24, 47–8, 74–7, *75*, 89–90,
    97–100, 141, 143–4, 147, 149,
    154–8, 166, 168–9, 171, 173,
    177–8, 187, 191, 193, 205
  as social construct   157–8
  fog   192, 194
  landscape and   13, 47, 55, 60,
    74–5, *75*, 90, 93–8, 163
  motorway speed limits   141, 154,
    171, 191–2, 194, 205
  safety   169, 192–3
Spence, Basil   85, 240n
Spencer, Herbert   101
Spurrier, Raymond   9, 77–80, *78*, *79*,
    85, 87, 93–7, *95*
statistics   9, 42, 63, 113, 120, 159,
    163–4, 185, 192–3, 195–6
  accidents   9, 35–6, 44, 159, 187–8,
    192–3, 195–6
  breakdowns   162, *163*
  cost-benefit Analysis   187–9
  M1 traffic flows   162–4
  Road Research Laboratory
    189–90
  subjectivity   187–8
Strathern, Marilyn   189
streets   16–17
subjectivity   5, 7–8, 142, 207
subtopia   74, 89, 93, 96, 101
suburbia   2, 21, 29–30, 46, 74, 121,
    211, 214
Sunday drives   163–4
Surrey   24, 29–30, 52
Sussex   29–30, 39

Tarmac Civil Engineering
    Company   105, 129, 131, 227n
taxes   9, 26
  petrol   26
Taylor, Eva G. R.   45, 67
Taylor, Sir George   81, 96

Taylor, Ted, *see* Ted Taylor Four
Team Spirit (Laing's magazine) 85, 111, 114–15, 120–3, 139
technologies 12, 127, 157, 160, 226n
  automotive 9, 11–12, 147–52, 155
  of construction 104–9, *107*, *108*, 119–20, 123, 128, 131–3, 135, 226n
  of government 3, 9, 97–8, 102, 115–16, 121, 142, 144, 160
  sound recording 127
  unconscious 8–9
Ted Taylor Four 164–6, *165*, 240n
teenagers 21
  Absolute Beginners 181
  consumption by 21, 180–1
  motorways 169, 180–1
  service areas 180–1
  Teenage Consumer, The 180–1
  see also children; youth
television 8, 13, 19, 126, 134, 156, 158, 163
  programmes about motorways 156, 158, 163
  The Services, by Peter Kay 217
theatre 1, 125, 127
  Banner Theatre of Actuality 127
  Freeway, The (play) 208–10
  Semi-Detached (play) 164, 240n
  Theatre Workshop 125
Thrift, Nigel 4, 6, 8, 10–12, 17, 145, 157, 211, 224n–6n
Thwaite, B. H. 25–6
time-space compression 156–7
Tingrith (Bedfordshire) 70
Toddington service area 91, 204
Todt, Fritz 31–2, 35–7, 39–40, 42
toys
  motorway toys 16, 240n
traffic accidents and crashes 9, 35–6, 43, 61, 156, 159–62, 168–9, 171, 180, 190–9, 208–10, 243n
  blame 180, 192–9
  bridges, collisions with 197–9

history and 192
investigations into 159–60, 168–9, 194–9, *195*, *196*
  on M1 159–62, 193–9, *195*, *196*
  sociology and 192
  see also Select Committee of the House of Lords on the Prevention of Road Accidents
traffic congestion 8, 36, 65–6, 216
  dystopian narratives about 208–10
  Freeway, The (play) 208–10
traffic signs, *see* signs
Transport Research Laboratory, *see* Road Research Laboratory
trees and shrubs 51–5, 74, 81, 90, 96–7, 157
  'exotic' species 53–5, 74
  garden species 55, 74, 89
  indigenous or native species 37–8, 81, 90, 97
  ornamental species 89–90, 97
  see also roadside planting
Trowell service area 204, 242n
trunk roads 35, 40
  Trunk Roads Bill and Act 1936/1937 35, 38, 54
Trunk Roads Joint Committee 54, 80, 230n
Tyme, John 70, 207
typography, *see* signs

urban 54–5, 90, 96, 172
  heritage 204–6
  motorways 1–2, 14–15, 18, 204–5, 208–10
Urry, John 5–8, 11–13, 121, 142, 147, 156, 189
USA 1–3, 12, 14–15, 17–19, 31, 49, 51, 76, 90–1, 98, 125–7, 190–1, 187, 191, 193, 205, 213, 226n, 229n
  Boston 2, 14, 49
  Conditional Aid 187
  construction machines from 106–9, *108*
  freeways 1–3, 18, 20

USA (*cont'd*)
  landscaping   31, 76, 226n
  Las Vegas   2
  Los Angeles   1–2, 12, 20
  New York   49, 51, 229n
  parkways   18, 31, 49, 51
  see also Americanization

Vaughan-Thomas, Wynford   166–7
Venturi, Robert   2, 18
Victoria Line (London
    Underground)   139–40
  cost-benefit analysis study   188
  Irish labourers   139–40
Virilio, Paul   12–13, 157, 192–3
visuality and vision   3, 12–16, 25, 47,
    57, 73, *75*, 76, 83, 85, 89, 159,
    208, 225n–6n
  eyesight standards   156
  mirrors   144, 155–6, 173, 212
  motorway driving   3, 12–16, *75*,
    76, 83, 145–9, *150*, *151*, 155–6,
    212, 218
  railway travel   13–14

Wales   66, 123, 166
walking   1, 6, 17, 47–8, *75*, 75, 159
war
  motorways and   34–5, 39
  transport in   43–4
  World War II   35, 43–6, 56–9, 70,
    73, 80, 168
Watford
  North–South divide   244n
Watford Gap   90, 204, 244n
  communications corridor   67–8,
    67
  Jay H. Appleton   67–8
  North–South divide   204, 244n
Watford Gap service area   91, *92*,
    96–7, 167, 178–80, 185–6, 204,
    206–7
  architecture   91–4, *95*
  criticisms of design   91–4
  footbridge design   91–2, *92*
  lorry drivers   185–6

petrol   178–9
Roy Harper song   206–7
see also Blue Boar; Harry W.
    Weedon and Partners
Watkinson, Harold   70, 104, 119–20,
    143, 152, 158, 239n
Way, John Michael   216
weather   149, 173
  accidents and   192–4
  motorway construction   117
  social studies of   194
  see also fog
Weedon, Harry, *see* Harry W. Weedon
    and Partners
Wells, H. G.   25
Westway   25, 244n
Williams, Sir (Evan) Owen   69–71,
    81–7, 89, 90, 104, 115, 120,
    129, 132–3, 170, 197, 220,
    233n
  avant-garde   69, 233n
  biography   69, 233n
  European architecture   69, 233n
  M1 motorway   70–1, 84–7, 104,
    115, 120, 129, 132–3, 170, 197,
    233n
  modernism   69, 89, 233n
  personality   70–1, 129
  see also Sir Owen Williams and
    Partners
Williams, Owen Tudor (son of Sir E.
    Owen Williams)   90–1, 129, 132,
    197
Williams-Ellis, Clough   48–9, 80, 90,
    92–3, *92*, 230n–1n, 234n
  and CPRW   80
  criticisms of M1   86, 91, 93, *92*,
    234n
  England and the Octopus   48
  Landscape Advisory
    Committee   80, 90–1, 93
  Watford Gap footbridge   91–2, *92*
Woburn Abbey   177
women
  BBC Radio Ballads   133–5,
    138–9

construction site   120–2
drivers   120, 145, 167–8, 241n
Worboys Committee   97, 99
work   3–4, 11, 25, 185–6, 213, 217
  AA   160–2, *161*
  construction   33, 104–24
  labouring   33, 106, 120, 122–4
  lorry driving   96, 185–6, 131
  policing   158–60
  see also corporate spaces
working classes   25, 96, 125–6, 128,
    130, 132–3, 137–8, 180, 185–6
  folk cultures   125–6, 128–30,
    132–4, 136–9

lorry drivers   96, 185–6
motorway labourers   128, 130,
    132–3, 137–8
  speech   138
Wylie, John   3, 224n

Yorkshire   45, 52, 204, 207, 209
youth   7, 21, 144–5, 169, 174, 180–1
  driving   7, 144–5, 169, 174
  masculinity   7, 144–5, 169, 174
  motorways   144–5, 166–71, 180–1
  service areas   180–1
  see also children; teenagers
Youth Hostels Association   65